AF831448

Antifreeze Proteins Volume 1

Hans Ramløv • Dennis Steven Friis
Editors

Antifreeze Proteins Volume 1

Environment, Systematics and Evolution

Springer

Editors
Hans Ramløv
Department of Natural Sciences
Roskilde University
Roskilde, Denmark

Dennis Steven Friis
Copenhagen, Denmark

ISBN 978-3-030-41931-8 ISBN 978-3-030-41929-5 (eBook)
https://doi.org/10.1007/978-3-030-41929-5

This Springer imprint is published by the registered company Springer Nature Switzerland AG.
The registered company address is: Gewerbestrasse 11, 6330 Cham, Switzerland

This book is dedicated to the two great minds and most creative researchers within the field of antifreeze proteins and physiology of cold tolerance: Professor Arthur L. DeVries and the late Professor Karl Erik Zachariassen.

Preface

When I (Hans Ramløv) was asked by Springer Verlag if I would consider editing a book on antifreeze proteins, I was much honored and gladly took on the project. I asked my good friend and previous PhD student (Dennis Steven Friis) if he would join me as editor, as we could complement each other with my background in comparative physiology and his in molecular biology. He was also excited by the idea and gladly teamed up with me.

Two books have been published focusing on antifreeze proteins, the last one in 2010. Since then considerable progress has been made on various aspects of the subject. There are an increasing number of laboratories around the world taking part in the research on antifreeze proteins, and there is a series of conferences dedicated specifically to this subject. Thus, our understanding of the diversity, the structures, the mechanisms by which the antifreeze proteins interact with ice, the evolution, and the adaptive value as well as the ideas for applications and the problems involved in this has increased considerably during the past decade. We therefore thought it timely to edit a book where all aspects of antifreeze proteins were included showing the "state of the art" of the subject at this time. Achieving this could not be possible without the help from leading experts within the different branches of the antifreeze protein research. We therefore contacted the respective scientists and could luckily gather a strong team of contributing authors within a short time. We would like to extend our sincerest gratitude to all of them.

We would like to dedicate this book to the two brilliant scientists within the field of antifreeze proteins and physiological adaptations to cold in ectotherms: Professor Arthur L. DeVries and the late Professor Karl Erik Zachariassen.

Art DeVries is the key figure in the discovery of antifreeze proteins in the late 1960s, and he has ever since worked relentlessly to shed light on all aspects of the biology, role, and mechanism of these proteins, especially in polar fishes. His discoveries have opened a whole new field of research, which is now investigated by a large number of scientists and which has opened the eyes of the public to the exciting adaptations harbored by cold-tolerant ectothermic organisms.

Karl Erik Zachariassen was one of the most nondogmatic, unorthodox, and creative scientists one could ever meet. Together with Ted Hammel he discovered the ice-nucleating agents in insects, and during his career he made several large contributions to the understanding of drought and cold tolerance in insects. Karl Erik Zachariassen worked for many years on cold tolerance in Norwegian beetles. He always tried to put this work into a wider and more generalized perspective often making unexpected observations and sometimes bold statements, which would create heated discussions in the scientific community.

We thank them both for all their efforts, contributions, and enlightenment.

Roskilde, Denmark Hans Ramløv
Copenhagen, Denmark Dennis Steven Friis

Contents

Abbreviations

AA	Anti-agglomeration
ABA	Abscisic acid
ABP	Aluminum-binding peptide
ACC	Antarctic circumpolar current
ADIS	Anti-icing and de-icing systems
AFGL	Antifreeze glycolipid
AFGP	Antifreeze glycoprotein
AF(G)P	Antifreeze protein or antifreeze glycoprotein
AFP/AP	Antifreeze protein (xAFP, x = species or type identifier)
AFPP	Antifreeze potentiating protein
ALT	Alanine aminotransferase
AQP	Aquaporin
ASW	Amorphous solid water
BAC	Bacterial chromosome arm vector
BLAST(P)	Basic local alignment search tools (protein)
BSA	Bovine serum albumin
CCP	Carrot concentrate protein
CD	Circular dichroism
CDS	Coding sequence
CF	Carboxyfluorescein
CNT	Classic nucleation theory
CPD	Cryoprotective dehydration
CRD	Carbohydrate recognition domain
CTLD(cps)	C-Type lectin-like domain (containing proteins)
CT_{min}	Critical thermal minimum
DDC	Double siamond cages
DEPC	Dielaidoylphosphatidylcholine
DEPE	Dielaidoylphosphatidylethanolamine
DEPG	Dielaidoylphosphatidylglycerol
DGDG	Digalactosyldiacylglycerol

DMPC	Dimyristoylphosphatidylcholine
DMSO	Di-methyl sulfoxide
DPC	Dodecylphosphocholine
DPPC	Dipalmitoylphosphatidylcholine
DSC	Differential scanning microcalorimeter/microcalorimetry
DT	Delay time
EAC	Escape from adaptive conflict
EPC	Egg phophatidylcholine
EST	Expressed sequence tag
FFS	Forward-flux sampling
F_i	Ice fraction
FIPA	Fluorescent-based ice plane affinity
FP	Freezing point
FTIR	Fourier transform infrared
GC-MS	Gas chromatography mass spectrometry
GFP	Green fluorescent protein
GH	Growth hormone
GI	Gastrointestinal
HC	Hexagonal cages
HDA	High-density amorphous ice
HDL	High-density liquid
HES	Hydroxyethyl starch
hFP	hysteresis freezing point
HGW	Hyperquenched glassy water
hMP	hysteresis melting point
HP-DSC	High-pressure differential scanning microcalorimeter/ microcalorimetry
HRP	Horse radish peroxidase
IBD	Ice-binding domain
IBF	Ice-binding face
IBP	Ice-binding protein
IBS	Ice-binding site
I_c	Cubic ice
IDTA	Infrared differential thermal analysis
I_h	Hexagonal ice
INA	Ice-nucleating agent
INLP	Ice-nucleator lipoprotein
INP	Ice-nucleating protein
IR	Infrared
IRI	Ice recrystallization inhibition
IRRINA	Ice recrystallization rate inhibition assay
ISP	Ice structuring protein
ITC	Isothermal calorimetry or calorimetric
JH	Juvenile hormone

KHI	Kinetic hydrate inhibition
LDA	Low-density amorphous ice
LDH	Lactate dehydrogenase
LDHI	Low-dosage hydrate inhibitor
LDL	Low-density liquid
LGT	Lateral gene transfer
LL	Ladderlectin
LLCP	Liquid-liquid critical point
LLT	Lower lethal temperature
LPIN	Lipoprotein ice nucleator
LT_{50}	Temperature with 50% mortality (survival studies)
LUV	Large unicellular vesicles
MD	Molecular dynamics
MGDG	Monogalactosyldiacylglycerol
MP	Melting point
MW	Monoatomic water
Mya	Million years ago
Myr	Million years
NANS	N-Acetylneuraminic acid synthase
ND	Nuclear diffraction
NIBF	Non-ice-binding face
NMR	Nuclear magnetic resonance
Nt	Nucleotide
ORF	Open reading frame
PBS	Phosphate-buffered saline
PCR	Polymerase chain reaction
PDB (XXXX)	Protein Data Bank (ID no.)
PDMS	Polydimethylsiloxane
PEG	Poly(Ethylene) glycol
PI	Phosphatidylinositol
PNAG	Poly-N-Acetylglucosamine
PNIPAAm	Poly(N-Isopropyl acrylamide)
PPII	Polyproline type II
PR	Pathogenesis-related
PSP	Pancreatic stone protein
PTFE	Polytetrafluoroethylene (Teflon)
PVA	Poly(Vinyl alcohol)
PVCap	Polyvinylcaprolactam
PVP	Polyvinylpyrrolidone
PVPip	Poly(N-Vinylpiperidone)
QAE	Quaternary aminoethyl
RCH	Rapid cold hardening
REG protein	Regenerating protein
RFU	Relative fluorescent unit

RIP	Recrystallization inhibition protein
ROS	Reactive oxygen species
SAS	Sialic acid synthase, eukaryotes (NeuB in prokaryotes). Also called NANS
SAXS	Small angle X-ray scattering
SCP	Supercooling point
SDM	Site-directed mutagenesis
SDS	Sodium dodecyl sulfate
SEM	Scanning electron microscopy
SHS	Superhydrophobic surface
sI, sII	Structures of ice types, I, II
SO	Southern Ocean
SP	Sulfopropyl
SPI	Soy protein isolate
SSR	Simple sequence repeats
STM	Scanning tunneling microscopy
TCA	Trichloroacetic acid
TEM	Transmitted electron microscopy
Tf	Freezing temperature
Tg	Glass transition temperature
TH/THA	Thermal hysteresis/thermal hysteresis activity
THF	Tetrahydrofuran/thermal hysteresis factor
THP	Thermal hysteresis protein
TL	Thaumatin-like
TLP	Trypsinogen-like protease
T_m	Melting temperature or transition temperature
TMB	Tetramethylbenzidine
TSS	Transcription start site
ULT	Upper lethal temperature
UTR	Untranslated region
VHDA	Very high-density amorphous ice
VSFG	Vibrational sum-frequency generation
WAP	West Antarctic Peninsula
XDM	X-ray diffraction
ZnO	Zinc oxide

Chapter 1
Introduction

Hans Ramløv and Dennis Steven Friis

Antifreeze proteins (AFPs), also called thermal hysteresis proteins (THPs) (thermal hysteresis factors (THFs)), ice-binding proteins (IBPs), and ice structuring proteins (ISPs) are a unique group of proteins that despite large differences in structure on all levels, share the common properties that they recognize, bind to, and inhibit (structure) the growth of ice crystals in cold-adapted ectothermic organisms. In the following, we will call these proteins as antifreeze proteins (AFPs) or, when specifically referring to the glycosylated variants, antifreeze glycoproteins (AFGPs). The binding of AFPs to ice surfaces has two main effects; thermal hysteresis and inhibition of recrystallization.

Thermal hysteresis is defined as the thermal difference between the equilibrium freezing point and the temperature where a minute ice crystal in the solution begins to grow. When AFPs bind to the surface of an ice crystal, they inhibit further growth of the crystal. This inhibition occurs even if the temperature is lowered 10 °C below the freezing point for the most potent AFPs. Thus, the solution is actually supercooled despite the fact that tiny ice crystals are present in the solution. However, when the temperature is lowered further, at some point the inhibition effect fails, and rapid ice growth occurs. Thus, the AFPs prevent the growth of ice, whether that being ice which has been ingested via the drinking of seawater in polar fishes or being ice inoculation insects and other organisms via body orifices. Antifreeze proteins are hence an important adaptation for many freeze-avoidant species living in areas where ice is present in the environment at some time during the annual cycle.

H. Ramløv (✉)
Department of Natural Sciences, Roskilde University, Roskilde, Denmark
e-mail: hr@ruc.dk

D. S. Friis
Copenhagen, Denmark

H. Ramløv, D. S. Friis (eds.), *Antifreeze Proteins Volume 1*,
https://doi.org/10.1007/978-3-030-41929-5_1

Recrystallization occurs in a polycrystalline ice solution. Due to thermodynamic forces, the larger ice crystals will slowly grow at the expense of small crystals, until the solution will contain one single large ice crystal. When AFPs are bound to the ice surfaces of the polycrystalline solution, they inhibit the recrystallization resulting in virtually no change of the sizes of the crystal. In organisms, the ice recrystallization can cause cell and tissue damage as the large ice crystals expand. Therefore, the AFPs are also an important asset for many freeze-tolerant species. However, in these species, the AFPs are not very potent in regard to thermal hysteresis, as undercooling is not favorable in freeze-tolerant organisms.

The AFPs thus have a large adaptive role in freeze-avoiding and freeze-tolerant species. The proteins are widespread among various species inhabiting the cold regions of our planet. The species include fish, insects, bacteria, plants, and many more, and in each case the proteins are adapted to provide a certain degree of thermal hysteresis. The most potent AFPs in regard to thermal hysteresis are found in freeze-avoiding insects, while plant AFPs evoke virtually no thermal hysteresis, but instead protect the organisms by inhibition of recrystallization. The vast difference among AFPs, in potency, size, and structure, arises from the fact that many AFPs have evolved convergently from various ancestral proteins and genes.

Though being part of many species arsenals against freezing temperatures, the discovery of AFPs is relatively new. The thermal hysteresis phenomenon in living organisms was first reported by Ramsay in 1964, during a study on water reabsorption in the hemolymph of the common mealworm, the larva of the beetle *Tenebrio molitor* (Ramsay 1964). Before Ramsay's observation, Scholander and co-workers and Gordon and co-workers had observed that in Arctic fish, the total amount of osmotically active substances in the body fluids did not account for their freezing point. Thus, the fish were supercooled even in water where ice was present (Gordon et al. 1962; Scholander et al. 1957). Interestingly, Scolander et al. already in 1957 mentioned the spicular ice growth, which is now known as a clear indication of the presence of fish antifreeze proteins (Scholander et al. 1957). It was, in fact, coincidental that Ramsay discovered the antifreeze effect. While investigating the water absorption via the cryptonephridial rectal complex in *T. molitor*, he was determining the osmolalities of the various compartments of the rectal complex. In a footnote in his paper he states "*the crystals which appear in fluid from the anterior perinephric space tend to have a jagged outline and large ice crystals do not grow on the expense of small ones. Furthermore, the system is not temperature-reversible. As the temperature is raised the crystals decrease in size, but as the temperature is lowered they do not increase in size. After the temperature has been lowered by a few degrees the crystals suddenly begin to grow rapidly*" (Ramsay 1964). From this citation, it can be seen that Ramsay observed recrystallization inhibition as well as antifreeze activity. In fact, in some cases, Ramsay observed an undercooling of the ice-containing sample at 10 °C. In a later paper on the cryptonephridial complex in *T. molitor*, Grimstone et al. again mention the observation of the strange behavior of ice crystals. In this chapter, there is actually a small but very lucid discussion of the reason for this behavior of the ice crystals as the authors try to explain such a behavior occurring in solutions which may never see ice (Grimstone et al. 1968).

Grimstone and co-workers are so curious about this phenomenon that they try to identify the substance responsible for the observations. They show that in supernatants of centrifuged homogenates of *T. molitor* they can abolish the activity of the unknown substance by trypsin, which indicates that it is a protein. They further perform paper chromatography as well as size exclusion chromatography and ion chromatography and find that a molecule with a molecular weight in the range of 10 kDa–12 kDa is responsible for the activity (Grimstone et al. 1968). The size range found by Grimstone et al. is somewhat higher than what is actually the case (see below) but still close to what was later found. Indeed the substance was a protein and not only one but several (Patterson and Duman 1979; Schneppenheim and Theede 1980) but it was not before many years later, that the proteins were sequenced (Graham et al. 1997; Liou et al. 1999).

In 1969, DeVries and Wohlschlag discovered the first proteins responsible for evoking thermal hysteresis in Antarctic fish (DeVries and Wohlschlag 1969). Later, in 1971, it was established that it was the AFGPs that were responsible for the thermal hysteresis (DeVries 1971). After the discovery of the AFGPs in Antarctic fish non-glycosylated proteins with a similar effect were discovered in the North American Flounder *Pseudopleuronectes americanus* in 1974 (Duman and Devries 1974, 1976). Antifreeze proteins in insects were first discovered by Duman in 1977 where he named them as macromolecular antifreezes (Duman 1977a, b). Since the first discoveries of antifreeze proteins in fish and insects, these proteins have been found in a large number of cold-tolerant ectothermic organisms such as plants, algae, lichens, and bacteria. In some, the actual antifreeze activity is of the largest importance inhibiting extracellular or intracellular ice growth or inoculation, in others it is recrystallization inhibition which seems to be the important property whereas others may excrete the antifreeze (ice structuring) proteins to their immediate environment thereby changing the ice structures surrounding the organisms or helping them to adhere to the ice.

The common notion is that AFPs, by their interaction with the ice crystal surface, cause a non-colligative depression of the freezing point of water whereas the melting point is only depressed according to the colligative freezing point depression. However, this notion can be disputed, from a physicochemical point of view. A solution already containing ice crystals per definition cannot be said to have its freezing point depressed as this would implicate that the solution is supercooled and thus without ice crystals [see also remarks by (Franks et al. 1987)]. The supercooling point is defined as the temperature at which the solution spontaneously freezes and thus the supercooled solution lacks ice crystals. Several authors have implied that AFPs can stabilize the supercooled state (Block 1991; Duman et al. 1993). Based on the observation that in a given volume of a solution-containing AFPs, there is a linear and inverse relationship between the antifreeze activity and the logarithm of the mass fraction of ice, Zachariassen and Husby suggested that AFPs might bind to microscopic (embryonic) ice crystals (Zachariassen and Husby 1982; Zachariassen et al. 2002) and thereby inhibiting crystallization completely stabilizing the supercooled condition. Despite some attempts, this issue has not been fully elucidated yet (see Wilson and Leader below). Zachariassen and Husby also claimed that

AFPs are unable to mask ice-nucleating agents (INAs) and thus, it is only solutions with no INAs which are stabilized by the AFPs (Zachariassen and Husby 1982). Franks et al. showed that in solutions containing AFGPs, the homogeneous nucleation temperature is depressed to the same extent as solutions containing a solute (PVP) lacking antifreeze activity, which was acting as a colligative freezing point depressant and thus also depressing the supercooling point (Franks et al. 1987). Later findings have shown that fish AFGPs as well as AFPs from insects are able to inhibit ice-nucleating activity of bacterial membrane-bound INAs and certain INAs (Duman et al. 1991; Parody-Morreale et al. 1988). Using AFGPs from fish, Wilson and Leader have shown that these molecules cannot stabilize the supercooled state in a Ringer solution as much as if the solution contains INAs. On the basis of these results, the authors conclude that the AFGPs and possibly all AFPs bind to nucleation sites rather than bind to heterogeneous ice nuclei in the supercooled solution (Wilson and Leader 1995). The reduction in supercooling temperature in the Ringer solution containing INAs was approximately 2.5 °C whereas in the solution devoid of INAs the reduction in supercooling temperature was approximately 1.2 °C. Taken the above-mentioned results into consideration it seems likely that AFPs mainly act on embryo ice crystals and thus as such do not stabilize a supercooled solution. However, the results also indicate that AFPs may mask INAs in freeze avoiding organisms and this can, of course, be considered as a stabilization of the supercooled state.

Since the discovery of the antifreeze proteins, many questions have haunted the investigators. When ice crystals enter the organisms and their growth is inhibited by the antifreeze proteins then what is their fate? How is the interaction between the proteins and the ice surface and with water? What drives this interaction? Which structural features are essential? What determines the potency of the AFPs? How have the antifreeze proteins evolved? Are there common properties in a physicochemical perspective? Is the observed interaction between antifreeze proteins and biological membranes adaptive or merely coincidental? And can we exploit the AFPs and use them in some industrial applications?

Some of the above questions have been elucidated quite extensively, i.e., the structures and structural properties, especially the fish antifreeze proteins and during the recent years also the insect antifreeze proteins. Although the structural description has been the basis for the understanding of the AFPs' interactions with water and ice, much still remains to be elucidated concerning the binding of the AFPs to ice. Recently, it has been possible, for the first time, to experimentally observe the interactions between the antifreeze and water molecules when the antifreeze protein was bound to the ice surface (Meister et al. 2019). However, to achieve a complete understanding of the interactions many more experiments with similar techniques as used here will be necessary.

Physicochemical properties have only been investigated in a few AFPs and only with respect to few properties so also here is there a large amount of work to be done by future investigators.

The interactions with ice and water in the intact organisms as well as the fate of ingested ice crystals, for example, those which are associated with the spleen of

Antarctic fishes, remain to be fully elucidated. Some of the interactions are so elusive that we only have vague ideas about the mechanisms determining the fate of the ice crystals in the organism.

The evolution of antifreeze proteins shows some of the best examples of molecular convergent evolution. Especially the evolution of fish antifreeze proteins has during the last couple of decades been a field of intense research and much has been discovered. The evolution of insect AFPs is much less elucidated and much remains to be investigated but it seems likely that AFPs in insects have originated several times. Certain molecular structures seem to be evolved convergently as they are likely to be very important for the observed strong antifreeze activity found in insects.

The interactions between AFPs and biological membranes were a matter of much debate about 20 years ago but since then there has been a paucity in studies concerning such interactions. Recently, new studies in this area have been published and it is clear that much is to be learned. However, it still remains uncertain if these interactions are coincidental or if they are adaptive. Interactions between pathogens and AFPs and the biological importance of this is a totally new area of investigation that has so far showed some promising results and given rise to some interesting hypotheses as to the adaptional value to the organisms by these interactions.

The last question concerning industrial applications of AFPs has become the focus of many investigations in recent years, and many potential users are in play. Despite the fact that a number of AFPs have been cloned and expressed in both prokaryotic and eukaryotic systems, it has until now not been possible to produce the most active AFPs in commercially interesting amounts. But much effort both to produce commercially available AFPs and to investigate the potential uses is being done. These efforts are making the field of the application of AFPs an intriguing field to follow.

This book on antifreeze proteins contains elaborate reviews on all the subjects described above by contributors who are leading experts within the field. The book is divided in two volumes, in main terms focusing on either the biology or the biochemistry of the antifreeze proteins.

Volume 1 is divided into an introductory section, Part I on ice formation and Part II on the biology of the antifreeze proteins. The introduction includes a short presentation of the antifreeze proteins as well as an outline of the book contents (this chapter), and a chapter with short résumés of the chapters included in volume 1.

Volume 2 also includes an introductory chapter with short résumés of the chapters included in the volume. The volume is divided into four parts concerning the biochemistry of the antifreeze proteins, their molecular mechanisms, some of the industrial applications of antifreeze proteins, and finally a closing part.

Each chapter contains the relevant introductory text needed to comprehend the chapter, so each chapter can be read separately.

References

Block W (1991) To freeze or not to freeze? Invertebrate survival of sub-zero temperatures. Funct Ecol 5:284–290

DeVries AL (1971) Glycoproteins as biological antifreeze agents in Antarctic fishes. Science 172:1152–1155

DeVries AL, Wohlschlag DE (1969) Freezing resistance in some Antarctic fishes. Science 163:1073–1075

Duman JG (1977a) The role of macromolecular antifreeze in the darkling beetle, *Meracantha contracta*. J Comp Physiol 115:279–286

Duman JG (1977b) Variations in macromolecular antifreeze levels in larvae of the darkling beetle, *Meracantha contracta*. J Exp Zool 201:85–92

Duman JG, Devries AL (1974) Freezing resistance in winter flounder *Pseudopleuronectes americanus*. Nature 247:237–238

Duman JG, DeVries AL (1976) Isolation, characterization, and physical properties of protein antifreezes from the winter flounder, *Pseudopleuronectes americanus*. Comp Biochem Physiol B 54:375–380

Duman JG, Xu L, Neven LG, Tursman D, Wu DW (1991) Hemolymph proeins involved in insect low temperature tolerance: ice nucleators and antifreeze proteins. In: Lee RE, Denlinger DL (eds) Insects at low temperatures. Chapman Hall, New York, pp 94–127

Duman JG, Wu DW, Yeung KL, Wolf EE (1993) Hemolymph proteins involved in the cold tolerance of terrestrial arthropods: antifeeze and ice nucleator proteins. In: Prisco GD (ed) Life under extreme conditions. Springer, Berlin, pp 282–300

Franks F, Darlington J, Schenz T, Mathias SF, Slade L, Levine H (1987) Antifreeze activity of Antarctic fish glycoprotein and a synthetic polymer. Nature 325:146–147

Gordon MS, Amdur BH, Scholander PF (1962) Freezing resistance in some northern fishes. Biol Bull 122:52–62

Graham LA, Liou Y-C, Walker VK, Davies PL (1997) Hyperactive antifreeze protein from beetles. Nature 388:727–728

Grimstone AV, Mullinger AM, Ramsay JA (1968) Further studies on the rectal complex of mealworm *Tenebrio molitor*, L. (Coleoptera, Tenebrioidae). Philos Trans R Soc Lond Ser B Biol Sci 253:343–382

Liou YC, Thibault P, Walker VK, Davies PL, Graham LA (1999) A complex family of highly heterogeneous and internally repetitive hyperactive antifreeze proteins from the beetle *Tenebrio molitor*. Biochemistry 38:11415–11424

Meister K, Moll CJ, Chakraborty S, Jana B, DeVries AL, Ramlov H, Bakker HJ (2019) Molecular structure of a hyperactive antifreeze protein adsorbed to ice. J Chem Phys 150:131101

Parody-Morreale A, Murphy KP, Di Cera E, Fall R, DeVries AL, Gill SJ (1988) Inhibition of bacterial ice nucleators by fish antifreeze glycoproteins. Nature 333:782–783

Patterson JL, Duman JG (1979) Composition of a protein antifreeze from larvae of the beetle, *Tenebrio molitor*. J Exp Zool 210:361–367

Ramsay JA (1964) The rectal complex of the mealworm *Tenebrio molitor*, L. (Coleoptera, Tenebrionidae). Philos Trans R Soc Lond Ser B Biol Sci 248:279–314

Schneppenheim R, Theede H (1980) Isolation and characterization of freezing point depressing peptides from larvae of *Tenebrio molitor*. Comp Biochem Physiol 67:561–568

Scholander PF, van Dam L, Kanwisher JW, Hammel HT, Gordon MS (1957) Supercooling and osmoregulation in arctic fish. J Cell Comp Physiol 49:5–24

Wilson PW, Leader JP (1995) Stabilization of supercooled fluids by thermal hysteresis proteins. Biophys J 68:2098–2107

Zachariassen KE, Husby JA (1982) Antifreeze effect of thermal hysteresis agents protects highly supercooled insects. Nature 298:865–867

Zachariassen KE, DeVries AL, Hunt B, Kristiansen E (2002) Effect of ice fraction and dilution factor on the antifreeze activity in the hemolymph of the cerambycid beetle *Rhagium inquisitor*. Cryobiology 44:132–141

Chapter 2
Contents of Volume 1—Antifreeze Proteins: Environment, Systematics, and Evolution

Hans Ramløv and Dennis Steven Friis

In this volume of the book, we present background on water and ice formation, the history of the antifreeze proteins (AFPs), their physiological role in freeze avoiding or freeze tolerating species, as well as the evolution and grouping of the antifreeze proteins. In the second volume, we delve into the functions and interactions of antifreeze proteins at the molecular level.

This volume is divided into two sections (Parts I and II of this work) as well as an introductory part. Part I focuses on ice formation, both in regards to physics as an isolated event as well as its biological impact in living organisms. Part II describes the AFPs characteristics, potency, and mode of action in the different groups of organisms they are found in, divided into either one of the main groups of fish, insects, or plants or in a miscellaneous group including the rest, as well as coverage of the interesting convergent evolution of the AFPs.

Below we give a short summary of the chapters that constitute this first volume of the book, besides the introductory Chap. 1 and this chapter.

2.1 Part I: Ice Formation

In Chap. 3, Prof. Dr. Amir Haji-Akbari gives an introduction to ice and its formation. The chapter concerns all the aspects of ice relevant for the later chapters in the book, including the driving forces of nucleation, the structure and the polymorphisms of ice as well as the amorphous glassy state. The nucleation is described in a detailed

H. Ramløv (✉)
Department of Natural Sciences, Roskilde University, Roskilde, Denmark
e-mail: hr@ruc.dk

D. S. Friis
Copenhagen, Denmark

H. Ramløv, D. S. Friis (eds.), *Antifreeze Proteins Volume 1*,
https://doi.org/10.1007/978-3-030-41929-5_2

manner with the thermodynamics variables that affect the event, and the different types and anomalies of ice are presented and those relevant for biological systems are discussed in-depth, along with the crystallography and planes or faces of ice crystals. The author also describes the event of recrystallization process and the thermal hysteresis phenomenon that are key processes in the investigation of antifreeze proteins.

In Chap. 4, Prof. Dr. Hans Ramløv and Dr. Dennis Steven Friis give a general overview of ice formation in living organisms. Here, we introduce the consequences of cold exposure "per se". Basic concepts with regard to cold adaption and consequences of freezing temperatures are introduced. In the chapter the organisms' responses to cold such as rapid cold hardening (RCH), the synthesis and actions and mechanisms of low- and high-molecular weight cryoprotectants are explained. In the final part of the chapter, antifreeze proteins are introduced and their occurrence in various organisms and their mode of action is mentioned.

This chapter describes the consequences of cold exposure and ice formation on the cellular, tissue, and the organismal level in cold-tolerant ectothermic organisms. Defensive/preventive adaptations and mechanisms of the organisms are described as well, along with a description of freeze tolerance and freeze avoidance. The synthesis and mechanisms of action of low molecular weight cryoprotectants as well as their distribution are discussed. High molecular weight cryoprotectants are also described concluding with the introduction of the AFPs.

2.2 Part II: The Biology of Antifreeze Proteins

In Chap. 5, Prof. Dr. Arthur L. DeVries presents the current investigation on antifreeze proteins in fish. This is the most thoroughly investigated group of antifreeze proteins, as it can readily be harnessed from the blood of fish expressing them. Most polar marine teleost fish express antifreeze proteins, as they live in waters of that are about 1 °C colder than the freezing point of their blood. Several types of antifreeze proteins are found in fish, covering both glycosylated and non-glycosylated types, of various sizes, which have arisen by convergent evolution.

The chapter also describes the physiological process of the introduction of ice in the blood or gut of the fish, the synthesis of antifreeze proteins, and the interaction with ice as well as the neutralization and removal of the ice/antifreeze protein complex.

In Chap. 6, Prof. Dr. John G. Duman and Dr. Samuel S. Newton cover the antifreeze proteins found in insects. They present and discuss the three main functions of the ice-binding proteins; their ability to produce a high thermal hysteresis (AFPs), the inhibition of recrystallization (RIPs), and nucleator proteins (INPs), with focus on the first. The AFPs are a key element for freeze avoiding insects during cold periods, but the insects have several other adaptions that are important as well, which are explained in detail as well. Also, the types and structures of the most studied insect AFPs are presented, as well as the prevalence of AFPs throughout several insect groups.

The RIPs and INPs are relevant for freeze-tolerant organisms, where they make sure the nucleation occur. Thermal hysteresis is achieved by several adaptions besides AFPs in the insects, and the degree of thermal hysteresis is much larger than seen in fish. The RIPs and INPs are found in freeze-tolerant insects, which are also covered in this chapter due to their ice-binding capabilities.

In Chap. 7, Prof. Dr. Michael Wiesniewski et al. present the current investigation on antifreeze proteins in plants. Plant AFPs are not as potent as other AFPs, as they produce only small thermal hysteresis gaps, and their main purpose is to inhibit ice recrystallization. For this reason, they are often just referred to as ice-binding protein or recrystallization inhibitors. The chapter covers the cold damage that plants are subjected to, and how and where the ice-binding proteins evoke their protection, from higher plants to algal species, as well as other adaptations improving cold hardiness. Transgenic crops with higher cold hardiness are a "hot potato" on the field, and the construction of these organisms is discussed in the chapter as well.

In Chap. 8, Prof. Dr. John G. Duman and Dr. Samuel S. Newton present the current knowledge of all species not included in fish, insects, or plants, where antifreeze proteins, or generally ice-binding proteins, are found. The chapter goes in-depth with the groups and species that express ice-binding proteins, covering potency, tissue localization, structure, and special characteristics to the extent it has been investigated. The species expressing ice-binding proteins include arthropods, mollusks, nemerteans, cestodes, sponges, nematodes, and even vertebrates, as well as fungi and archaea. The proteins are found in virtually every taxonomic family inhabiting the cold regions of the world, and certainly in many more species than we have uncovered so far.

In Chap. 9, Prof. Dr. Chris Cheng and Dr. Xuan Zhuang dive into the evolution of antifreeze proteins. The antifreeze proteins are very interesting in regards to evolution, as they have evolved convergently in several families and species.

The chapter explains the different ways new genes arise and the significance of the selective pressure. The research that has been made on the evolution of antifreeze proteins mainly concerns fish antifreeze proteins, which are also the focus of this chapter, detailed down to the molecular processes of the evolution, and discussing the evidence on a genetic level. The selective pressure driving the evolution of fish antifreeze proteins is presented, followed by the different occurrences of fish antifreeze protein evolution which not surprisingly is seen between the different AFP types, but also even within the same AFP type found in different species.

Part I
Ice Formation

Chapter 3
Ice and Its Formation

Amir Haji-Akbari

3.1 Introduction

Freezing of water into ice is a ubiquitous phenomenon that has fascinated mankind for millennia. In addition to biological cells (Padayachee et al. 2009), ice formation can occur in a wide variety of other environments, including atmospheric clouds (Baker 1997; Carslaw et al. 2002), soil (Chamberlain and Gow 1979), freshwater resources (Beltaos and Prowse 2009), aircraft (Potapczuk 2013), and aviation and transportation infrastructure (Ye et al. 2013). The ubiquity and relative ease of freezing water have, since antiquity, been used in applications such as traditional medicine (Yang and Mochizuki 2003), transportation (Li et al. 2013a), mummification (Cockburn et al. 1998), and food preservation (Varshney and Singh 2015) and processing (Arbuckle 1986). Ice formation also plays a pivotal role in many natural processes, such as cloud microphysics (Fowler et al. 1996; Herbert et al. 2015). For instance, freezing of supercooled microdroplets of water in atmospheric clouds triggers a cascade of events that eventually result in precipitation. Predicting the spatiotemporal distribution of ice droplets within a cloud is, therefore, an important—and yet challenging—component of climate modeling (Wilson and Ballard 1999).

Freezing is usually disruptive in biological systems, and can result in cell death. Not only can ice particles damage cell membranes (Muldrew and McGann 1990, 1994), but they can also result in cold denaturation of proteins (Privalov 1990). A large portion of this book is dedicated to understanding how antifreeze proteins (AFPs) prevent and/or control ice formation within cells and organisms. In order to understand how AFPs function, however, it is first necessary to have a fundamental

A. Haji-Akbari (✉)
Department of Chemical & Environmental Engineering, Yale University, New Haven, CT, USA
e-mail: amir.hajiakbaribalou@yale.edu

H. Ramløv, D. S. Friis (eds.), *Antifreeze Proteins Volume 1*,
https://doi.org/10.1007/978-3-030-41929-5_3

understanding of the microscopic structure of ice, and the thermodynamics and kinetics of its formation.

Considering the ubiquity of freezing and its important ramifications in many disciplines, the question of ice formation has been extensively studied, and this current chapter offers only a brief introduction. In this context, the term "ice" refers to any condensed phase of water that exhibits mechanical properties of a solid. This notion encompasses both *crystalline ices*, i.e., structures in which the oxygen atoms of constituent water molecules are arranged into periodic lattices, and *amorphous ices*, or kinetically arrested glassy states of liquid water. Like almost everything else about water, crystalline and glassy ices exhibit various interesting anomalies some of which will be discussed in this chapter.

This chapter is organized as follows. Section 3.2 is dedicated to a brief discussion of the thermodynamics of crystallization. Section 3.3 discusses ice anomalies, including water's complicated phase diagram and its ability to form open-network crystals. Section 3.4 is dedicated to ice I, the thermodynamically stable crystalline phase of water under ambient conditions, and discusses the thermodynamics and kinetics of its formation in pure water and aqueous solutions. The chapter is concluded with a discussion of amorphous ices in Sect. 3.5.

3.2 Physics of Crystallization

Before discussing ice, its anomalies, and the thermodynamics and kinetics of its formation, it is useful to provide a quick precept of some basic thermodynamic concepts that will be frequently referred to in this chapter. According to classical thermodynamics, a system will reach thermodynamic equilibrium by minimizing its *free energy*. The precise notion of what constitutes free energy, i.e., the particular thermodynamic function that needs to be minimized at equilibrium, depends on the thermodynamic *ensemble* or the set of thermodynamic variables controlled during an experiment. In most experiments, it is the temperature and pressure that are kept constant, and therefore the quantity that needs to minimize at equilibrium is the *Gibbs free energy*, G, which is related to internal energy E, volume V, and entropy S by:

$$G = E + PV - TS \tag{3.1}$$

Here, P and T are pressure and temperature, respectively. Therefore, under isothermal isobaric conditions, a crystal becomes more stable when its molar Gibbs free energy—also known as *chemical potential*, μ—is smaller than that of its competitors.

Crystallization is a *first-order phase transition* in the sense that it involves discontinuities in first derivatives of Gibbs free energy, such as enthalpy, H, and volume. For instance, a latent heat of fusion is released during crystallization, which is due to the fact that the solid has a lower enthalpy than the liquid. Similarly, the crystal and the liquid of the same substance usually have different densities. Despite this discontinuity in first derivatives, a crystal and a liquid can coexist under conditions at which their chemical potentials are equal. The loci of temperatures and pressures at which this can happen are called the *coexistence line*, with its slope given by the *Clausius–Clapeyron* equation:

$$\frac{dP}{dT} = \frac{\Delta h_{\mathrm{m}}}{T\Delta v_{\mathrm{m}}} \tag{3.2}$$

Here, Δh_{m} is the latent enthalpy of fusion and Δv_{m} is the difference between specific volumes of liquid and the solid. Each point on a solid–liquid coexistence line is technically a *melting point*, but the term usually refers to the coexistence temperature at ambient pressure, i.e., at 1 bar. The temperature and pressure at which a solid–liquid coexistence line intersects the solid–gas and liquid–gas coexistence lines are called a *triple point*. Also, coexistence lines that separate two disordered phases, such as gas and liquid, usually end in a *critical point* at which the transition between the two phases becomes *second order* and the macroscopic distinction between them, e.g., in terms of density and enthalpy difference, disappears. Beyond the critical point, a transformation between the two phases will become continuous, i.e., without any discontinuities in thermodynamic functions.

3.3 Anomalies of Crystalline Ices

Water is a material with numerous anomalous properties, the origins of which are far from fully understood (Debenedetti 2003; Gallo et al. 2016). In order to understand anomalies associated with crystalline ices, it is first necessary to discuss what we expect from the crystallization of a typical single-component liquid. For most materials, liquid–solid transition will result in the formation of a *single* thermodynamically stable crystalline solid, which will always be *denser* than the liquid at coexistence. Therefore, according to Eq. (3.2), the coexistence line will have a positive slope. From a molecular perspective, this behavior emanates from the fact that intermolecular interactions between most small molecules are nonspecific and nondirectional in nature. Therefore, only the densely packed crystalline structures can be energetically favored, and since there are not very many such structures (Hales 2006; Damasceno et al. 2012), the phase diagram will be simple, with one—or a handful of—crystalline phases only.

Water is a material that deviates from this generic picture as it not only can exist in several distinct crystalline forms, but it can also form crystals that are less dense than the liquid. These two anomalies will be discussed at length in Sects. 3.3.1 and 3.3.2, respectively. A simple conceptual framework to understand them, however, is to note the importance of hydrogen bonding in water. As strong and highly directional specific interactions, hydrogen bonds contribute so strongly to the internal energy of condensed phases of water that such phases can only be stable if their constituent molecules form as many hydrogen bonds with their neighbors as possible. As will be discussed later, this constraint can be satisfied by a wide variety of permitted—including some open network—crystalline structures. This is unlike simple liquids, which, as discussed above, mostly form densely packed crystals.

3.3.1 Polymorphism in Water

In materials science, *polymorphism* refers to a material's ability to form distinct crystalline structures, i.e., structures with distinct sets of point symmetry operations. Each such structure is typically called a *polymorph*. As explained above, most single-component systems only form one thermodynamically stable polymorph, and therefore have simple phase diagrams with one solid/liquid/gas triple point and one liquid/gas critical point. Water is a notable exception, as evidenced by its very complex phase diagram (Fig. 3.1), and has 18 experimentally observed crystalline polymorphs, denoted by Roman numerals according to their chronological order of discovery. Some of these polymorphs, including five thermodynamically

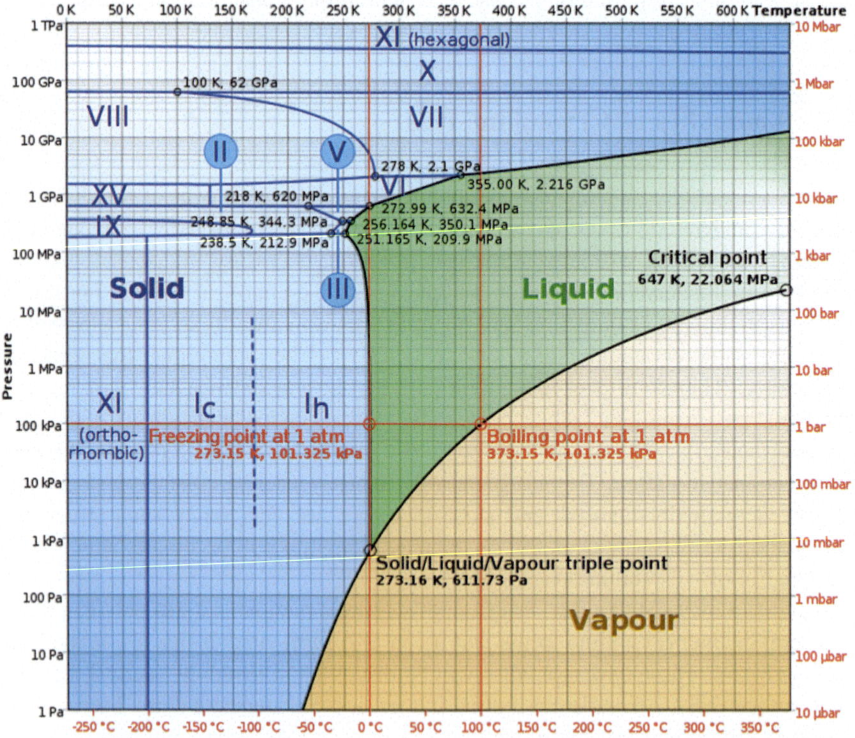

Fig. 3.1 Equilibrium phase diagram of water. Note that ice can exist in many different polymorphs. Reproduced from Zhang et al. (2015)

stable ones, can directly form within—and coexist with—liquid water, while the remaining ones are only accessible through solid–solid transformations. Due to this rich polymorphism, the phase diagram of water has five solid–solid–liquid (Bridgman 1912; Schwager et al. 2004; Goncharov et al. 2005, 2009), eight solid–solid–solid (Bridgman 1912; Kauzmann and Eisenberg 1969; Fletcher 1970; Mercury et al. 2001; Song et al. 2003; Salzmann et al. 2009; Yen and Chi 2015), and one gas–solid–solid (Petrenko and Whitworth 1999) triple points, in addition to the generic gas–liquid–solid triple point commonly observed for other materials. Despite their structural differences, all these crystalline ices are similar in the sense that every water molecule in them forms exactly four hydrogen bonds with its neighbors, two as a proton donor and two as a proton acceptor (Zheligovskaya and Malenkov 2006). These constraints are typically referred to as *Bernal Fowler rules* or *ice rules* (Bernal and Fowler 1933).

Ice I is the most common polymorph of ice, and forms at ambient pressures. It has two polymorphs that are stacking variants of one another: the thermodynamically stable *hexagonal ice* (I_h) and the thermodynamically metastable *cubic ice* (I_c). The structure of ice I and the thermodynamics and kinetics of its formation will be discussed in detail in Sect. 3.4. Other ice polymorphs can only form under non-ambient conditions, such as very high pressures, and/or very low temperatures. Historically, the first non-ambient polymorph of ice is ice II, which was discovered in the beginning of the twentieth century, and has a rhombohedral unit cell composed of 12 water molecules (Kamb 1964). Ice II can be obtained from ice I at 198 K and 300 MPa (Fuentes-Landete et al. 2015). Since then, 16 more ice polymorphs have been discovered. The most recent polymorphs, ice XVII and XVIII, were only discovered in 2016 (del Rosso et al. 2016a, b) and 2019 (Millot et al. 2019), respectively. Ice polymorphs span a wide range of densities (from 0.85 g cm^{-3} in the case of ice XVII (del Rosso et al. 2016a, b) to 2.785 g cm^{-3} in the case of ice X (Hemley et al. 1987) and ~3 g cm^{-3} in the case of ice XVIII (Millot et al. 2019)) and have unit cells with as few as two molecules—e.g., in the case of ice VII (Kamb and Davis 1964) and ice X (Hemley et al. 1987)—to 136 molecules—e.g., in the case of ice XVI (Falenty et al. 2014).

Polymorphism in water is very rich and unprecedented even in comparison to other tetrahedral liquids such as silicon (Jones and Stevanovíc 2017), which also have multiple polymorphs. One important factor that contributes to this richness is the asymmetric nature of a hydrogen bond. In other words, a hydrogen atom that bridges two oxygen atoms is covalently bonded—and is, therefore, closer—to one of those oxygens. This asymmetry allows for degeneracy in assigning protons to oxygen that are located on a periodic lattice, with the number of all possible assignments scaling exponentially with the number of water molecules in the lattice.[1] There are usually small energetic differences between these distinct arrangements, but such differences are usually too small for any particular arrangement to be

[1]Note that a vanishingly small fraction of such arrangements will have nonzero electrostatic dipole or quadrupole moments. Such structures will have a vanishingly small probability of emerging in a proton-disordered phase.

favored unless at very low temperatures. Therefore, ice polymorphs at higher temperatures are usually *proton-disordered*, and the entropy associated with such proton disorder lowers their free energy and enhances their thermodynamic stability. When temperatures are sufficiently low, however, the entropic gain associated with proton disorder will become smaller, and eventually a transition into a proton-ordered phase will take place. Without such proton disorder–order transitions, far fewer ice polymorphs would have existed, as twelve of the eighteen known polymorphs come in pairs that only differ in their proton order (Fuentes-Landete et al. 2015). For instance, ice I transforms into its proton-ordered counterpart, ice XI, at 72 K (Fukazawa et al. 1998). Ice X is the only polymorph in which this hydrogen bonding asymmetry is broken, and each proton is equally spaced between the two oxygens that it bridges (Hirsch and Holzapfel 1984).

In addition to these experimentally observed polymorphs, there are many more polymorphs whose existence has been predicted using computational methods such as density functional theory (Gross and Dreizler 2013) or classical force fields (Guillot 2002; Vega and Abascal 2011). Such polymorphs are typically known as *computer ices*, and are usually difficult to isolate in experiments. One important class of computer ices constitutes those that are predicted to form at experimentally inaccessible pressures. The most notable example is *metallic ice*, which is predicted to form at pressures above 4.8 TPa. Metallic ice has the crystallographic space group C_2/m, and its electrons are delocalized like a metal (Hermann et al. 2012). [A possible experimental candidate for metallic ice has only been observed recently, though its electronic properties vis-a-vis the diffusion of protons are yet to be confirmed (Millot et al. 2019).]

Not all computer ices are expected to form at high pressures. For instance, ice 0, which was recently proposed as a precursor for the formation of ice I, has an open tetragonal structure with a unit cell composed of 12 water molecules (Russo et al. 2014). In principle, any crystalline structure that is tetrahedral, satisfies ice rules, and is mechanically stable, can at least be thermodynamically metastable. As to which of these computer ices can form in reality is an open question that can only be addressed experimentally.

The polymorphs discussed so far can form—or be stable—in the bulk. In ultra-confined geometries, such as nanotubes and nanopores, however, water's hydrogen bond network can be disrupted, which can, in turn, culminate in the formation of novel ice polymorphs. Such polymorphs are typically called *two-dimensional ices* and can only exist under the geometrical constraints induced as a result of confinement. Historically, such two-dimensional ices were originally discovered in molecular simulations, but the existence of some of them has been recently confirmed in experiments (Zhao et al. 2014b). 2D ices can exist in many different forms, including monolayers (Zangi and Mark 2003; Kumar et al. 2005; Bai et al. 2010; Kaneko et al. 2013; Zhao et al. 2014a; Chen et al. 2016; Corsetti et al. 2016; Barati Farimani and Aluru 2016), bilayers (Koga et al. 1997; Kastelowitz et al. 2010; Johnston et al. 2010; Algara-Siller et al. 2015; Zhu et al. 2016; Chen et al. 2017), and ice nanotubes (Koga et al. 2001; Bai et al. 2006; Takaiwa et al. 2008). Unconventional 2D ice polymorphs can, in theory, form in biological systems due to the presence of ultra-confined geometries within biological cells. Such possibilities have, however, not been thoroughly investigated.

3.3.2 Formation of Open-Network Crystals

Another interesting consequence of hydrogen bonding in water is its ability to form open-network crystalline structures, or phases that are less dense than the liquid. This rather striking feature is only shared by a handful of other small molecules, e.g., Group IV elements such as carbon and germanium (Brygoo et al. 2007; Jayaraman et al. 1963). Open-network ices typically reside in the low-P region of the phase diagram. Among thermodynamically stable ices, only ice I has an open structure and is $\approx 8\%$ less dense than liquid water at coexistence. The remaining four ice polymorphs that can coexist with liquid water, i.e., ices III, V, VI, and VII, are all denser than the liquid.

The fact that ice I is less dense than liquid water has important physiological consequences. First of all, one of the reasons that makes ice formation so disruptive to biological cells is the volumetric expansion of liquid water upon freezing. Therefore, ice particles that form intracellularly or extracellularly will occupy a larger volume than the liquid and can, therefore, exert mechanical stress on cell membranes (Meryman 1956). Ecologically, however, this density anomaly is critical to the sustenance of life in lakes and rivers during cold winters, since such freshwater resources freeze from the top and the marine life can still survive within the liquid entrapped below the ice cap.

Another important consequence of this density anomaly is what is known as *pressure melting*. Indeed, for a liquid that is denser than its crystal, such as liquid water and ice I, $\Delta v_m < 0$ and according to Eq. (3.2), the solid–liquid coexistence line will have a negative slope. This will imply that ice I can become unstable upon both heating and pressurization. At temperatures above 251 K, ice will eventually melt at some pressure below ~210 MPa (Fuentes-Landete et al. 2015). Pressure–melting of ice is believed to play a pivotal role in glacier mobility (Weertman 1957).

Ice I is not the only open-network crystal of pure water. In recent years, several other such polymorphs have been discovered. These ices are metastable, can usually form at negative pressures only, and are obtained by exposing two-component crystals of water and small gaseous molecules to vacuum. For instance, ice XVI is obtained by vacuum pumping a clathrate hydrate of neon (Falenty et al. 2014). Gas hydrates are crystalline solids in which water molecules form cages that encompass small guest molecules such as methane, carbon dioxide, hydrogen, and neon (Maslin et al. 2010; Knott et al. 2012), and will be discussed in further detail in Chap. 12 of Vol. 2. At ambient pressures, such cages will collapse in the absence of guest molecules. Applying negative pressure to such hydrates, however, can generate a driving force for the removal of the guest molecules and the subsequent formation of *guest free clathrates*. Ice XVI is the first experimentally observed guest-free clathrate of water, even though guest-free clathrates had been previously observed in other systems such as germanium (Guloy et al. 2006). Not all open-network ice polymorphs are obtained from applying negative pressures to clathrate hydrates. For instance, ice XVII (del Rosso et al. 2016a, b) is the guest-free variant of a two-component crystal composed of water and hydrogen, which has a quartz-like structure (Strobel et al. 2016).

3.4 Ice I: The Ambient and Biologically Relevant Form of Ice

As mentioned above, ice I is the only ice polymorph that forms at ambient pressures, and is, therefore, the only form of crystalline ice relevant to biological systems. This section is dedicated to a discussion of its structure and the thermodynamics and kinetics of its formation in pure water and aqueous solutions.

3.4.1 Structure of Ice I

As mentioned in Sect. 3.3.1, every water molecule in a crystalline ice forms hydrogen bonds with four of its neighbors in accordance with Bernard–Fowler rules. In ice I, each molecule has exactly four neighbors in its first hydration shell, and forms hydrogen bonds with all of them. These nearest neighbors are all arranged tetrahedrally around the central molecule with an average O–O distance of 0.276 nm. As will be discussed in Sect. 3.5, this local tetrahedral arrangement is not unique to ice I and can exist in amorphous ices as well. What distinguishes ice I, however, is a hierarchy of structures that culminate in a periodic lattice with long-range translational order. There are different ways of describing this hierarchy, two of which will be discussed here.

The first approach is based on viewing ice I as a layered structure with water molecules within each layer organizing into hexagons. The two-dimensional projection of each such layer will thus be the *honeycomb* lattice (Fig. 3.2a, e). Due to the tetrahedral nature of hydrogen bonding, however, these layers are not flat, and their constituent hexagons adopt chair-like configurations. In Fig. 3.2b, f, the oxygens that are above and below the average plane of a hexagon are depicted in purple and green, respectively. Among the four hydrogen bonds that each molecule forms, three are with molecules that are within the same layer, while the fourth is with a molecule in an adjacent layer. This constraint is, however, not sufficient for uniquely determining the arrangement of adjacent layers, and ice I can, therefore, exist in different stacking variants, also known as *polytopes*.

The most commonly known ice I polytope is *hexagonal ice*, which is denoted by I_h and has the space group $P6_3/mmc$ (Fig. 3.2a–d). In hexagonal ice, adjacent layers are mirror images of one another, and each upward oxygen (purple) is adjacent to a

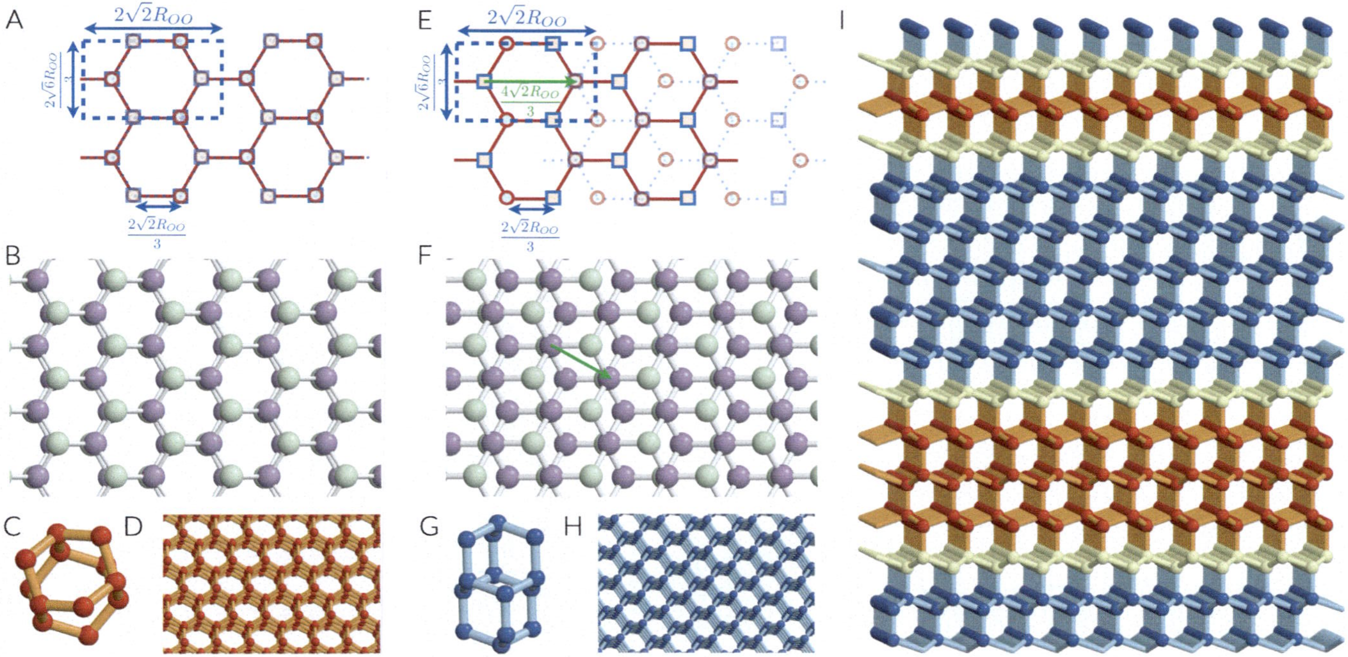

Fig. 3.2 Polymorphs of ice I. (**a–d**) Hexagonal ice, I_h, in which adjacent layers are reflected with respect to one another, with their up (square) and down (circular) oxygens sitting on top of one another (**a, b**). This culminates in a layered structure (**d**) composed of hexagonal cages (**c**). (**e–h**) Cubic ice, I_c, in which adjacent layers have the same orientation, but are shifted with respect to one another by $4\sqrt{2}R_{OO}/3$. (**e, f**) The corresponding layered structure (**h**) is composed of double-diamond cages (**g**). (**i**) Stacking disordered ice, with hexagonal and cubic layers depicted in red and blue, respectively. Stacking faults are shown in light yellow

downward oxygen (green) in the layer above it (Fig. 3.2b). Ice I_h, therefore, has a stacking sequence of $ABAB\cdots$, with A and B corresponding to layers related to one another via reflection. The second idealized polytope of ice I is cubic ice (Fig. 3.2e–h), which is denoted by I_c and has the space group $Fd3m$ (König 1943). In I_c, adjacent layers are not mirror images of one another but are instead related to one another via a translation by a distance $4\sqrt{2}r_{OO}/3$. Since each layer will be recovered after three such translations, I_c has a stacking sequence of $ABCABC\cdots$. Both cubic and hexagonal ices are proton disordered, with a zero-point entropy of 3.41 J mol^{-1} K^{-1} (Kolafa 2014).

These two polytopes can also be understood in terms of their constituent topological building blocks (Haji-Akbari and Debenedetti 2015), as hexagonal and cubic ices are composed of hexagonal and double-diamond cages, respectively. Hexagonal cages (HCs) (Fig. 3.2c) are anisotropic motifs composed of five hexagons. Two of these hexagons participate in stacking layers, have chair like conformations, and are known as *basal* faces, while the remaining three have boat-like conformations and are referred to as *prismatic* faces. Double-diamond cages (DDCs) (Fig. 3.2g) are, however, structurally compact and are composed of six identical chair-like hexagons surrounding an internal chair-like hexagon that participates in the stacking layer. A DDC can, therefore, be viewed as two tetrahedra sitting on top of one another, and rotated with respect to each other by 60°.

Since these two polytopes differ in their stacking sequence only, their lattice energies are almost identical. Hexagonal ice is, however, thermodynamically more stable (Handa et al. 1986), a fact that has recently been attributed to its lower anharmonic nuclear vibrational energy (Engel et al. 2015). Cubic ice, which is thermodynamically metastable, has never been isolated in its single crystalline form. It is instead an idealized representation of ices that form at very low temperatures (Dowell and Rinfret 1960; Kohl et al. 2000; Murray et al. 2005; Shilling et al. 2006) and/or in ultra-confined geometries such as nano-droplets (Johari 2005) and nano-pores (Steytler et al. 1983; Morishige and Uematsu 2005; Morishige et al. 2009). Such metastable ices are usually a mixture of cubic and hexagonal stacks with a high ($\approx$60–70%) fraction of cubic stacks, and are collectively referred to as *stacking disordered ice* I_{sd} (Fig. 3.2i). Geometrically, I_{sd} can be constructed in a layer-by-layer fashion by randomly applying reflection or translation to any new layer that is added on top of the existing layers. The importance of cubic ice and stacking disorder in ice nucleation will be discussed in detail in Sect. 3.4.2. The kinetics of transformation from I_{sd} into the thermodynamically stable I_h is, however, very slow at low temperatures (Murray et al. 2005; Gainaru et al. 2018). Further information about stacking disordered ice can be found in a review by Malkin et al. (2015).

3.4.1.1 Crystallography of Hexagonal Ice

Considering the importance of I_h in biological systems, it is worthwhile to discuss its crystallography in further detail. I_h belongs to the hexagonal family of crystals, and has two crystallographic axes (Fig. 3.3a). Its first major axis is denoted by c and is perpendicular to the basal faces of HCs. Its second major axis—denoted by a—lies on a plane perpendicular to c and points toward the vertices of individual hexagons. Note that there are three equivalent directions for a, denoted by $\mathbf{a}_1$, $\mathbf{a}_2$, and $\mathbf{a}_3$ in Fig. 3.3a. These three degenerate vectors, along with $\mathbf{c}$, uniquely determine the unit cell of hexagonal ice. Using such a mathematically overdetermined basis set might seem strange at first, but has important mathematical advantages explained in Allen (2010).

In principle, it is possible to define countably infinite number of distinct crystallographic planes of any crystal by constructing integer valued linear combinations of its basis set. Most such planes will be experimentally uninteresting, either due to their high surface energy, or because of their similarity to simpler crystallographic planes with smaller integer indices. Hexagonal ice has three major crystallographic planes. The *basal plane* (or the 0001 plane) is perpendicular to the c-axis and correspond to basal faces of HCs (Fig. 3.3b). The basal plane is the lowest energy plane of I_h and has a surface enthalpy of 5.57 μJ cm^{-2} (Shultz et al. 2015). The *primary prismatic plane*—or the $01\bar{1}0$ plane—is perpendicular to the a-axis (Fig. 3.3c), and has a surface enthalpy of 5.95 μJ cm^{-2} (Shultz et al. 2015), while

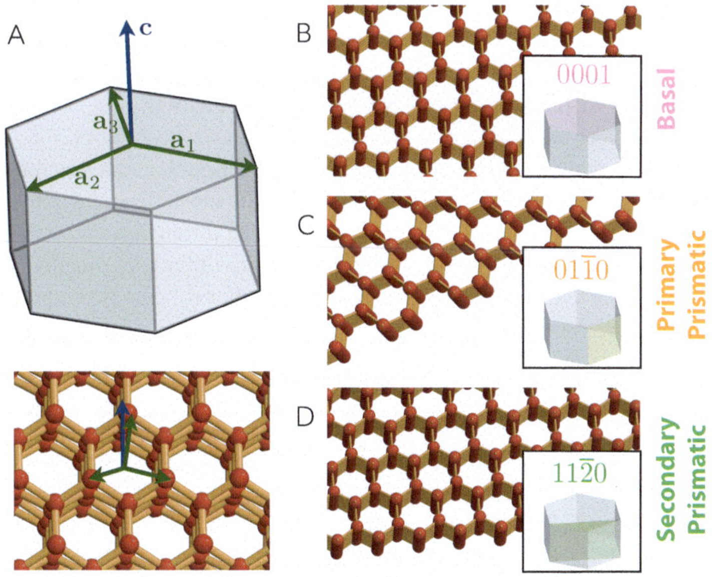

Fig. 3.3 Major crystallographic planes of hexagonal ice. (**a**) Major crystallographic axes overlaid on the structure of I_h. Note that there are three equivalent a axes. (**b**) Basal, (**c**) primary prismatic and (**d**) secondary prismatic planes

the *secondary prismatic plane*—or the $11\bar{2}0$ plane—is perpendicular to a prismatic face and has a surface enthalpy of 6.90 μJ cm^{-2} (Shultz et al. 2015). These crystallographic planes are important in understanding the function of antifreeze proteins, as different AFPs can preferentially attach to different crystallographic planes of ice, and affect the mechanism of ice formation—or lack thereof—accordingly (Olijve et al. 2016; Haji-Akbari 2016).

3.4.2 *Thermodynamics and Kinetics of Ice Formation in Pure Water*

3.4.2.1 Thermodynamics of Ice Formation

In Sect. 3.2, a brief overview is provided on the basics of the thermodynamics of crystallization, and concepts such as melting point and triple point are introduced. Table 3.1 summarizes the thermodynamic properties of ice I and its corresponding liquid–solid transition. Ice I has a melting temperature of $T_m = 273.15$ K, and has an enthalpy of fusion of 1.436 kcal mol^{-1} at T_m. The gas–liquid–solid triple point of water is at 273.16 K and 61 Pa. Note that the triple point temperature is slightly higher than T_m due to the negative slope of the ice I–liquid coexistence line. Also, water has an unusually high melting point in comparison to similar molecules, a fact that has been attributed to the dominant role that hydrogen bonding plays in condensed phases of water.

3.4.2.2 Kinetics of Ice Formation: Nucleation and Growth

According to classical thermodynamics, a *supercooled* liquid, i.e., a liquid below its equilibrium melting temperature, is thermodynamically unstable and will eventually freeze into its respective crystal. But the fact that such a transformation is thermodynamically favored does not imply that it will occur within experimentally accessible timescales. Indeed, supercooled water—or any other supercooled liquid for that

Table 3.1 Thermodynamic and structural properties of hexagonal ice (Petrenko 1993)

Property	Value
Density	917 kg m^{-3}
Melting point	273.15 K
Latent heat of fusion	1.436 kcal mol^{-1}
Heat capacity	8.79 cal mol^{-1} K^{-1}
Thermal conductivity	0.234 W m^{-1} K^{-1}
Solid/liquid surface energy	33 mJ m^{-2}
Solid/vapor surface energy	109 mJ m^{-2}
Triple Point (ice I/liquid/gas)	273.16 K, 61 Pa
Dielectric permittivity	95

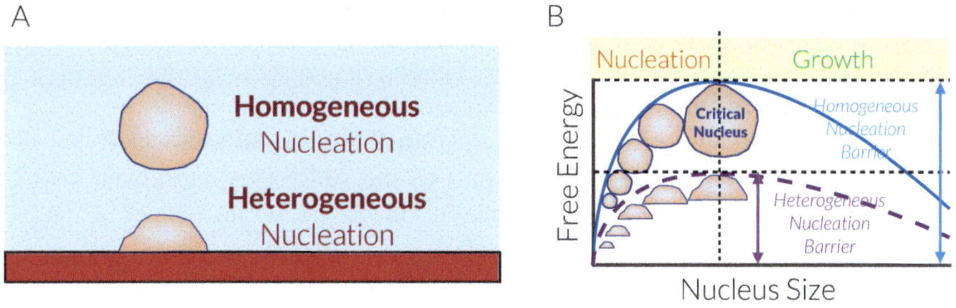

Fig. 3.4 (**a**) Ice can nucleate either homogeneously (e.g., in pure water in the absence of a facilitating surface) or heterogeneously in the vicinity of an interface. (**b**) Schematic free energy of the system as a function of nucleus size for both homogeneous and heterogeneous nucleation

matter—can remain liquid, sometimes for very long times. Such a delay arises from the fact that crystallization—like any other first-order phase transition—proceeds through a mechanism known as *nucleation and growth* (Fig. 3.4). Nucleation is an activated process and involves the formation of a sufficiently large crystalline nucleus within the supercooled liquid, while growth refers to the subsequent enlargement of that nucleus. The activated nature of nucleation arises from the free energetic penalty associated with the formation of a solid–liquid interface. This penalty, which exists due to structural mismatch between an ordered crystal and a disordered liquid, scales linearly with the surface area of the crystalline nucleus, and is eventually overcome by the thermodynamic driving force for crystallization, which scales linearly with the volume of the nucleus. For small nuclei, however, this surface contribution is dominant and makes nuclei that are smaller than a critical size thermodynamically unstable. This makes the formation of a critical nucleus a stochastic activated process unless its free energy of formation is comparable to thermal fluctuations in the system.

The qualitative argument provided above is made more rigorous in *classical nucleation theory (CNT)*. Proposed in the early twentieth century (Volmer and Flood 1934; Becker and Döring 1935), CNT is the most widely used theoretical framework for crystal nucleation (Oxtoby 1992). In the simplest variant of CNT, it is assumed that crystalline nucleus is spherical in shape, and the properties of the solid and the liquid are bulk-like and are not affected by the presence of an interface. The free energy of formation of a crystalline nucleus composed of N molecules will then be given by (Debenedetti 1996):

$$G(N) = -\frac{N|\Delta\mu|}{N_\mathrm{A}} + \left(\frac{36\pi}{\rho_\mathrm{s}^2}\right)^{1/3} \sigma_\mathrm{ls} N^{2/3} \quad (3.3)$$

Here, $\Delta\mu$ is the chemical potential difference between the liquid and the solid, N_A is the Avogadro number, and ρ_s and σ_ls are the solid number density and the solid–liquid surface tension, respectively. Equation (3.3) has a maximum at:

$$N_{\mathrm{c}} = \frac{\sigma_{\mathrm{ls}}}{|\Delta\mu|}\left(\frac{32\pi}{3\rho_{\mathrm{s}}^2}\right)^{2/3} \tag{3.4}$$

The free energy of formation for a nucleus with N_c molecules is called the *nucleation barrier* and is given by:

$$\Delta G_{\mathrm{nuc}} = G(N_{\mathrm{c}}) = \frac{N_{\mathrm{c}}}{2N_{\mathrm{A}}}|\Delta\mu| = \frac{16\pi\sigma_{\mathrm{ls}}^3 N_{\mathrm{A}}^2}{3\rho_{\mathrm{s}}^2|\Delta\mu|^2} \tag{3.5}$$

Due to this non-monotonicity in $G(N)$, forming a critical nucleus is an uphill battle thermodynamically. Assuming that a steady-state nucleus size distribution is established at timescales much shorter than the nucleation time, it can be demonstrated that a fluctuation resulting in barrier crossing will have an occurrence probability proportional to $\exp[-\Delta G_{\mathrm{nuc}}/kT]$ with k the Boltzmann constant. The *nucleation rate*, or the expected number of nucleation events per unit time per unit volume, will then be given by:

$$R = \frac{24\rho_{\mathrm{liq}} D N_{\mathrm{c}}^{3/2}}{l^2}\left(\frac{|\Delta\mu|}{6\pi kTN_{\mathrm{c}}}\right)^{\frac{1}{2}} \exp\left[-\frac{\Delta G_{\mathrm{nuc}}}{kT}\right] \tag{3.6}$$

with ρ_{liq} and D the density and the self-diffusivity of the liquid, and l the atomic jump distance in the liquid. Note that the stochastic nature of nucleation will imply that the nucleation time τ will be a stochastic random variable and will have a mean given by $\langle\tau\rangle = 1/(RV)$, with V the volume of the supercooled liquid. It is generally assumed that nucleation is a Poisson process[2] and nucleation time is exponentially distributed (Koop 2004). Upon the formation of a critical nucleus, however, there is no longer any thermodynamic resistance for its further growth, and the growth timescale is governed by the pace of structural relaxation in the system.

Depending on the magnitude of the nucleation barrier, either nucleation or growth can be the rate-limiting step in crystallization. When the nucleation barrier is very large, i.e., when $\Delta G_{\mathrm{nuc}} \gg kT$, crystallization will be *nucleation limited* and nucleation times will be orders of magnitude larger than growth times. For $\Delta G_{\mathrm{nuc}} \approx kT$, however, crystallization will be *growth limited* as a large number of nucleation events will occur simultaneously in the system, and the crystallization timescale will be determined by the intrinsic dynamics of the system, most importantly by its self-diffusivity.

In order to identify the conditions under which crystallization is nucleation limited or growth limited, it is first necessary to examine how $|\Delta\mu|$ and σ_{ls} change with thermodynamic variables such as temperature. In general, solid–liquid surface tension is extremely difficult to measure in the deeply supercooled regime, and even

[2]To be more precise, it is the number of nucleation events per unit time within an ensemble of independent nucleation experiments, e.g., isolated microdroplets, that is a Poisson process.

though it is known that σ_{ls} is an increasing function of temperature, its precise dependence on T is not known due to technical difficulties of measuring it far from equilibrium (Gŕańasy et al. 2002). $|\Delta\mu|$, however, can be rigorously calculated from integrating the equations of state of the supercooled liquid and the crystal:

$$|\Delta\mu| = T\int_{T}^{T_m} \frac{\left|h_{I_h} - h_{liq}\right|}{\overline{T}^2} d\overline{T} \tag{3.7}$$

with h_{I_h} and h_{liq} the molar enthalpies of ice I_h and the supercooled liquid, respectively. Assuming that molar enthalpies are not strong functions of temperature, it can be easily shown that:

$$\Delta\mu = \Delta h_m \left[1 - \frac{T}{T_m}\right] \tag{3.8}$$

where Δh_m is the enthalpy of fusion at T_m. Combining Eqs. (3.5) and (3.8) yields:

$$\frac{\Delta G_{nuc}}{kT} = \frac{16\pi\sigma_{ls}^3 N_A^2 T_m^2}{3kT\rho_s^2 \Delta h_m^2 (T_m - T)^2} = \frac{B}{kT(T - T_m)^2} \tag{3.9}$$

Therefore, the nucleation barrier depends on $\Delta T = T_m - T$, which is typically referred to as *the degree of supercooling*.[3] Equation (3.9) predicts that ΔG_{nuc} is a decreasing function of ΔT, an assertion that is generally true even if Eq. (3.8) is not valid. Therefore, nucleation rate should increase exponentially upon decreasing temperature. Such an increase in rate is, however, eventually offset by the slowdown of intrinsic dynamics at low temperatures. Assuming that transport properties such as self-diffusivity have an Arrhenius-type dependence[4] on temperature, i.e., $D \sim \exp[-E_a/kT]$, the overall crystallization rate will be given by:

$$R = A\,\exp\left\{-\frac{1}{kT}\left[\underbrace{E_a}_{\text{diffusion}} + \underbrace{\frac{B}{\Delta T^2}}_{\text{nucleation}}\right]\right\} \tag{3.10}$$

Therefore, nucleation-limited crystallization will occur at small supercoolings where the rate of increase in $\exp[-\Delta G_{nuc}/kT]$ is much faster than the slowdown in D.

[3] An alternative definition for the degree of supercooling is $\theta = T/T_m$. See Gianetti et al. (2016) for discussion.

[4] For strong liquids, D has an Arrhenius-like dependence on temperature. In fragile liquids, however, D drops much more abruptly as temperature decreases. Depending on the extent of supercooling water can be strong or fragile. In the nucleation-limited regime, i.e., at temperatures close to T_m, water behaves as a fragile liquid, while amorphous ices act as strong liquids. Therefore, it has been speculated that a fragile-to-strong transition occurs in the no man's land.

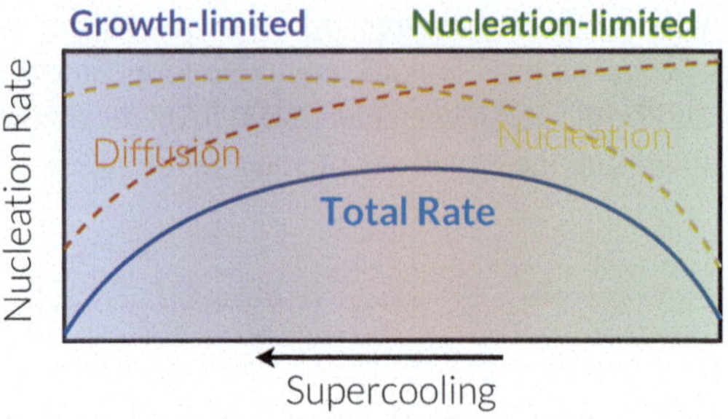

Fig. 3.5 Schematic representation of the dependence of the rate of ice nucleation on temperature. Nucleation- and growth-limited regimes are depicted in green and blue, respectively

Growth-limited crystallization, however, will take place under conditions at which Eq. (3.10) is dominated by the E_a/kT term, i.e., at large supercoolings. In the middle, there will always be a temperature at which the nucleation rate will be the largest. Figure 3.5 depicts the schematic dependence of nucleation rate on ΔT, with the nucleation- and growth-limited regimes depicted in green and blue, respectively.

The discussion presented so far only applies to *homogeneous nucleation* in which a critical nucleus forms in the absence of an external surface. In the case of water, barriers for homogeneous nucleation are so large that it can only occur at temperatures below ≈235 K, and is practically impossible for temperatures above –20 °C (Sanz et al. 2013). With extra care, water microdroplets can be supercooled for temperatures as low as 227 K (Sellberg et al. 2014). All our daily experience of freezing, therefore, involve *heterogeneous nucleation* in which an external surface facilitates freezing by decreasing nucleation barriers. There are a wide variety of materials that can heterogeneously nucleate ice, including mineral dust (Chen et al. 2008; Zimmermann et al. 2008; Hoose et al. 2010; Atkinson et al. 2013; Kiselev et al. 2017), volcanic ash (Schumann et al. 2011), soot (Jensen and Toon 1997; Chen et al. 2008; Hoose et al. 2010), organic materials (Wilson et al. 2015), ammonium sulfate particles (Hoose and Möhler 2012), and biological materials such as pollen and bacteria (Chen et al. 2008; Hoose et al. 2010; Joly et al. 2013; Wilson et al. 2015; O'Sullivan et al. 2015) and diatoms (Knopf et al. 2011). Interestingly biological materials are among the most potent ice-nucleating agents (Murray et al. 2012).

An extension of CNT for heterogeneous nucleation was formulated by Turnbull (1950), in which it is assumed that the crystalline nucleus is a spherical cap and its shape does not change during nucleation. According to CNT for heterogeneous nucleation, the nucleation barrier will be given by $\Delta G_{\text{nuc}}^{\text{het}} = f(\theta_c)\Delta G_{\text{nuc}}^{\text{homo}}$. Here, θ_c is the three-phase contact angle between the crystal, the supercooled liquid and the external surface, and $f(\theta_c) = \frac{1}{4}(1 - \cos\theta_c)^2(2 + \cos\theta_c)$ is called the *potency factor*. Note that whenever $\theta_c < \pi$, i.e., when the external surface has a higher propensity for the crystal than for the supercooled liquid, the barrier for heterogeneous nucleation will always be smaller than the barrier for homogeneous nucleation at the same temperature. Therefore, the temperature at which the nucleation rate becomes maximum increases in heterogeneous nucleation, but the overall qualitative picture depicted in Fig. 3.5 remains unchanged. Also, the nucleation rate in heterogeneous nucleation should be expressed as the number of nucleation events per time per area, and nucleation time will have a mean given by $\langle\tau\rangle = 1/(RS)$ with S the total surface area of all ice-nucleating particles in the system. It is, however, usually possible to obtain an effective volumetric nucleation rate by accounting for the volumetric concentration of ice-nucleating particles, and the surface area of each such particle.

Another remarkable feature about heterogeneous nucleation is that it does not occur on solid-like surfaces only, but can also take place at a fluid–fluid interface. For instance, it has been demonstrated that amphiphilic alcohol monolayers (Gavish et al. 1990; Popovitz-Biro et al. 1994; Seeley and Seidler 2001) and terpenes (Rosinski 1980; Rosinski and Lecinski 1981; Rosinski et al. 1990) can facilitate ice nucleation. A more interesting possibility, which has been studied extensively both in experiments (Djikaev et al. 2002; Tabazadeh et al. 2002; Kuhn et al. 2011) and simulations (Gianetti et al. 2016; Vrbka and Jungwirth 2006; Pluhárová et al. 2010; Li et al. 2013b; Haji-Akbari et al. 2014; Haji-Akbari and Debenedetti 2017a), is that water–vapor interfaces can facilitate ice nucleation at their vicinity. This latter scenario is called *surface freezing* and has been predicted to become dominant in submicron microdroplets (Kuhn et al. 2011). Due to the difficulty of generating monodispersed submicron droplets, however, whether surface freezing occurs or not is still controversial (Sigurbjörnsson and Signorell 2008) and is regarded as one of the ten biggest open questions about ice and snow (Bartels-Rausch 2013). A detailed discussion of surface freezing can be found in an earlier review (Haji-Akbari and Debenedetti 2017b).

3.4.2.3 Ice Nucleation and Ice I Polytopes

As mentioned earlier, hexagonal ice is the thermodynamically stable form of ice I under ambient conditions, and the type of ice that forms at temperatures close to T_m. However, this does not necessarily imply that the crystalline nuclei that emerge during nucleation are exclusively hexagonal. Indeed, the ice that nucleates homogeneously at low temperatures (Dowell and Rinfret 1960; Mayer and Hallbrucker 1987; Kohl et al. 2000; Murray et al. 2005; Shilling et al. 2006) or in ultra-confined geometries (Johari 2005; Steytler et al. 1983; Morishige and Uematsu 2005; Morishige et al. 2009) is stacking disordered, with a considerable fraction of cubic stacks (Malkin et al. 2012; Amaya et al. 2017). If temperatures are sufficiently high, stacking disordered ice will eventually anneal to the thermodynamically stable hexagonal ice (Murray et al. 2005; Gainaru et al. 2018).

The fact that a liquid would nucleate into a thermodynamically metastable crystal is not surprising and has been observed in other systems (ten Wolde and Frenkel 1999). It has indeed been argued that if that metastable crystal has a lower solid–liquid surface tension than the thermodynamically stable crystal, it will be the first to nucleate in the liquid. The emerging nucleus will then nucleate the thermodynamically stable phase from within when it becomes sufficiently large (ten Wolde et al. 1996). This is typically called *Ostwald's step rule*, and stipulates that crystallization of a liquid into the thermodynamically stable crystal will usually occur in stages

(Van Santen 1984). Ostwald's step rule has been invoked to explain the formation of I_{sd} during ice nucleation (Takahashi 1982). It has, however, not been established whether I_c and I_{sd} have lower solid–liquid surface tensions than the thermodynamically stable hexagonal ice. Furthermore, the validity of Ostwald's step rule at a molecular level cannot be confirmed using existing experimental techniques due to their insufficient spatiotemporal resolution.

In recent decades, molecular simulations have shed more light on the molecular origin of stacking disorder in ice nucleation. Molecular simulations utilizing different water models and simulation techniques have indeed revealed the formation of stacking disordered ice during nucleation (Brukhno et al. 2008; Moore and Molinero 2010, 2011a, b; Johnston and Molinero 2012; Reinhardt and Doye 2012; Reinhardt et al. 2012; Haji-Akbari and Debenedetti 2015, 2017a; Lupi et al. 2017). In unbiased molecular dynamics simulations of the monoatomic water (mW) potential (Molinero and Moore 2009), a coarse-grained model of water, it has been observed that homogeneous nucleation in the deeply supercooled regime results in the formation of nuclei that are stacking disordered, and their fraction of cubic stacks increases when they become larger, eventually reaching a plateau of $\sim$60% (Moore and Molinero 2010, 2011a, b). The formation of I_{sd} is, however, not confined to mW and has been observed for atomistic models of water as well (Brukhno et al. 2008; Reinhardt et al. 2012; Haji-Akbari and Debenedetti 2015, 2017a).

So far, two distinct explanations have been proposed for the emergence of stacking disorder during ice nucleation. The first explanation is kinetic in nature, and was provided by Haji-Akbari and Debenedetti (2015), who utilized atomistic simulations and forward-flux sampling (FFS) (Allen et al. 2006) to compute the rate and explore the mechanism of homogeneous ice nucleation for TIP4P/Ice (Abascal et al. 2005), one of the most accurate atomistic models of water. They monitored the presence of HCs and DDCs, the topological building blocks of I_h and I_c, and observed that DDC-rich nuclei have smaller surface-to-volume ratios and are therefore more likely to grow and become post-critical. HC-rich nuclei, however, emerge mostly as a result of the attachment of new HCs to prismatic faces of existing HCs. The dominance of prismatic growth culminates in HC-rich nuclei that are composed of chains of hexagonal cages, and that have large surface-to-volume ratios. The fact that new HCs are added prismatically has a kinetic origin, and results in disfavoring HC-rich crystallites and favoring the formation of stacking disordered ice with a high fraction of cubic stacks. This mechanism has been confirmed in a later study utilizing a more accurate variant of the FFS technique (Haji-Akbari 2018). Computational studies of bulk ice also confirm that ice growth is the fastest at prismatic planes, presumably due to their higher surface energies (Seo et al. 2012; Rozmanov and Kusalik 2012).

The second explanation is thermodynamic in nature and was proposed by Lupi et al. (2017), who used molecular dynamics simulations and the aimless shooting method (Mullen et al. 2015) to compute the free energy landscape for homogeneous nucleation in the mW system, as a function of nucleus size and cubicity. They concluded that the entropy associated with mixing cubic and hexagonal layers makes stacking disordered nuclei more stable than their pure cubic and hexagonal counterparts. This entropic contribution is proportional to the number of stacks, and scales as $N^{1/3}$, with N the number of water molecules in the nucleus. In bulk ice, i.e., for $N \to \infty$, this entropic contribution will be negligible in comparison to the free energy difference between I_c and I_h, which will scale as $O(N)$. For small nuclei with a few hundred water molecules, however, this entropic contribution will be significant and will decrease the nucleation barrier by enhancing the stability of stacking disordered nuclei. It is necessary to note that these explanations are not mutually exclusive, and the preference for stacking disordered ice is likely a consequence of both thermodynamic and kinetic effects.

3.4.2.4 Melting of Ice I: Gibbs–Thomson Equation and Recrystallization

Similar to freezing, melting of crystals, such as ice, is a first-order transition, proceeds through nucleation and growth, and can be delayed at temperatures higher than T_m for kinetic reasons. In other words, similar to liquid water that can be supercooled and remain liquid at temperatures below T_m, ice can also be *superheated* and remain crystalline at temperatures above T_m (Iglev et al. 2006). What is different about melting, however, is that every crystal has an intrinsic stability limit, i.e., a temperature above which it becomes mechanically unstable due to high-amplitude thermal vibrations of its constituent atoms and/or molecules (Lindemann 1910). Homogeneous nucleation of the liquid can, however, occur before this stability limit is reached, as confirmed in several experimental (Wang et al. 2012, 2015) and computational (Mochizuki et al. 2013; Samanta et al. 2014) studies. In reality, most crystals melt heterogeneously at temperatures that are only slightly higher than T_m and are thus much lower than this stability limit. Heterogeneous melting can be induced at external surfaces, or at internal defects such as grain boundaries. What is remarkable about heterogeneous melting is the ability of most external surfaces to destabilize crystals even at temperatures below T_m, resulting in the formation of a thin pre-melted quasi-liquid layer in their vicinity. This phenomenon—also known as *premelting*—has been extensively studied for ice (Dash et al. 1995; Bluhm et al. 2002; Sánchez et al. 2017) and other materials (Ronchi and Hiernaut 1996; Kristensen and Cotterill 1977) in recent decades. A crystal that is in contact with a surface that induces premelting cannot be superheated, as at any $T > T_m$, the pre-melted layer will immediately destabilize the bulk crystal at its vicinity and will result in its melting without a need to cross a nucleation barrier.

Considering the preponderance of surface premelting, superheating is not very common, and in order for a crystal to be superheated, it needs to be engulfed by surfaces that do not induce premelting. This can, for instance, be achieved by embedding nanocrystals of the target crystal within a matrix of a second crystal with a higher melting point, a method that has been used to superheat metals (Banhart et al. 2003) and Group IV elements such as germanium (Xu et al. 2006). Crystals that have been frozen within solutions with kinetic inhibitors such as antifreeze proteins can also be superheated, as such inhibitors can bind to crystalline particles and provide the engulfment discussed above in an irregular manner. Therefore, the ice that forms in the presence of antifreeze proteins can be superheated (Knight and DeVries 1989; Celik et al. 2010; Cziko et al. 2014), a process that will be discussed in further detail in other chapters.

It is necessary to emphasize that confinement does not only affect the kinetics of melting, but can also impact its thermodynamics. For instance, the equilibrium melting temperature can change considerably for water entrapped in nanopores (Findenegg et al. 2008; Sliwinska-Bartkowiak et al. 2008; Moore et al. 2012). One of the most notable examples of how confinement can affect the thermodynamics of melting is the depression of equilibrium melting point for small crystalline particles. In order to understand the origin of this depression, consider a spherical ice particle of radius r, which, due to its surface curvature, will experience a heightened internal pressure. The pressure difference between the crystal interior and the surrounding liquid is called the *Laplace pressure* and can be estimated using the Young–Laplace equation (Chen et al. 2006):

$$\Delta P = \frac{2\sigma_{\mathrm{ls}}}{r} \tag{3.11}$$

This ΔP will increase the chemical potential of the crystal by $\Delta\mu = v_{\mathrm{ice}}\Delta P = 2\sigma_{\mathrm{ls}}/\rho_{\mathrm{ice}}r$ and will make it thermodynamically unstable at the bulk melting temperature T_{m}. Instead, thermodynamic equilibrium will be achieved at a lower temperature $T_{\mathrm{m}}(r)$ where this curvature-induced penalty is offset by the bulk thermodynamic driving force given by Eq. (3.8). It can be shown using simple thermodynamic reasoning that $T_{\mathrm{m}}(r)$ will be given by:

$$T_{\mathrm{m}}(r) = T_{\mathrm{m,bulk}} - \frac{2\sigma_{\mathrm{ls}}T_{\mathrm{m,bulk}}}{\rho_{\mathrm{s}}r\Delta h_{\mathrm{m}}} \tag{3.12}$$

Equation (3.12) is known as the *Gibbs–Thomson equation* (Jackson and McKenna 1990; Johari 1997). A similar argument can be made to obtain the depressed T_{m} for arbitrarily shaped surfaces (Johnson 1965; Jones 1974), including a spherical cap of radius r sitting on top of an ice-nucleating surface. A closely related equation is the *Kelvin equation*, which describes the dependence of the equilibrium vapor pressure of a droplet as a function of its size (Thomson 1872):

$$P(r) = P(\infty)\exp\left[\frac{2\sigma_{\mathrm{lv}}}{\rho_{\mathrm{l}}rkT}\right] \tag{3.13}$$

with $P(r)$ and $P(\infty)$ the equilibrium vapor pressure of a droplet of radius r, and bulk liquid, respectively, σ_{lv}, the liquid–vapor surface tension, and ρ_{l} the number density of the liquid.

An important consequence of this size-dependent stability is a phenomenon known as *recrystallization* or *Ostwald ripening* (Voorhees 1985), which refers to the growth of larger crystalline particles at the expense of smaller ones. This occurs when the nucleation barrier is so small that a large number of post-critical nuclei emerge within the liquid. However, not all such nuclei will be identical in shape and size. Such polydispersity will generate a thermodynamic driving force for the growth of larger nuclei at the expense of smaller ones, which, as explained above, are less stable. One of the important functions of AFPs is to inhibit recrystallization, i.e., to prevent small crystalline nuclei—that are still larger than their surrounding nuclei—from growing any larger. This is important from a biological perspective since smaller crystallites are less harmful to biological cells (Knight et al. 1984). The ability of AFPs to prevent recrystallization will be discussed in detail in Chap. 7 of Vol. 2.

3.4.2.5 Thermal Hysteresis

Nucleation is, in essence, an out-of-equilibrium process in which a thermodynamically metastable system, such as a supercooled liquid, turns into a thermodynamically stable structure, such as a crystal. Therefore, when the coexistence line is crossed, e.g., by cooling a liquid, freezing does not occur immediately. Instead, the system remains in the metastable supercooled state until nucleation occurs. However, when nucleation is complete, the system reaches thermodynamic equilibrium, and remains there until the coexistence line is crossed in the opposite direction. Therefore, the behavior of a system that crosses a coexistence line depends on the direction of such crossing. This is known as *hysteresis* and is an important feature of all first-order phase transitions, including crystallization. As explained in Sect. 3.4.2, hysteresis can occur for both ice formation and ice melting, but is less common in the case of melting. An important manifestation of this is *thermal hysteresis* in biological systems, which refers to a situation in which a biological solution can exist as a liquid at temperatures below T_m (Finger and Bischof 2018) or as a solid at temperatures above T_m (Knight and DeVries 1989). In other words, at finite timescales, temperature is not sufficient for determining whether a system is frozen or not, and the system might behave differently based on its initial state and processing history. One measure of an AFP's potency is its ability to induce the largest possible thermal hysteresis, which will enable cells and/or organisms to survive freezing at lower temperatures.

3.4.3 Ice Formation in Solutions

So far, only the thermodynamics and kinetics of ice formation in pure water has been discussed. Biological water, however, is never pure, and in order to explore freezing in biological systems, it is important to understand how soluble impurities affect ice formation. The conceptual approach for analyzing freezing in aqueous solutions will depend on whether any of the solutes participates in the thermodynamic stable form of the aqueous crystal. Most solutes are excluded from the crystal, and therefore the freezing of the corresponding solution results in the formation of pure ice I. As mentioned in Sect. 3.3.2, however, there are notable exceptions to this rule including aqueous solutions of small gaseous molecules that can crystallize into gas hydrates (Maslin et al. 2010; Knott et al. 2012). Here, the focus will be on the former scenario, i.e., on solutes that do not participate in the ice structure, and on their impact on the thermodynamics and kinetics of freezing.

3.4.3.1 Thermodynamics of Ice Formation in Solutions

The effect of soluble entities on the thermodynamics of ice formation is well understood, and is discussed in detail in all standard thermodynamics textbooks. In summary, adding a solute to water results in a decrease in its chemical potential given by:

$$\mu_{\mathrm{w,liq}}\left(T,P,x_{\mathrm{w,liq}}\right)=\mu^0_{\mathrm{w,liq}}(T,P)+RT\ln a_{\mathrm{w,liq}} \tag{3.14}$$

Here, $x_{\mathrm{w,liq}}$ is the mole fraction of water and $a_{\mathrm{w,liq}}<1$ is a dimensionless quantity known as *activity*. As mentioned earlier, thermodynamic coexistence is achieved when the chemical potentials of water in the liquid and the crystal become identical. Assuming that the solute does not participate in the thermodynamically stable crystal structure, the condition for thermodynamic coexistence between the solution and ice I will be given by:

$$\ln a_{\mathrm{w,liq}}(T_{\mathrm{m}})=\frac{\mu^0_{\mathrm{w,ice}}(T_{\mathrm{m}},P)-\mu^0_{\mathrm{w,liq}}(T_{\mathrm{m}},P)}{RT_{\mathrm{m}}}=-\frac{|\Delta\mu|}{RT_{\mathrm{m}}} \tag{3.15}$$

Since $a_{\mathrm{w,liq}}<1$, Eq. (3.15) can only be satisfied if $T_{\mathrm{m}}<T_{\mathrm{m}}^{\mathrm{pure}}$. This decrease in equilibrium freezing temperature is typically referred to as *freezing point depression*, which is defined as $\Delta T_{\mathrm{m}}:=T_{\mathrm{m}}^{\mathrm{pure}}-T_{\mathrm{m}}(x_{\mathrm{w}})$. Assuming the availability of $|\Delta\mu(T)|$, which can be obtained by integrating the equation of state, and the activity data versus temperature, which can, for instance, be readily obtained from vapor pressure—or fugacity—measurements, the depressed freezing point can be accurately calculated from Eq. (3.15). A simpler approach can, however, be employed for dilute solutions, i.e., solutions in which x_{S}, the total mole fraction of *all* solutes in the system, is much smaller than unity, i.e., when $x_{\mathrm{S}}\ll 1$. Under such circumstances, water activity will approach its mole fraction, and its chemical potential can be expressed as:

$$\begin{aligned}\mu_{\mathrm{w,liq}}\left(T,P,x_{\mathrm{w,liq}}\right)&=\mu^0_{\mathrm{w,liq}}(T,P)+RT\ln\left(1-x_{\mathrm{S}}\right)\\&\overset{x_{\mathrm{S}}\ll 1}{\approx}\mu^0_{\mathrm{w,liq}}(T,P)-x_{\mathrm{S}}RT\end{aligned} \tag{3.16}$$

Combining Eqs. (3.8) with Eqs. (3.15 and 3.16) yields:

$$T_{\mathrm{m}}(x_{\mathrm{S}}) = \frac{T_{\mathrm{m}}^{\mathrm{pure}}}{1 + \frac{x_{\mathrm{S}} R T_{\mathrm{m}}^{\mathrm{pure}}}{\Delta h_{\mathrm{m}}}} \approx T_{\mathrm{m}}^{\mathrm{pure}} \left[1 - \frac{x_{\mathrm{S}} R T_{\mathrm{m}}^{\mathrm{pure}}}{\Delta h_{\mathrm{m}}} \right] \tag{3.17}$$

Therefore, freezing point depression in dilute solutions is a *colligative property* as it only depends on the overall molal concentration of the solutes and not on their chemical identities. Other notable colligative properties include vapor pressure depression, boiling point elevation, and osmotic pressure. Due to their relatively large molar masses, aqueous solutions of most biomolecules can be usually considered dilute, and Eq. (3.17) can usually be used for computing ΔT_{m}.

3.4.3.2 Kinetics of Ice Formation in Solutions

The presence of solutes in liquid water will alter its molecular structure, such as its hydrogen bond network. Such changes will inevitably affect the kinetics—and possibly the mechanism—of ice nucleation, and considering the out-of-equilibrium nature of nucleation and its strong sensitivity to even modest changes in thermodynamic properties, it is not a priori clear how the nucleation kinetics will be affected by the identities and concentrations of the corresponding solutes. Even if we assume the validity of the classical nucleation theory, it is not trivial to determine how properties such as solid–liquid surface tension will depend on solute concentration. The situation is even more hopeless in the case of heterogeneous nucleation, since an ice-nucleating surface can interact differently with water molecules and the solutes, and such differences can induce compositional anisotropies that can, in turn, affect the kinetics and mechanism of ice nucleation in nontrivial ways.

Despite these complexities, there seems to be overwhelming experimental evidence for the colligative nature of nucleation kinetics. In 1977, Kanno and Angel discovered that the amount of decrease in the kinetic freezing temperature[5] of an aqueous alkali halide solution is a colligative property and depends on cation concentration only (Kanno and Angell 1977). This idea was later generalized in the pioneering work of Koop et al. (2000) who studied sixteen different aqueous solutions, and demonstrated that water activity is sufficient for predicting the nucleation kinetics, irrespective of the identity of the solute. The qualitative argument for this empirical observation is that water activity is a proxy for how much its hydrogen bond network is perturbed in the presence of solute(s). This empirical framework is known as the *activity shift model* and works remarkably well for most simple solutes. Small deviations from it, however, have been observed in several systems (Swanson 2009; Knopf and Rigg 2011). Antifreeze proteins are among the most remarkable example of solutes that deviate from this colligative picture, resulting in kinetic suppressions considerably larger than what would otherwise be expected from their concentrations (Dolev et al. 2016; Olijve et al. 2016; Haji-

[5]Here, the kinetic freezing temperature refers to the highest temperature at which freezing occurs within the observation timescale of an experiment.

Akbari 2016). The activity shift model was originally proposed for homogeneous nucleation. Interestingly, the kinetics of heterogeneous nucleation has also been shown to follow the activity shift model (Zobrist et al. 2008; Wilson and Haymet 2009; Knopf and Alpert 2013; Zeng et al. 2015) with minor deviations also reported in some systems (Cantrell and Robinson 2006).

Despite its success in predicting the kinetics of ice nucleation in aqueous solutions, the activity shift model does not provide any information about how the nucleation mechanism is affected by a solute. For instance, the presence of soluble impurities has been shown to affect polymorph selection (i.e., the ratio of cubic to hexagonal ice) in deeply supercooled water (Murray and Bertram 2007; Murray 2008). In addition, there is a wide gap in our understanding of when and why the activity shift model is violated, and studying such violations can be a potent area of research.

3.4.3.3 Solute Exclusion and Its Consequences

As mentioned above, most aqueous solutions crystallize into ice I, and their constituent solutes do not participate in the crystal structure. This results in a phenomenon known as *solute exclusion* or *solute rejection* in which the freezing of an aqueous solution results in the formation of ice crystals that coexist with regions composed of more concentrated—and possibly saturated—aqueous solutions. Solute rejection has very important consequences in biological systems. Not only such concentrated regions can be toxic simply due to their high solute concentrations, but there will also be a steep osmotic gradient between them and less concentrated solutions, e.g., in the cytoplasm. Such a gradient will induce a strong water flow, which can, in turn, exert mechanical damage to cell membranes. This *osmotic shock* scenario is thought to be an important mechanism of ice toxicity. Interestingly, osmosis-induced ice toxicity usually occurs when freezing is extracellular, in which case water flow will be from the cytoplasm to the extracellular region (Muldrew and McGann 1990, 1994).

3.5 Amorphous Ices

The focus of Sect. 3.4 was to discuss water's equilibrium behavior at ambient pressures and subfreezing temperatures. This section, however, will deal with out-of-equilibrium glassy states of water that share the molecular structure of liquid water but, due to their slow dynamics, exhibit mechanical properties of a solid. Like most other liquids, water can be turned into a glass via a number of means including—but not limited to—rapidly quenching the liquid. Such kinetically arrested states are known as *amorphous ices* and the process employed for making them is called *vitrification*. Amorphous ices can avoid crystallization within experimentally accessible timescales as long as they are kept at temperatures below the glass transition temperature T_g, which, for liquid water, is $\sim$136 K at ambient pressures (Fuentes-Landete et al. 2015). Note that crystallization can be delayed for relatively long times even at temperatures slightly above T_g, since in order for the viscous water to crystallize, a large number of molecular rearrangements will still need to occur. At ambient pressures, for instance, crystallization will become sufficiently fast to be considered instantaneous at or above $T_x \sim 150$ K (Fuentes-Landete et al. 2015). In the interstellar space, temperatures are usually much lower than T_g. Consequently, amorphous ice is the predominant form of ice in the outer space (Mitchell et al. 2017). This is in contrast to earth where subfreezing temperatures are much higher than T_x, and therefore it is the crystalline ice that is prevalent.

Amorphous ices are out of equilibrium in nature, and their properties depend heavily on their processing history. Consequently, classifying amorphous ices based on their structural and physical properties is not trivial, and there are a large number of names and acronyms currently in use in the literature to describe different types of amorphous ices. In terms of processing history, amorphous ices are broadly classified into five categories as depicted in Fig. 3.6a. Historically, glassy water was first obtained through a process in which water vapor was physically deposited onto the surface of a cold copper rod at temperatures below 130 K (Burton and Oliver 1935). Vapor-deposited amorphous ice is typically referred to as *amorphous solid water* (*ASW*), is usually very porous, and can have densities as low as 0.3 g cm^{-3}. Upon annealing, however, ASW can undergo pore collapse and its specific surface area can decrease by as much as 2–3 orders of magnitude (Baragiola 2003). Vapor deposition is the primary mechanism of vitrification in comets and in the interstellar space (Mayer and Pletzer 1986).

Hyperquenched glassy water (*HGW*) is a type of glassy water obtained via rapid quenching of water microdroplets, which are then deposited onto the surface of a cold "cryoplate" (Mayer 1985). In order for HGW to form, the quenching rate must be $\sim 10^6$ K/s or higher. *High-density amorphous ice (HDA)* is another type of amorphous ice, obtained via pressure melting of hexagonal ice, e.g., at 77 K and 1.0 GPa (Mishima et al. 1984). Annealing HDA at low and high pressures results in the formation of *low-density amorphous ice* (*LDA*) (Mishima et al. 1985) and *very high-density amorphous ice* (*VHDA*) (Loerting et al. 2001), respectively.

Three of these five classes, i.e., ASW, HGW, and LDA, are low-density amorphous ices, and share the same molecular structure. The local environment of each water molecule within LDA (and other related low-density glasses) is remarkably similar to that of ice I in the sense that each molecule has four tetrahedrally arranged nearest neighbors and there is a clear separation between the first and second hydration shells, as manifested in the oxygen–oxygen radial distribution functions depicted in Fig. 3.6b. Unlike ice I, however, LDA lacks long-range translational order. The local tetrahedrality of the first hydration shell is disrupted in both HDA and VHDA in which the first hydration shell is penetrated by one and two water molecules from the second shell, respectively, as evident from the emergence of a weak second peak at $\approx$0.4 nm in HDA (Finney et al. 2002b) and $\approx$0.35 nm in VHDA (Finney et al. 2002a). Therefore, from a structural perspective, amorphous ices are classified into three categories: LDA, HDA, and VHDA.

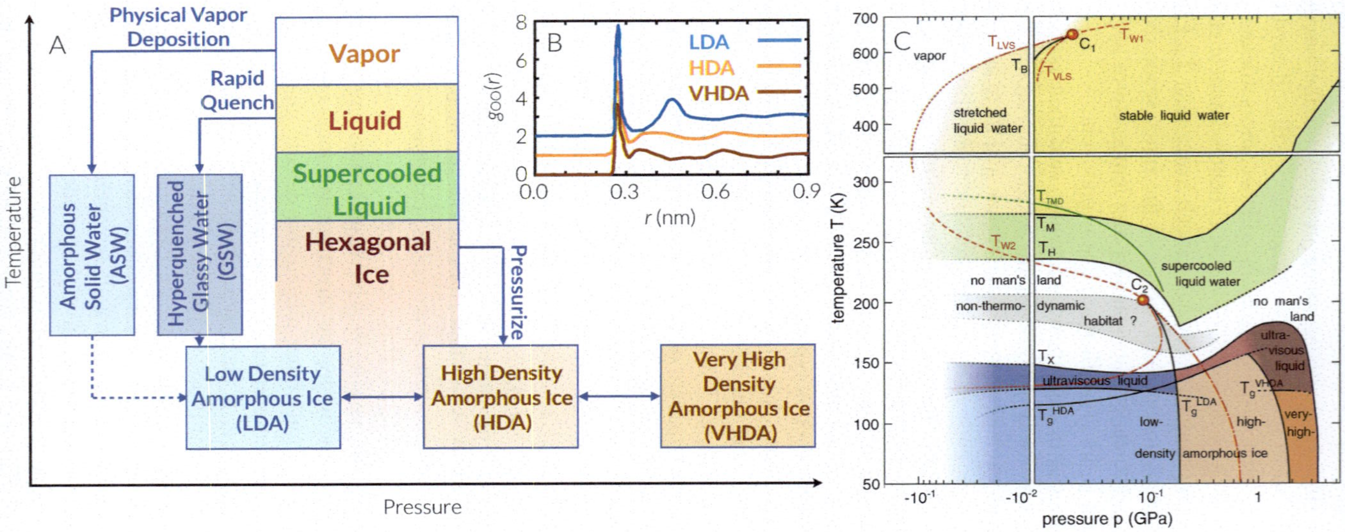

Fig. 3.6 Metastable phase diagram of liquid water. (**a**) Procedures for preparing different types of amorphous ices, adapted and simplified from Fig. 8 of Fuentes-Landete et al. (2015). (**b**) O–O radial distribution functions for LDA, HDA, and VHDA adapted and replotted from Finney et al. (2006). (**c**) Metastable phase diagram of liquid water (Reproduced from Amann-Winkel et al. 2016)

Another remarkable feature of these three structurally distinct amorphous ices is that they can be reversibly (Winkel et al. 2008) transformed into one another without crystallizing. The transition from LDA to the other two types is particularly sharp and is believed to be first order in nature (Finney et al. 2002a; Lemke et al. 2017). The notion that a material can exist in multiple amorphous phases that can reversibly transform into one another via first-order transitions is called *polyamorphism*. Studying glassy water polyamorphism is particularly interesting in the context of low-temperature water anomalies (Ball 2008; Brovchenko and Oleinikova 2008). There are a large number of such anomalies in the literature, including density maximum at 4 °C, isothermal compressibility minimum 46 °C and isobaric heat capacity minimum at and 35 °C. The origin of such anomalies are, however, yet to be fully understood, and four distinct scenarios have been proposed to explain them (Stanley and Teixeira 1980; Stanley et al. 1981; Speedy 1982; Poole et al. 1992; Sastry et al. 1996; Angell 2008; Stokely et al. 2009). Among these scenarios, the liquid–liquid transition hypothesis (Poole et al. 1992) is inspired by amorphous ice polyamorphism and attributes water anomalies to the existence of two distinct liquids in the supercooled regime, namely the low-density liquid (LDL) and the high-density liquid (HDL). According to this hypothesis, HDL and LDL are separated by a coexistence line that culminates in a liquid–liquid critical point (LLCP), and are the liquid counterparts of LDA and HDA, respectively. As can be seen in the metastable phase diagram of supercooled water in Fig. 3.6c (Amann-Winkel et al. 2016), the predicted liquid–liquid critical point is located below the homogeneous nucleation line and is therefore not accessible to existing experimental techniques. In recent years, there have been numerous experimental efforts to access this LLCP (Sellberg et al. 2014; Laksmono et al. 2015). Also, molecular simulations have been extensively utilized to understand liquid–liquid transition in water (Poole et al. 1992; Errington and Debenedetti 2001; Sastry et al. 1996; Palmer et al. 2014, 2018a) and other tetrahedral liquids (Saika-Voivod et al. 2000; Sastry and Angell 2003; Xu et al. 2005; Smallenburg et al. 2014). The topic of liquid–liquid transition in supercooled water is, however, far from resolved and is an active area of research (Palmer et al. 2018b).

Due to their metastable nature, amorphous ices should, in principle, crystallize at sufficiently long times. Crystallization, however, becomes practically implausible below T_g, and only becomes instantaneous at temperatures considerably higher than T_g. The process in which an amorphous solid turns into a crystal is called *devitrification* or *recrystallization*. Note that the general meaning of the term "recrystallization" is already introduced in Sect. 3.4.2. Crystallization within a glass is a growth-limited process resulting in the formation of a large number of nuclei, and therefore falls into the general framework of recrystallization introduced in Sect. 3.4.2.

Different amorphous ices can freeze into different crystalline phases. LDA crystallizes within experimentally accessible timescales at around 150 K and transforms into stacking disordered ice with a large fraction of cubic stacks (Johari et al. 1996; Johari 1998). High-density amorphous ices, however, crystallize into other ice polymorphs depending on pressure and processing history. For instance, HDA has been shown to crystallize into ice VII and ice VIII (Hemley et al. 1989), and ice IX

(Seidl et al. 2013). VHDA, however, can transform into a wide variety of ice polymorphs at different pressures (Klotz et al. 2005). Amorphous ices can also be directly transformed into liquid water if they are heated so quickly that crystallization is avoided. The heating rates needed for this to occur are, however, around two orders of magnitude larger than the cooling rates needed for vitrification.

Similar to pure water, rapid quenching of an aqueous solution can result in the formation of an amorphous solid. As outlined in Sect. 3.4.3, the presence of solutes results in melting point depression, and a considerable decline in the nucleation rate. Meanwhile, aqueous solutions usually have higher viscosities in comparison to pure liquid, and their viscosities are stronger functions of temperature (Angell 2002). This deceleration of dynamics can become very pronounced in concentrated aqueous solutions to the extent that their T_g can considerably exceed that of pure water. For instance, concentrated divalent electrolytic solutions such as $Mg(NO_3)_2$ and $Ca(NO_3)_2$ can have T_g's that are tens of degrees Kelvin higher than that of pure water (Angell and Sare 1970), and aqueous solutions of sugars such as trehalose can have T_g's at or above room temperature (Green and Angell 1989). For most solutes, however, the extent of increase in T_g is not that large, but it can be generally stated that aqueous solutions have lower T_m's and higher T_g's. Consequently, homogeneous ice nucleation can occur over a narrower range of temperatures. This will imply that vitrifying an aqueous solution can be achieved at cooling rates considerably lower than what would be needed for vitrifying pure water. Similarly, it will be easier to bypass crystallization during the thawing of such glasses, as the required heating rates will also be considerably lower.

3.5.1 Biological Applications of Vitrification

Vitrification is, in essence, a process in which the liquid is kinetically arrested without considerable change to its microscopic structure. Therefore, whenever a biological cell is vitrified, its structure and the spatial distribution of its composition remain unchanged, unlike freezing that can damage it through a number of means, such as volumetric expansion, solute exclusion, and osmotic shock. Vitrification is therefore of considerable interest in *cryopreservation* applications (Fahy et al. 1984). Vitrified cells and tissues can, in principle, be preserved for years, if not decades, and considering that biological cells are composed of concentrated aqueous solutions of electrolytes, carbohydrates, and other biomolecules, they can be vitrified (and thawed) at considerably lower cooling (heating) rates in comparison to what will be needed for pure water. Yet, it is still difficult to vitrify entire tissues and organs due to possible heat transfer limitations. In other words, when a tissue or organ is rapidly cooled, its interior will be at a higher temperature than its surface at all times, and such temperature differences can result in the freezing of the interior while the surface is being vitrified. A similar problem can arise during thawing, and the tissue or organ can freeze from within due to inadequate heat transfer. Therefore, vitrification is generally used for preserving individual cells, such as sperms (Isachenko et al. 2004), oocytes (Kuwayama et al. 2005; Cobo et al. 2008) and embryos (Rall and Fahy 1985; Rall 1987), or small tissues, such as cryo-electron microscopy specimen (Dubochet et al. 1988).

3.6 Conclusions

Water is a material that exhibits numerous anomalies, and as discussed in detail in this chapter, its crystallization is also anomalous in many different ways. In particular, water can form a large number of crystalline polymorphs, some of which are less dense than the liquid water at coexistence. In all these ice polymorphs, each water molecule is hydrogen bonded to four of its neighbors, with two as a proton donor, and two as a proton acceptor. The polymorph of ice that forms under ambient pressures is ice I, which is a layered structure in which each water molecule has four neighbors within its first hydration shell. Ice I has two stacking variants—or polytopes—known as hexagonal (I_h) and cubic (I_c) ice. Hexagonal ice is the thermodynamically stable polytope, and is the form that emerges upon freezing under physiologically relevant conditions. Structurally, I_h is anisotropic and has three major crystallographic planes with differing surface energies. Cubic ice, which is structurally less anisotropic, is metastable and only becomes relevant during crystallization at very low temperatures.

As a first-order phase transition, ice formation proceeds through a nucleation and growth mechanism, and can occur both homogeneously and heterogeneously. At temperatures close to T_m, the thermodynamic driving force for crystallization is small, and freezing is nucleation-limited. At deep supercoolings, however, nucleation becomes almost instantaneous and crystallization becomes growth-limited. What is particularly interesting about ice I is that its homogeneous nucleation can be delayed for temperatures as low as 235 K. Freezing of liquid water into ice is therefore primarily heterogeneous under physiologically relevant conditions.

Freezing is generally slower in aqueous solutions, in comparison to pure water. Not only the thermodynamic freezing temperature is depressed in the presence of soluble impurities, but also the nucleation rate decreases considerably. For most solutions, the deceleration of nucleation kinetics can be predicted using the activity shift model, which is based on the empirical observation that solutions with identical water activities will freeze at the same rate at any given temperature.

Finally, water can be transformed into an amorphous solid via a number of means, including rapid quenching of the liquid, or depositing water vapor onto the surface of a cold substrate. Such glassy states are known as amorphous ices, and exhibit the mechanical properties of a solid despite being structurally liquid-like. Amorphous ices form during vitrification, which is widely used in cryopreservation applications.

Acknowledgments A.H.-A. gratefully acknowledges the support of the National Science Foundation CAREER Award (Grant No. CBET-1751971).

References

Abascal JLF, Sanz E, Fernández RG, Vega C (2005) A potential model for the study of ices and amorphous water: TIP4P/Ice. J Chem Phys 122:234511

Algara-Siller G, Lehtinen O, Wang F, Nair R, Kaiser U, Wu H, Geim A, Grigorieva I (2015) Square ice in graphene nanocapillaries. Nature 519:443

Allen PB (2010) Interpreting the 4-index notation for hexagonal systems. arXiv preprint arXiv:1006.2842

Allen RJ, Frenkel D, ten Wolde PR (2006) Forward flux sampling type schemes for simulating rare events: efficiency analysis. J Chem Phys 124:194111

Amann-Winkel K, Böhmer R, Fujara F, Gainaru C, Geil B, Loerting T (2016) Colloquium: water's controversial glass transitions. Rev Mod Phys 88:011002

Amaya AJ, Pathak H, Modak VP, Laksmono H, Loh ND, Sellberg JA, Sierra RG, McQueen TA, Hayes MJ, Williams GJ et al (2017) How cubic can ice be? J Phys Chem Lett 8:3216–3222

Angell C (2002) Liquid fragility and the glass transition in water and aqueous solutions. Chem Rev 102(8):2627–2650

Angell CA (2008) Insights into phases of liquid water from study of its unusual glass-forming properties. Science 319:582–587

Angell C, Sare E (1970) Glass-forming composition regions and glass transition temperatures for aqueous electrolyte solutions. J Chem Phys 52:1058–1068

Arbuckle W (1986) Development of the ice cream industry. In: Ice cream. Springer, Boston, MA, pp 1–8

Atkinson JD, Murray BJ, Woodhouse MT, Whale TF, Baustian J, Carslaw KS, Dobbie S, O'Sullivan D, Malkin TL (2013) The importance of feldspar for ice nucleation by mineral dust in mixed-phase clouds. Nature 498:355–358

Bai J, Wang J, Zeng XC (2006) Multiwalled ice helixes and ice nanotubes. Proc Natl Acad Sci U S A 103:19664–19667

Bai J, Angell CA, Zeng XC (2010) Guest-free monolayer clathrate and its coexistence with two-dimensional high-density ice. Proc Natl Acad Sci U S A 107:5718–5722

Baker MB (1997) Cloud microphysics and climate. Science 276:1072–1078

Ball P (2008) Water: water—an enduring mystery. Nature 452:291

Banhart F, Hernandez E, Terrones M (2003) Extreme superheating and supercooling of encapsulated metals in fullerene-like shells. Phys Rev Lett 90(18):185502

Baragiola R (2003) Water in confining geometries. Springer, Berlin, p 359

Barati Farimani A, Aluru NR (2016) Existence of multiple phases of water at nanotube interfaces. J Phys Chem C 120:23763–23771

Bartels-Rausch T (2013) Chemistry: ten things we need to know about ice and snow. Nature 494:27–29

Becker R, Döring W (1935) Kinetische behandlung der keimbildung in übersättigten dämpfen. Ann Phys 416:719–752

Beltaos S, Prowse T (2009) River-ice hydrology in a shrinking cryosphere. Hydrol Process 23:122–144

Bernal J, Fowler R (1933) A theory of water and ionic solution, with particular reference to hydrogen and hydroxyl ions. J Chem Phys 1:515–548

Bluhm H, Ogletree DF, Fadley CS, Hussain Z, Salmeron M (2002) The premelting of ice studied with photoelectron spectroscopy. J Phys Condens Matter 14:L227

Bridgman PW (1912) Water, in the liquid and five solid forms, under pressure. Proc Am Acad Arts Sci 47:441–558

Brovchenko I, Oleinikova A (2008) Multiple phases of liquid water. ChemPhysChem 9:2660–2675

Brukhno AV, Anwar J, Davidchack R, Hande R (2008) Challenges in molecular simulation of homogeneous ice nucleation. J Phys Condens Matter 20:494243

Brygoo S, Henry E, Loubeyre P, Eggert J, Koenig M, Loupias B, Benuzzi-Mounaix A, Le Gloahec MR (2007) Laser-shock compression of diamond and evidence of a negative-slope melting curve. Nat Mater 6:274
Burton E, Oliver W (1935) X-ray diffraction patterns of ice. Nature 135:505
Cantrell W, Robinson C (2006) Heterogeneous freezing of ammonium sulfate and sodium chloride solutions by long chain alcohols. Geophys Res Lett 33:L07802
Carslaw KS, Harrison RG, Kirkby J (2002) Cosmic rays, clouds, and climate. Science 298:1732–1737
Celik Y, Graham LA, Mok Y-F, Bar M, Davies PL, Braslavsky I (2010) Superheating of ice crystals in antifreeze protein solutions. Proc Natl Acad Sci U S A 107:5423–5428
Chamberlain EJ, Gow AJ (1979) Effect of freezing and thawing on the permeability and structure of soils. Eng Geol 13:73–92
Chen T, Chiu M-S, Weng C-N (2006) Derivation of the generalized Young-Laplace equation of curved interfaces in nanoscaled solids. J Appl Phys 100(7):074308
Chen J-P, Hazra A, Levin Z (2008) Parameterizing ice nucleation rates using contact angle and activation energy derived from laboratory data. Atmos Chem Phys 8:7431–7449
Chen J, Schusteritsch G, Pickard CJ, Salzmann CG, Michaelides A (2016) Two dimensional ice from first principles: structures and phase transitions. Phys Rev Lett 116:025501
Chen J, Schusteritsch G, Pickard CJ, Salzmann CG, Michaelides A (2017) Double-layer ice from first principles. Phys Rev B 95:094121
Cobo A, Kuwayama M, Pérez S, Ruiz A, Pellicer A, Remohí J (2008) Comparison of concomitant outcome achieved with fresh and cryopreserved donor oocytes vitrified by the cryotop method. Fertil Steril 89:1657–1664
Cockburn A, Cockburn E, Reyman TA (1998) Mummies, disease and ancient cultures. Cambridge University Press, London
Corsetti F, Matthews P, Artacho E (2016) Structural and configurational properties of nanoconfined monolayer ice from first principles. Sci Rep 6:18651
Cziko PA, DeVries AL, Evans CW, Cheng C-HC (2014) Antifreeze protein-induced superheating of ice inside Antarctic notothenioid fishes inhibits melting during summer warming. Proc Natl Acad Sci USA 111:14583–14588
Damasceno PF, Engel M, Glotzer SC (2012) Predictive self-assembly of polyhedra into complex structures. Science 337:453–457
Dash J, Fu H, Wettlaufer J (1995) The premelting of ice and its environmental consequences. Rep Prog Phys 58(1):115
Debenedetti PG (1996) Metastable liquids: concepts and principles. Princeton University Press, Princeton, NJ
Debenedetti PG (2003) Supercooled and glassy water. J Phys Condens Matter 15:R1669
del Rosso L, Celli M, Ulivi L (2016a) New porous water ice metastable at atmospheric pressure obtained by emptying a hydrogen-filled ice. Nat Commun 7:13394
del Rosso L, Grazzi F, Celli M, Colognesi D, Garcia-Sakai V, Ulivi L (2016b) Refined structure of metastable ice XVII from neutron diffraction measurements. J Phys Chem C 120 (47):26955–26959
Djikaev Y, Tabazadeh A, Hamill P, Reiss H (2002) Thermodynamic conditions for the surface-stimulated crystallization of atmospheric droplets. J Phys Chem A 106(43):10247–10253
Dolev MB, Braslavsky I, Davies PL (2016) Ice-binding proteins and their function. Annu Rev Biochem 85:515–542
Dowell LG, Rinfret AP (1960) Low-temperature forms of ice as studied by X-ray diffraction. Nature 188:1144
Dubochet J, Adrian M, Chang J-J, Homo J-C, Lepault J, McDowall AW, Schultz P (1988) Cryo-electron microscopy of vitrified specimens. Q Rev Biophys 21:129–228
Engel EA, Monserrat B, Needs RJ (2015) Anharmonic nuclear motion and the relative stability of hexagonal and cubic ice. Phys Rev X 5:021033

Errington JR, Debenedetti PG (2001) Relationship between structural order and the anomalies of liquid water. Nature 409(6818):318

Fahy GM, MacFarlane D, Angell C, Meryman H (1984) Vitrification as an approach to cryopreservation. Cryobiology 21(4):407–426

Falenty A, Hansen TC, Kuhs WF (2014) Formation and properties of ice XVI obtained by emptying a type sII clathrate hydrate. Nature 516:231

Findenegg GH, Jähnert S, Akcakayiran D, Schreiber A (2008) Freezing and melting of water confined in silica nanopores. ChemPhysChem 9:2651–2659

Finger EB, Bischof JC (2018) Cryopreservation by vitrification: a promising approach for transplant organ banking. Curr Opin Organ Tran 23(3):353–360

Finney J, Bowron D, Soper A, Loerting T, Mayer E, Hallbrucker A (2002a) Structure of a new dense amorphous ice. Phys Rev Lett 89(20):205503

Finney J, Hallbrucker A, Kohl I, Soper A, Bowron D (2002b) Structures of high and low density amorphous ice by neutron diffraction. Phys Rev Lett 88(22):225503

Finney J, Hallbrucker A, Kohl I, Loerting T, Soper A (2006) The local and intermediate range structures of the five amorphous ices at 80 K and ambient pressure: a Faber-Ziman and Bhatia-Thornton analysis. J Chem Phys 125:194502

Fletcher NH (1970) Other forms of ice. In: Cambridge monographs on physics. Cambridge University Press, Cambridge, pp 49–72

Fowler LD, Randall DA, Rutledge SA (1996) Liquid and ice cloud microphysics in the CSU general circulation model. Part 1: Model description and simulated microphysical processes. J Clim 9:489–529

Fuentes-Landete V, Mitterdorfer C, Handle P, Ruiz G, Bernard J, Bogdan A, Seidl M, Amann-Winkel K, Stern J, Fuhrmann S et al (2015) Crystalline and amorphous ices. In: Proceedings of the International School of Physics “Enrico Fermi”, vol 187, pp 173–208

Fukazawa H, Ikeda S, Mae S (1998) Incoherent inelastic neutron scattering measurements on ice XI; the proton-ordered phase of ice Ih doped with KoH. Chem Phys Lett 282:215–218

Gainaru C, Vynokur E, Köster K, Fuentes-Landete V, Spettel N, Zollner J, Loerting T, Böhmer R (2018) Dynamic signatures of the transition from stacking disordered to hexagonal ice: dielectric and nuclear magnetic resonance studies. J Chem Phys 148:134502

Gallo P, Amann-Winkel K, Angell CA, Anisimov MA, Caupin F, Chakravarty C, Lascaris E, Loerting T, Panagiotopoulos AZ, Russo J et al (2016) Water: a tale of two liquids. Chem Rev 116:7463–7500

Gavish M, Popovitz-Biro R, Lahav M, Leiserowitz L (1990) Ice nucleation by alcohols arranged in monolayers at the surface of water drops. Science 250:973–975

Gianetti MM, Haji-Akbari A, Longinotti MP, Debenedetti PG (2016) Computational investigation of structure, dynamics and nucleation kinetics of a family of modified stillinger–weber model fluids in bulk and freestanding thin films. Phys Chem Chem Phys 18(5):4102–4111

Goncharov AF, Goldman N, Fried LE, Crowhurst JC, Kuo I-FW, Mundy CJ, Zaug JM (2005) Dynamic ionization of water under extreme conditions. Phys Rev Lett 94(12):125508

Goncharov AF, Sanloup C, Goldman N, Crowhurst JC, Bastea S, Howard W, Fried LE, Guignot N, Mezouar M, Meng Y (2009) Dissociative melting of ice vii at high pressure. J Chem Phys 130:124514

Gŕańasy L, Pusztai T, James PF (2002) Interfacial properties deduced from nucleation experiments: a Cahn–Hilliard analysis. J Chem Phys 117:6157

Green JL, Angell CA (1989) Phase relations and vitrification in saccharide-water solutions and the trehalose anomaly. J Phys Chem 93:2880–2882

Gross EK, Dreizler RM (2013) Density functional theory, vol 337. Springer Science & Business Media, Berlin

Guillot B (2002) A reappraisal of what we have learnt during three decades of computer simulations on water. J Mol Liq 101:219–260

Guloy AM, Ramlau R, Tang Z, Schnelle W, Baitinger M, Grin Y (2006) A guest-free germanium clathrate. Nature 443:320

Haji-Akbari A (2016) Rating antifreeze proteins: not a breeze. Proc Natl Acad Sci U S A 113:3714–3716
Haji-Akbari A (2018) Forward-flux sampling with jumpy order parameters. J Chem Phys 149 (7):072303
Haji-Akbari A, Debenedetti PG (2015) Direct calculation of ice homogeneous nucleation rate for a molecular model of water. Proc Natl Acad Sci U S A 112:10582
Haji-Akbari A, Debenedetti PG (2017a) Computational investigation of surface freezing in a molecular model of water. Proc Natl Acad Sci U S A 114:3316–3321
Haji-Akbari A, Debenedetti PG (2017b) Perspective: surface freezing in water: a nexus of experiments and simulations. J Chem Phys 147:060901
Haji-Akbari A, DeFever RS, Sarupria S, Debenedetti PG (2014) Suppression of sub-surface freezing in free-standing thin films of monoatomic water. Phys Chem Chem Phys 16:25916–25927
Hales TC (2006) Historical overview of the Kepler conjecture. Discret Comput Geom 36:5–20
Handa YP, Klug D, Whalley E (1986) Difference in energy between cubic and hexagonal ice. J Chem Phys 84:7009–7010
Hemley R, Jephcoat A, Mao H, Zha C, Finger L, Cox D (1987) Static compression of H_2O-ice to 128 GPa (1.28 Mbar). Nature 330(6150):737
Hemley R, Chen L, Mao H (1989) New transformations between crystalline and amorphous ice. Nature 338:638
Herbert RJ, Murray BJ, Dobbie SJ, Koop T (2015) Sensitivity of liquid clouds to homogenous freezing parameterizations. Geophys Res Lett 42:1599–1605
Hermann A, Ashcroft N, Hoffmann R (2012) High pressure ices. Proc Natl Acad Sci U S A 109:745–750
Hirsch K, Holzapfel W (1984) Symmetric hydrogen bonds in ice X. Phys Lett A 101:142–144
Hoose C, Möhler O (2012) Heterogeneous ice nucleation on atmospheric aerosols: a review of results from laboratory experiments. Atmos Chem Phys 12:9817–9854
Hoose C, Kristjánsson JE, Chen J-P, Hazra A (2010) A classical theory-based parameterization of heterogeneous ice nucleation by mineral dust, soot, and biological particles in a global climate model. J Atmos Sci 67:2483–2503
Iglev H, Schmeisser M, Simeonidis K, Thaller A, Laubereau A (2006) Ultrafast superheating and melting of bulk ice. Nature 439:183
Isachenko V, Isachenko E, Katkov II, Montag M, Dessole S, Nawroth F, van der Ven H (2004) Cryoprotectant-free cryopreservation of human spermatozoa by vitrification and freezing in vapor: effect on motility, dna integrity, and fertilization ability. Biol Reprod 71:1167–1173
Jackson CL, McKenna GB (1990) The melting behavior of organic materials confined in porous solids. J Chem Phys 93:9002–9011
Jayaraman A, Klement W Jr, Kennedy G (1963) Melting and polymorphism at high pressures in some group IV elements and III-V compounds with the diamond/zincblende structure. Phys Rev 130:540
Jensen EJ, Toon OB (1997) The potential impact of soot particles from aircraft exhaust on cirrus clouds. Geophys Res Lett 24:249–252
Johari G (1997) The Gibbs–Thomson effect and intergranular melting in ice emulsions: interpreting the anomalous heat capacity and volume of super cooled water. J Chem Phys 107:10154–10165
Johari G (1998) Liquid state of low-density pressure-amorphized ice above its T_g. J Phys Chem B 102:4711–4714
Johari G (2005) Water's size-dependent freezing to cubic ice. J Chem Phys 122:194504
Johari G, Hallbrucker A, Mayer E (1996) Two calorimetrically distinct states of liquid water below 150 kelvin. Science 273:90–92
Johnson CA (1965) Generalization of the Gibbs-Thomson equation. Surf Sci 3:429–444
Johnston JC, Molinero V (2012) Crystallization, melting, and structure of water nanoparticles at atmospherically relevant temperatures. J Am Chem Soc 134:6650–6659

Johnston JC, Kastelowitz N, Molinero V (2010) Liquid to quasicrystal transition in bilayer water. J Chem Phys 133:154516

Joly M, Attard E, Sancelme M, Deguillaume L, Guilbaud C, Morris CE, Amato P, Delort A-M (2013) Ice nucleation activity of bacteria isolated from cloud water. Atmos Environ 70:392–400

Jones D (1974) The free energies of solid-liquid interfaces. J Mater Sci 9:1–17

Jones EB, Stevanovíc V (2017) Polymorphism in elemental silicon: probabilistic interpretation of the realizability of metastable structures. Phys Rev B 96:184101

Kamb B (1964) Ice. II. A proton-ordered form of ice. Acta Crystallogr 17:1437–1449

Kamb B, Davis BL (1964) Ice VII, the densest form of ice. Proc Natl Acad Sci U S A 52:1433–1439

Kaneko T, Bai J, Yasuoka K, Mitsutake A, Zeng XC (2013) New computational approach to determine liquid–solid phase equilibria of water confined to slit nanopores. J Chem Theory Comput 9(8):3299–3310

Kanno H, Angell CA (1977) Homogeneous nucleation and glass formation in aqueous alkali halide solutions at high pressures. J Phys Chem 81:2639–2643

Kastelowitz N, Johnston JC, Molinero V (2010) The anomalously high melting temperature of bilayer ice. J Chem Phys 132(12):124511

Kauzmann W, Eisenberg D (1969) The structure and properties of water. Clarendon Press, Oxford

Kiselev A, Bachmann F, Pedevilla P, Cox SJ, Michaelides A, Gerthsen D, Leisner T (2017) Active sites in heterogeneous ice nucleation—the example of K-rich feldspars. Science 355:367–371

Klotz S, Strässle T, Saitta A, Rousse G, Hamel G, Nelmes R, Loveday J, Guthrie M (2005) In situ neutron diffraction studies of high density amorphous ice under pressure. J Phys Condens Matter 17:S967

Knight C, DeVries A (1989) Melting inhibition and superheating of ice by an antifreeze glycopeptide. Science 245:505–507

Knight CA, De Vries AL, Oolman LD (1984) Fish antifreeze protein and the freezing and recrystallization of ice. Nature 308:295

Knopf DA, Alpert PA (2013) A water activity based model of heterogeneous ice nucleation kinetics for freezing of water and aqueous solution droplets. Faraday Discuss 165:513–534

Knopf DA, Rigg YJ (2011) Homogeneous ice nucleation from aqueous inorganic/organic particles representative of biomass burning: water activity, freezing temperatures, nucleation rates. J Phys Chem A 115(5):762–773

Knopf DA, Alpert PA, Wang B, Aller JY (2011) Stimulation of ice nucleation by marine diatoms. Nat Geosci 4(2):88–90

Knott BC, Molinero V, Doherty MF, Peters B (2012) Homogeneous nucleation of methane hydrates: unrealistic under realistic conditions. J Am Chem Soc 134:19544–19547

Koga K, Zeng XC, Tanaka H (1997) Freezing of confined water: a bilayer ice phase in hydrophobic nanopores. Phys Rev Lett 79:5262

Koga K, Gao G, Tanaka H, Zeng XC (2001) Formation of ordered ice nanotubes inside carbon nanotubes. Nature 412:802

Kohl I, Mayer E, Hallbrucker A (2000) The glassy water–cubic ice system: a comparative study by X-ray diffraction and differential scanning calorimetry. Phys Chem Chem Phys 2:1579–1586

Kolafa J (2014) Residual entropy of ices and clathrates from Monte Carlo simulation. J Chem Phys 140:204507

König H (1943) Eine kubische eismodifikation. Z Kristallogr 105:279–286

Koop T (2004) Homogeneous ice nucleation in water and aqueous solutions. Z Phys Chem 218:1231–1258

Koop T, Luo B, Tsias A, Peter T (2000) Water activity as the determinant for homogeneous ice nucleation in aqueous solutions. Nature 406:611–614

Kristensen J, Cotterill R (1977) On the existence of pre-melting and after melting effects a neutron scattering investigation. Philos Mag 36:437–452

Kuhn T, Earle ME, Khalizov AF, Sloan JJ (2011) Size dependence of volume and surface nucleation rates for homogeneous freezing of supercooled water droplets. Atmos Chem Phys 11:2853–2861

Kumar P, Buldyrev SV, Starr FW, Giovambattista N, Stanley HE (2005) Thermodynamics, structure, and dynamics of water confined between hydrophobic plates. Phys Rev E 72:051503
Kuwayama M, Vajta G, Kato O, Leibo SP (2005) Highly efficient vitrification method for cryopreservation of human oocytes. Reprod BioMed Online 11(3):300–308
Laksmono H, McQueen TA, Sellberg JA, Loh ND, Huang C, Schlesinger D, Sierra RG, Hampton CY, Nordlund D, Beye M, Martin AV, Barty A, Seibert MM, Messerschmidt M, Williams GJ, Boutet S, Amann-Winkel K, Loerting T, Pettersson LGM, Bogan MJ, Nilsson A (2015) Anomalous behavior of the homogeneous ice nucleation rate in "no-man's land". J Phys Chem Lett 6:2826–2832
Lemke S, Handle PH, Plaga LJ, Stern JN, Seidl M, Fuentes-Landete V, Amann-Winkel K, Köster KW, Gainaru C, Loerting T et al (2017) Relaxation dynamics and transformation kinetics of deeply super cooled water: temperature, pressure, doping, and proton/deuteron isotope effects. J Chem Phys 147(3):034506
Li J, Chen H, Stone HA (2013a) Ice lubrication for moving heavy stones to the forbidden city in 15th- and 16th-century China. Proc Natl Acad Sci U S A 110:20023–20027
Li T, Donadio D, Galli G (2013b) Ice nucleation at the nanoscale probes no man's land of water. Nat Commun 4:1887
Lindemann F (1910) Fa lindemann. Phys Z 11:609
Loerting T, Salzmann C, Kohl I, Mayer E, Hallbrucker A (2001) A second distinct structural "state" of high-density amorphous ice at 77 k and 1 bar. Phys Chem Chem Phys 3:5355–5357
Lupi L, Hudait A, Peters B, Grünwald M, Mullen RG, Nguyen AH, Molinero V (2017) Role of stacking disorder in ice nucleation. Nature 551:218–222
Malkin TL, Murray BJ, Brukhno AV, Anwar J, Salzmann CG (2012) Structure of ice crystallized from supercooled water. Proc Natl Acad Sci U S A 109:1041–1045
Malkin TL, Murray BJ, Salzmann CG, Molinero V, Pickering SJ, Whale TF (2015) Stacking disorder in ice I. Phys Chem Chem Phys 17:60–76
Maslin M, Owen M, Betts R, Day S, Jones TD, Ridgwell A (2010) Gas hydrates: past and future geohazard? Philos Trans R Soc A 368:2369–2393
Mayer E (1985) New method for vitrifying water and other liquids by rapid cooling of their aerosols. J Appl Phys 58:663–667
Mayer E, Hallbrucker A (1987) Cubic ice from liquid water. Nature 325:601
Mayer E, Pletzer R (1986) Astrophysical implications of amorphous ice—a microporous solid. Nature 319:298
Mercury L, Vieillard P, Tardy Y (2001) Thermodynamics of ice polymorphs and 'ice-like' water in hydrates and hydroxides. Appl Geochem 16:161–181
Meryman HT (1956) Mechanics of freezing in living cells and tissues. Science 124:515–521
Millot M, Coppari F, Rygg JR, Barrios AC, Hamel S, Swift DC, Eggert JH (2019) Nanosecond X-ray diffraction of shock-compressed superionic water ice. Nature 569:251
Mishima O, Calvert LD, Whalley E (1984) Melting ice' I at 77 K and 10 kbar: a new method of making amorphous solids. Nature 310:393–395
Mishima O, Calvert L, Whalley E (1985) An apparently first-order transition between two amorphous phases of ice induced by pressure. Nature 314:76
Mitchell EH, Raut U, Teolis BD, Baragiola RA (2017) Porosity effects on crystallization kinetics of amorphous solid water: implications for cold icy objects in the outer solar system. Icarus 285:291–299
Mochizuki K, Matsumoto M, Ohmine I (2013) Defect pair separation as the controlling step in homogeneous ice melting. Nature 498:350
Molinero V, Moore EB (2009) Water modeled as an intermediate element between carbon and silicon. J Phys Chem B 113:4008–4016
Moore EB, Molinero V (2010) Ice crystallization in water's "no-man's land". J Chem Phys 132:244504
Moore EB, Molinero V (2011a) Is it cubic? Ice crystallization from deeply supercooled water. Phys Chem Chem Phys 13:20008–20016

Moore EB, Molinero V (2011b) Structural transformation in supercooled water controls the crystallization rate of ice. Nature 479:506–508
Moore EB, Allen JT, Molinero V (2012) Liquid-ice coexistence below the melting temperature for water confined in hydrophilic and hydrophobic nanopores. J Phys Chem C 116:7507–7514
Morishige K, Uematsu H (2005) The proper structure of cubic ice confined in mesopores. J Chem Phys 122:044711
Morishige K, Yasunaga H, Uematsu H (2009) Stability of cubic ice in mesopores. J Phys Chem C 113:3056–3061
Muldrew K, McGann LE (1990) Mechanisms of intracellular ice formation. Biophys J 57 (3):525–532
Muldrew K, McGann LE (1994) The osmotic rupture hypothesis of intracellular freezing injury. Biophys J 66:532–541
Mullen RG, Shea J-E, Peters B (2015) Easy transition path sampling methods: flexible-length aimless shooting and permutation shooting. J Chem Theory Comput 11:2421–2428
Murray BJ (2008) Enhanced formation of cubic ice in aqueous organic acid droplets. Environ Res Lett 3(2):025008
Murray BJ, Bertram AK (2007) Strong dependence of cubic ice formation on aqueous droplet ammonium to sulphate ratio. In: O'Dowd CD, Wagner PE (eds) Nucleation and atmospheric aerosols. Springer, Dordrecht, pp 432–435
Murray BJ, Knopf DA, Bertram AK (2005) The formation of cubic ice under conditions relevant to earth's atmosphere. Nature 434:202–205
Murray BJ, O'Sullivan D, Atkinson JD, Webb ME (2012) Ice nucleation by particles immersed in supercooled cloud droplets. Chem Soc Rev 41:6519–6554
O'Sullivan D, Murray BJ, Ross JF, Whale TF, Price HC, Atkinson JD, Umo NS, Webb ME (2015) The relevance of nanoscale biological fragments for ice nucleation in clouds. Sci Rep 5:8082
Olijve LLC, Meister K, Devries AL, Duman JG, Guo S, Bakker HJ, Voets IK (2016) Blocking rapid ice crystal growth through non-basal plane adsorption of antifreeze proteins. Proc Natl Acad Sci U S A 113:3740–3745
Oxtoby DW (1992) Homogeneous nucleation: theory and experiment. J Phys Condens Matter 4:7627
Padayachee K, Watt MP, Edwards N, Mycock DJ (2009) Cryopreservation as a tool for the conservation of Eucalyptus genetic variability: concepts and challenges. South Forests 71:165–170
Palmer JC, Martelli F, Liu Y, Car R, Panagiotopoulos AZ, Debenedetti PG (2014) Metastable liquid–liquid transition in a molecular model of water. Nature 510:385–388
Palmer JC, Haji-Akbari A, Singh RS, Martelli F, Car R, Panagiotopoulos AZ, Debenedetti PG (2018a) Comment on "The putative liquid-liquid transition is a liquid-solid transition in atomistic models of water" [I and II: J. Chem. Phys. 135, 134503 (2011); J. Chem. Phys. 138, 214504 (2013)]. J Chem Phys 148:137101
Palmer JC, Poole PH, Sciortino F, Debenedetti PG (2018b) Advances in computational studies of the liquid–liquid transition in water and water-like models. Chem Rev 118:9129–9151
Petrenko VF (1993) Structure of ordinary Ice Ih. Part 1: Ideal structure of ice. Technical report, Thayer School of Engineering, Hanover NH
Petrenko VF, Whitworth RW (1999) Physics of ice. Oxford University Press, Oxford
Pluhárová E, Vrbka L, Jungwirth P (2010) Effect of surface pollution on homogeneous ice nucleation: a molecular dynamics study. J Phys Chem C 114:7831–7838
Poole PH, Sciortino F, Essmann U, Stanley E (1992) Phase behaviour of metastable water. Nature 360:324–328
Popovitz-Biro R, Wang J, Majewski J, Shavit E, Leiserowitz L, Lahav L (1994) Induced freezing of supercooled water into ice by self-assembled crystalline monolayers of amphiphilic alcohols at the air-water interface. J Am Chem Soc 116:1179–1191
Potapczuk MG (2013) Aircraft icing research at NASA Glenn Research Center. J Aerosp Eng 26:260–276

Privalov PL (1990) Cold denaturation of protein. Crit Rev Biochem Mol 25:281–306
Rall W (1987) Factors affecting the survival of mouse embryos cryopreserved by vitrification. Cryobiology 24:387–402
Rall WF, Fahy GM (1985) Ice-free cryopreservation of mouse embryos at −196°c by vitrification. Nature 313(6003):573
Reinhardt A, Doye JPK (2012) Free energy landscapes for homogeneous nucleation of ice for a monatomic water model. J Chem Phys 136:054501
Reinhardt A, Doye JPK, Noya EG, Vega C (2012) Local order parameters for use in driving homogeneous ice nucleation with all-atom models of water. J Chem Phys 137:194504
Ronchi C, Hiernaut J (1996) Experimental measurement of pre-melting and melting of thorium dioxide. J Alloys Compd 240:179–185
Rosinski J (1980) Heterogeneous nucleation of ice on surfaces of liquids. J Phys Chem 84:1829–1832
Rosinski J, Lecinski A (1981) Further studies of heterogeneous nucleation of ice at the liquid-liquid interface. J Phys Chem 85:2993–2997
Rosinski J, Kopcewicz B, Sandoval N (1990) Heterogeneous nucleation of ice at the liquid-liquid interface. J Aerosol Sci 21(1):87–96
Rozmanov D, Kusalik PG (2012) Anisotropy in the crystal growth of hexagonal ice, I-h. J Chem Phys 137:094702
Russo J, Romano F, Tanaka H (2014) New metastable form of ice and its role in the homogeneous crystallization of water. Nat Mater 13:733–739
Saika-Voivod I, Sciortino F, Poole PH (2000) Computer simulations of liquid silica: equation of state and liquid–liquid phase transition. Phys Rev E 63(1):011202
Salzmann CG, Radaelli PG, Mayer E, Finney JL (2009) Ice XV: a new thermodynamically stable phase of ice. Phys Rev Lett 103:105701
Samanta A, Tuckerman ME, Yu T-Q, Weinan E (2014) Microscopic mechanisms of equilibrium melting of a solid. Science 346:729–732
Sánchez MA, Kling T, Ishiyama T, van Zadel M-J, Bisson PJ, Mezger M, Jochum MN, Cyran JD, Smit WJ, Bakker HJ et al (2017) Experimental and theoretical evidence for bilayer-by-bilayer surface melting of crystalline ice. Proc Natl Acad Sci U S A 114:227–232
Sanz E, Vega C, Espinosa JR, Caballero-Bernal R, Abascal JLF, Valeriani C (2013) Homogeneous ice nucleation at moderate supercooling from molecular simulation. J Am Chem Soc 135:15008–15017
Sastry S, Angell CA (2003) Liquid–liquid phase transition in supercooled silicon. Nat Mater 2:739
Sastry S, Debenedetti PG, Sciortino F, Stanley HE (1996) Singularity free interpretation of the thermodynamics of supercooled water. Phys Rev E 53:6144
Schumann U, Weinzierl B, Reitebuch O, Schlager H, Minikin A, Forster C, Baumann R, Sailer T, Graf K, Mannstein H, Voigt C, Rahm S, Simmet R, Scheibe M, Lichtenstern M, Stock P, Rüba H, Schäuble D, Tafferner A, Rautenhaus M, Gerz T, Ziereis H, Krautstrunk M, Mallaun C, Gayet J-F, Lieke K, Kandler K, Ebert M, Weinbruch S, Stohl A, Gasteiger J, Grob S, Freudenthaler V, Wiegner M, Ansmann A, Tesche M, Olafsson H, Sturm K (2011) Airborne observations of the Eyjafjalla Volcano ash cloud over Europe during air space closure in April and May 2010. Atmos Chem Phys 11:2245–2279
Schwager B, Chudinovskikh L, Gavriliuk A, Boehler R (2004) Melting curve of H_2O to 90 gpa measured in a laser-heated diamond cell. J Phys Condens Matter 16:S1177
Seeley L, Seidler G (2001) Two-dimensional nucleation of ice from super cooled water. Phys Rev Lett 87:055702
Seidl M, Amann-Winkel K, Handle PH, Zifferer G, Loerting T (2013) From parallel to single crystallization kinetics in high-density amorphous ice. Phys Rev B 88:174105
Sellberg JA, Huang C, McQueen TA, Loh ND, Laksmono H, Schlesinger D, Sierra RG, Nordlund D, Hampton CY, Starodub D, DePonte DP, Beye M, Chen C, Martin AV, Barty A, Wikfeldt KT, Weiss TM, Caronna C, Feldkamp J, Skinner LB, Seibert MM, Messerschmidt M,

Williams GJ, Boutet S, Pettersson LGM, Bogan MJ, Nilsson A (2014) Ultrafast X-ray probing of water structure below the homogeneous ice nucleation temperature. Nature 510:381–384
Seo M, Jang E, Kim K, Choi S, Kim JS (2012) Understanding anisotropic growth behavior of hexagonal ice on a molecular scale: a molecular dynamics simulation study. J Chem Phys 137:154503
Shilling J, Tolbert M, Toon O, Jensen E, Murray BJ, Bertram AK (2006) Measurements of the vapor pressure of cubic ice and their implications for atmospheric ice clouds. Geophys Res Lett 33:L17801
Shultz MJ, Brumberg A, Bisson PJ, Shultz R (2015) Producing desired ice faces. Proc Natl Acad Sci U S A 112:E6096–E6100
Sigurbjörnsson OF, Signorell R (2008) Volume versus surface nucleation in freezing aerosols. Phys Rev E 77:051601
Sliwinska-Bartkowiak M, Jazdzewska M, Huang L, Gubbins KE (2008) Melting behavior of water in cylindrical pores: carbon nanotubes and silica glasses. Phys Chem Chem Phys 10:4909–4919
Smallenburg F, Filion L, Sciortino F (2014) Erasing no-man's land by thermodynamically stabilizing the liquid–liquid transition in tetrahedral particles. Nat Phys 10:653
Song M, Yamawaki H, Fujihisa H, Sakashita M, Aoki K (2003) Infrared investigation on ice VIII and the phase diagram of dense ices. Phys Rev B 68:014106
Speedy RJ (1982) Limiting forms of the thermodynamic divergences at the conjectured stability limits in superheated and supercooled water. J Phys Chem 86:3002–3005
Stanley HE, Teixeira J (1980) Interpretation of the unusual behavior of H_2O and D_2O at low temperatures: tests of a percolation model. J Chem Phys 73:3404–3422
Stanley HE, Teixeira J, Geiger A, Blumberg R (1981) Interpretation of the unusual behavior of H_2O and D_2O at low temperature: are concepts of percolation relevant to the "puzzle of liquid water"? Physica A 106:260–277
Steytler D, Dore J, Wright C (1983) Neutron diffraction study of cubic ice nucleation in a porous silica network. J Phys Chem 87:2458–2459
Stokely K, Mazza MG, Stanley HE, Franzese G (2009) Effect of hydrogen bond cooperativity on the behavior of water. Proc Natl Acad Sci U S A 107:1301–1306
Strobel TA, Somayazulu M, Sinogeikin SV, Dera P, Hemley RJ (2016) Hydrogen-stuffed, quartz-like water ice. J Am Chem Soc 138:13786–13789
Swanson BD (2009) How well does water activity determine homogeneous ice nucleation temperature in aqueous sulfuric acid and ammonium sulfate droplets? J Atmos Sci 66:741–754
Tabazadeh A, Djikaev YS, Reiss H (2002) Surface crystallization of supercooled water in clouds. Proc Natl Acad Sci U S A 99:15873–15878
Takahashi T (1982) On the role of cubic structure in ice nucleation. J Cryst Growth 59:441–449
Takaiwa D, Hatano I, Koga K, Tanaka H (2008) Phase diagram of water in carbon nanotubes. Proc Natl Acad Sci U S A 105:39–43
ten Wolde PR, Frenkel D (1999) Homogeneous nucleation and the Ostwald step rule. Phys Chem Chem Phys 1:2191–2196
ten Wolde PR, Ruiz-Montero MJ, Frenkel D (1996) Numerical calculation of the rate of crystal nucleation in a Lennard-Jones system at moderate undercooling. J Chem Phys 104:9932–9947
Thomson W (1872) 4. On the equilibrium of vapour at a curved surface of liquid. Proc R Soc Edin 7:63–68
Turnbull D (1950) Kinetics of heterogeneous nucleation. J Chem Phys 18:198–203
Van Santen R (1984) The Ostwald step rule. J Phys Chem 88:5768–5769
Varshney D, Singh M (2015) History of lyophilization. In: Varshney D, Singh M (eds) Lyophilized biologics and vaccines. Springer, New York, pp 3–10
Vega C, Abascal JLF (2011) Simulating water with rigid non-polarizable models: a general perspective. Phys Chem Chem Phys 13:19663–19688
Volmer M, Flood H (1934) Tröpfchenbildung in dämpfen. Z Phys Chem 170:273–285
Voorhees PW (1985) The theory of Ostwald ripening. J Stat Phys 38:231–252

Vrbka L, Jungwirth P (2006) Homogeneous freezing of water starts in the subsurface. J Phys Chem B 110:18126–18129
Wang Z, Wang F, Peng Y, Zheng Z, Han Y (2012) Imaging the homogeneous nucleation during the melting of superheated colloidal crystals. Science 338:87–90
Wang Z, Wang F, Peng Y, Han Y (2015) Direct observation of liquid nucleus growth in homogeneous melting of colloidal crystals. Nat Commun 6:6942
Weertman J (1957) On the sliding of glaciers. J Glaciol 3(21):33–38
Wilson DR, Ballard SP (1999) A microphysically based precipitation scheme for the UK Meteorological Office unified model. Q J R Meteor Soc 125:1607–1636
Wilson PW, Haymet ADJ (2009) Effect of solutes on the heterogeneous nucleation temperature of supercooled water: an experimental determination. Phys Chem Chem Phys 11:2679–2682
Wilson TW, Ladino LA, Alpert PA, Breckels MN, Brooks IM, Browse J, Burrows SM, Carslaw KS, Huffman JA, Judd C, Kilthau WP, Mason RH, McFiggans G, Miller LA, Nájera JJ, Polishchuk E, Rae S, Schiller CL, Si M, Temprado JV, Whale TF, Wong JPS, Wurl O, Yakobi-Hancock JD, Abbatt JPD, Aller JY, Bertram AK, Knopf DA, Murray BJ (2015) A marine biogenic source of atmospheric ice-nucleating particles. Nature 525:234–238
Winkel K, Elsaesser MS, Mayer E, Loerting T (2008) Water polyamorphism: reversibility and (dis) continuity. J Chem Phys 128:044510
Xu L, Kumar P, Buldyrev SV, Chen S-H, Poole PH, Sciortino F, Stanley HE (2005) Relation between the widom line and the dynamic crossover in systems with a liquid–liquid phase transition. Proc Natl Acad Sci U S A 102:16558–16562
Xu Q, Sharp I, Yuan C, Yi D, Liao C, Glaeser AM, Minor A, Beeman J, Ridgway MC, Kluth P et al (2006) Large melting-point hysteresis of Ge nanocrystals embedded in SiO_2. Phys Rev Lett 97:155701
Yang WJ, Mochizuki S (2003) Low temperature and cryogenic applications in medicine and surgery. In: Kakac S, Smirnov H, Avelino MR (eds) Low temperature and cryogenic refrigeration, NATO Science Series II: Mathematics, physics and chemistry. Springer, Dordrecht, pp 295–308
Ye Z, Wu J, Ferradi NE, Shi X (2013) Anti-icing for key highway locations: fixed automated spray technology. Can J Civ Eng 40:11–18
Yen F, Chi Z (2015) Proton ordering dynamics of H_2O ice. Phys Chem Chem Phys 17 (19):12458–12461
Zangi R, Mark AE (2003) Monolayer ice. Phys Rev Lett 91:025502
Zeng Q, Li K, Fen-Chong T (2015) Heterogeneous nucleation of ice from supercooled NaCl solution confined in porous cement paste. J Cryst Growth 409:1–9
Zhang X, Sun P, Yan T, Huang Y, Ma Z, Zou B, Zheng W, Zhou J, Gong Y, Sun CQ (2015) Water's phase diagram: from the notion of thermodynamics to hydrogen-bond cooperativity. Prog Solid State Ch 43:71–81
Zhao W-H, Bai J, Yuan L-F, Yang J, Zeng XC (2014a) Ferroelectric hexagonal and rhombic monolayer ice phases. Chem Sci 5:1757–1764
Zhao W-H, Wang L, Bai J, Yuan L-F, Yang J, Zeng XC (2014b) Highly confined water: two-dimensional ice, amorphous ice, and clathrate hydrates. Acc Chem Res 47:2505–2513
Zheligovskaya EA, Malenkov GG (2006) Crystalline water ices. Russ Chem Rev 75:57
Zhu W, Zhao W-H, Wang L, Yin D, Jia M, Yang J, Zeng XC, Yuan L-F (2016) Two-dimensional interlocked pentagonal bilayer ice: how do water molecules form a hydrogen bonding network? Phys Chem Chem Phys 18:14216–14221
Zimmermann F, Weinbruch S, Schütz L, Hofmann H, Ebert M, Kandler K, Worringen A (2008) Ice nucleation properties of the most abundant mineral dust phases. J Geophys Res 113:D23204
Zobrist B, Marcolli C, Peter T, Koop T (2008) Heterogeneous ice nucleation in aqueous solutions: the role of water activity. J Phys Chem A 112:3965–3975

Chapter 4
Ice Formation in Living Organisms

Hans Ramløv and Dennis Steven Friis

4.1 Introduction

One of the most important determinants of the geographical distribution of animals is temperature. Thus, a large number of adaptations to high or low temperatures are found in nature. The adaptations range from behavioural, i.e. migration at certain times of the year, shelter seeking and specific movements via morphological and anatomical to physiological and biochemical. Also, some temperatures may lead to changes in the physical environment, which are of great consequence to the organisms, i.e. ice formation within the environment or within the animals. Many of the adaptations are integrated with each other and thus there is usually not a single specific adaptation that determines the organism's appearance within a certain temperature range. However, it is possible to single out certain adaptations, which clearly are of importance within a certain range of temperatures.

In this chapter, we will be looking at the consequences and problems posed on the organisms exposed to low temperatures and ultimately ice formation, followed by an outline of the various adaptations cold tolerant organisms have evolved to deal with these challenges.

H. Ramløv (✉)
Department of Natural Sciences, Roskilde University, Roskilde, Denmark
e-mail: hr@ruc.dk

D. S. Friis
Copenhagen, Denmark

H. Ramløv, D. S. Friis (eds.), *Antifreeze Proteins Volume 1*,
https://doi.org/10.1007/978-3-030-41929-5_4

4.2 Problems of Cold "per se"

Even before the temperature drops below the equilibrium freezing point (melting point) of an animal's body fluids various problems can occur just as a consequence of the low temperature; as examples can be mentioned fluidity and phase changes in membranes, denaturation and changes in solubility of proteins, changes in solute gradients across membranes either as a function of possible phase changes or as a function of changes in membrane pump and channel conductivity and changes in pH as a function of temperature.

In many exothermic animals exposed to cold, cellular functions slow down or cease if the temperature decreases only a few degrees below the activity temperature and they may enter chill coma where all movement ceases (Mellanby and Gardiner 1939; Semper 1883). The temperature at which this happens is called the critical thermal minimum (CT_{min}). Chill coma can to a large extent be attributed to an inability to regulate ion balance that leads to a decreased membrane potential and reduced excitability of the neuromuscular system. In animals not adapted to cold, this condition is reversible if the cold exposure is not prolonged or if it is mild (Overgaard and MacMillan 2017).

However, in cold tolerant exothermic animals, various adaptations mean that even when the cold exposure is severe and/or prolonged cold exposure may only at the extreme lead to coma and this is a reversible condition. However, some cold tolerant exothermic animals remain active even at quite low temperatures, for example *Rhagium inquisitor* (−15 °C, pers. obs. Hans Ramløv) or Polar fishes that are active at the lowest temperatures to which they are exposed (−1.9 °C), the freezing point of full-strength seawater.

When the temperature gets low and the cold persists for a longer time, problems related to the cold "per se" may occur. Some physico-chemical parameters changes just as a function of the changing temperature; at the temperature range between 0 °C and 10 °C pH changes approximately 0.02 pH units/°C temperature change (Franks 1985; Hochachka and Somero 2002) and O_2 solubility increase approximately by 0.15 mg/l in the range from 10 °C to 0 °C in, for example seawater (Lange et al. 1972). This value may be somewhat different for body fluids but the trend and value range are the same. These conditions lead to an unbalanced state that may, in the end, cause coma or death if the organism is not adapted to cold, i.e. temperatures below the normal activity temperature of the organism (Ramlov 1999; Teets and Denlinger 2013; Teets et al. 2013).

At not too severe temperatures indirect chilling injury sets in. Here cellular metabolism is reduced, which may lead to irreversible changes in ion balance, depletion of cellular ATP and increasing toxic metabolic end products (Overgaard and MacMillan 2017). If the temperature becomes even lower, the animals may be subjected to direct chilling injury (Bayley et al. 2018) which includes actin depolymerization with a following reorganization of the cytoskeleton (Garlick and Robertson 2007; Kim et al. 2006; Lytvyak et al. 2016), protein denaturation (Ben-Naim 2013; Morris and Clarke 1987; Tantos et al. 2009), changes in ion homeostasis based upon changes in thermodynamic enzymatic or membrane pump activity (Overgaard and MacMillan 2017), change of fluidity of cell membranes (Morris and Clarke 1987; Waagner et al. 2013), phospholipids undergoing phase transitions from the liquid crystalline phase to the gel phase (Hazel 1995) leading to different diffusion rates, loss of intracellular metabolites and leakage of various potentially damaging substances into the cytosol (Clerc and Thompson 1995; Hazel and Williams 1990), increased hydrophobic thickness of the lipid membrane affecting the activity of integral proteins (Lee 2004), aggregation of integral proteins (Lee 2004) as a consequence of exclusion from domains of gel-phase lipids (Hazel 1995) and pressure/stretch forces within the membrane (Anishkin et al. 2014) affecting the function of integral proteins, which again leads to slower nerve signals and changes in the membrane potentials throughout the body of the individual.

The problems of cold that are listed above are all due to the cold "per se" and these problems are only accelerating when the temperature falls below the equilibrium freezing point (the melting point) of the body fluids of the organism, where freezing may or may not occur (see the previous chapter for an explanation on nucleation and freezing).

4.3 Rapid Cold Hardening

A range of adaptations collectively called the rapid cold hardening (RCH) response occurs as a response to moderate cold exposure, in both cold and non-cold adapted organisms.

Rapid cold hardening protects a range of insects as well as other arthropods (Bahrndorff et al. 2009; Worland and Convey 2001), both tropical, temperate (Elnitsky and Lee 2009; Lee Jr. et al. 2006b) and polar/alpine (Lee Jr. et al. 2006a; Sinclair et al. 2003; Worland and Convey 2001) from non-freezing injury, even after exposure to a modest cold stress over a short period; minutes to hours (Lee and Denlinger 2010). This response was first demonstrated in non-diapausing stages of the flesh fly *Sarcophaga crassipalpis*, the elm leaf beetle *Xanthogaleruca luteola* and the milkweed bug *Oncopeltus faciatus* (Lee et al. 1987). Rapid cold hardening is a response to small decreases in temperature and it protects against the kind of non-freezing injury called cold-shock or direct chilling injury mentioned above. Initially, it was described in chill susceptible and chill tolerant insects but later it has also been shown to be of importance in, at least certain, freeze tolerant insects such as the Antarctic midge *Belgica antarctica* (Lee Jr. et al. 2006a). In *B. antarctica* freeze tolerance was increased significantly after larvae and adults had been frozen for 1 h at −5 °C (Lee Jr. et al. 2006a). Interestingly the RCH response in *B. antarctica* was shown to confer a greater level of cryoprotection in larvae frozen to a certain temperature (even down to −12 °C) than in larvae supercooled to the same temperature (Kawarasaki et al. 2013). In the same study, Kawarasaki et al. also showed that only 15 min frozen at −5 °C increased the larval cold tolerance. Several studies have reported that the RCH response is not only elicited very quickly as seen above, but it is also lost very quickly, in *B. antarctica* it was lost after 2 h at 2 °C (Kawarasaki et al. 2013).

4.4 Cold Tolerance in Ectothermic Animals Adapted to Cold

As a contrast to the rapid cold hardening response is the seasonal cold hardening response. This response is directed towards a more long-term adaptation to longer-lasting cold conditions and is elicited as a response to decreasing temperature and change in photoperiod (Lee Jr. 2010). Ectothermic organisms living in cold climates such as polar, alpine, and temperate, where they at various periods of time are exposed to temperatures below the melting point of their body fluids, have developed a number of biochemical and physiological adaptations to survive (Ramlov 2000). These adaptations primarily involve the control of ice formation either within the organisms or in such a way that ice formation in the organisms is avoided. However, also adaptations to stabilize macromolecules and biological membranes at low temperatures are among the physiological and biochemical changes taking place during seasonal cold hardening (Cossins et al. 2002; Hazel and Williams 1990; Holmstrup et al. 1999). Overall cold tolerant organisms can be divided into two groups; *freeze tolerant* and *freeze avoiding*. Freeze tolerant organisms survive crystallization of various amounts of their body fluids and to a large extent only extracellularly with some exceptions (Wharton and Ferns 1995). Freeze avoiding organisms, on the other hand, do not survive ice formation in their tissues and their body fluids are supercooled at temperatures below the melting point (Zachariassen 1985). The supercooled condition is metastable, which means that the organism may experience ice formation in its tissues at any time (a nucleation is a stochastic event, see the previous chapter and Wilson et al. (2003)) and in the case of freeze avoiding organisms such an event is lethal (Zachariassen 1985). A special case of freeze avoidance is cryoprotective dehydration (CPD). This is an adaptation to temperatures below the melting point of the body fluids found in small soil-dwelling invertebrates with high cuticular permeability (Holmstrup and Westh 1994). Due to the high cuticular water permeability, animals adapted to cryoprotective dehydration lose a significant amount of the body water to the surroundings when ice forms in the soil (Holmstrup et al. 2002). This continues until the vapour pressure of the body fluids equals that of the surrounding ice. Thus, these animals do not survive the cold in a supercooled state which would be impossible due to the high water permeability of cuticle, when in close contact with moist soil, but in a dehydrated state (Holmstrup 2014). The animals are not frozen when cryoprotectively dehydrated and can thus be considered freeze avoiding. However, dehydration of the animal and hence the cells to some extent makes CPD resembling freeze tolerance.

Freeze avoiding organisms have developed a number of mechanisms to avoid nucleation and inoculation from the surroundings and to stabilize the supercooled body fluids (Ramlov 2000), i.e. adaptations that enable freeze avoiding organisms to stay undercooled for long time periods—even entire lives as, for example the Antarctic toothfish *Dissostichus mawsoni*, which may live for up to 30 years in the ice laden Antarctic waters (Mesa and Vacchi 2001) or to survive extensive dehydration during cold periods (Holmstrup 2014).

4.5 Freeze Tolerance

This section on freeze tolerance is an overview of the various challenges that freeze tolerant organisms face when the water in the body fluids turns into ice at the supercooling point (SCP).

The ice content is of importance to both the extent of the dehydration of the cells and the freeze concentration of solutes in the body fluids, which may impact membrane structure, protein denaturation, distribution of ions as well as changes from aerobic to anaerobic respiration. Freeze concentrations may also lead to the accumulation of reactive oxygen species, ROS. Ice content is controlled by low molecular weight cryoprotectants and ice nucleation, the control of the supercooling point and freezing rate is controlled by ice-nucleating agents as well as low molecular weight cryoprotectants.

In the frozen state, some anaerobic metabolism takes place and this leads to a number of problems for the organisms such as depletion of energy resources, accumulation of metabolic end products which may reach toxic concentrations that leads to acidification of the body fluids. However, some freeze tolerant organisms benefit from the low metabolism in the frozen state, so that survival is higher and fecundity is higher in cohorts which have been frozen than in cohorts which have remained unfrozen but at low temperature during the same time period.

As indicated above, the adaptations to the frozen state encompass the synthesis of ice-nucleating agents, low molecular weight cryoprotectants as well as increase in molecules that aid in the redistribution of water and the low molecular weight cryoprotectants.

4.5.1 *Ice Content*

Some ectothermic animals survive high ice contents in their tissues, for example some insects, i.e. the New Zealand alpine weta *Hemideina maori* (Fig. 4.1) where ice content may be as high as 85% of the body water. This equilibrium ice content is reached already at a temperature of −10 °C and the lower lethal temperature is in the vicinity of −15 °C (Ramlov et al. 1992; Ramlov and Westh 1993). Other insects, for example *Heleomyza borealis* survive 81% of its body water frozen, but in contrast to

Fig. 4.1 The freeze tolerant alpine weta from New Zealand *Hemideina maori*; (**a**) alpha male active during summer, (**b**) Frozen alpha male sitting beneath rock at −6 °C during winter in the Rock and Pillar Range (Photos: Hans Ramløv)

H. maori this is reached at −60 °C (Worland et al. 2000). Other invertebrates such as molluscs (Ansart and Vernon 2003), enchytraeid worms and earthworms also survive freezing of their body fluids many with ice contents of about 65%–70% (Calderon et al. 2009; Fisker et al. 2014; Holmstrup et al. 1999). However, some vertebrates such as various salmanders and frogs survive a fraction of their bodywater crystallized, i.e. the North American wood frog *Lithobates sylvaticus* (formerly *Rana sylvatica*), the Tree Frog *Hyla versicolor* (Layne Jr. and Lee Jr. 1987, 1989), the European brown frog *Rana arvalis* (Voituron et al. 2009) and the Siberean salamander *Salamandrella keyserlingi* (Berman et al. 2016). These vertebrates tolerate about 65% of their body water being crystallized (Layne Jr. and Lee Jr. 1987, 1989), usually reached at higher temperatures than insects and depending on the season and geographical distribution. However, *S. keyserlingi* survive temperatures down to at least −35 °C (Berman et al. 2016). Most freeze tolerant animals only survive extracellular ice formation although intracellular ice formation has been demonstrated in some species, i.e. nematodes (Wharton and Ferns 1995), cells from the fat body of the wheat stem sawfly *Cephus cinctus* (Salt 1961), the goldenrod gall fly *Eurosta solidaginis* (Lee et al. 1993) and in midgut cells from the alpine cockroach *Celattoblatta quinquemaculata* (Worland et al. 2004).

4.5.2 Ice Nucleation and Inoculation in Organisms

Freezing occurs usually as a consequence of heterogeneous nucleation (Zachariassen et al. 2004) where the nucleating agent either is exogenous, i.e. ingested as bacteria, fungi, algae or other substances such as dust particles or is endogenous as synthesized macromolecules that acts as ice nucleating agents (INAs) at relatively high temperatures (Duman and Horwath 1983; Duman et al. 1985; Wilson and Ramlov 1995). These substances arrange water molecules into an ice-like pattern on their surface (Zachariassen et al. 2004), decreasing the free energy of ice formation leading to embryo ice crystals, which upon cooling will reach a critical size and nucleate the supercooled solution (body fluid) (for an overview see papers in Duman et al. (1995); Lee et al. (1995) and the previous chapter). The size of the ice-like embryos is temperature dependent so that at a lower temperature these are larger. At the same time, the critical size of the embryo crystals decreases due to the slower thermal movement of the water molecules. This increases the probability of ice nucleation.

Freezing of the body fluids may also be initiated by inoculation due to ice occurring in the surrounding environment where ice may penetrate through moist skin or through body orifices such as the mouth, anus or trachea (Costanzo and Lee Jr. 2013; Gehrken et al. 1991; Shimada and Riihimaa 1988). This is probably the most common mechanism inhibiting supercooling in ectothermic vertebrates. Cold tolerant amphibians usually select moist hibernacula and the induction of ice crystallization across the moist and highly permeable skin (transintegumentary ice nucleation) of the animals may occur as a consequence of contact with environmental ice (Costanzo and Lee Jr. 1995). This limits the supercooling of these animals (Lee Jr. and Costanzo 1998). It is believed that supercooling is not a problem in large animals because of their large fluid volumes (Costanzo and Claussen 1990)—nucleation temperature has by other investigators been shown to be dependent on volume (Zachariassen et al. 2004). Exogenous ice nucleation occurring due to the ingestion of various microorganisms has been shown in freeze tolerant *Lithobates sylvaticus* (formerly *Rana sylvatica*) (Costanzo and Lee Jr. 1995). The retainment of bacteria with ice nucleating activity in the gut, even after feeding ceases during autumn, may ensure that protective freezing occurs in the hibernating animals even if ice nucleation is not initiated via inoculation across the skin (Costanzo and Lee Jr. 1995).

Ice nucleation occurs at various sites in insects and other organisms where the haemolymph is the most commonly reported site of nucleation (Duman et al. 1980; Wilson and Ramlov 1995) but also sites such as malpighian tubules (Mugnano et al. 1996), fat body cells (Mugnano et al. 1996), muscle cell membranes and epidermis cell membranes (Tsumuki and Konno 1991) have been reported as sites of nucleation. In the malpighian tubules, nucleation is due to the occurrence of calcium phosphate crystals (Mugnano et al. 1996). It is tempting to ask the question, whether nucleation in particular sites is adaptive? In the case of the freeze tolerant cockroach *C. quinquemaculata* the site of extracellular nucleation is not known. However, *C. quinquemaculata* is fortified with antifreeze proteins in the head and gut regions, both in the gut tissue and in the gut content so that ice crystals do not grow in these sites (Wharton et al. 2009). This together with the above-mentioned observation points towards a mechanism where the nucleation may be directed to some sites that tolerate ice formation whereas other areas in the organism are protected against ice growth. Tentatively, it could be speculated that if ice forms in the gut due to INAs in the gut content water will be drawn out of the rest of the body and join the growing ice in the gut, which may be damaging if the ice content here gets too large.

In plants, nucleation may also be limited to certain tissues (Griffith and Yaish 2004). These are thought to control the situation of the ice so that intracellular ice formation is avoided and that recrystallization is inhibited. An interesting case is the ice-nucleating activity of the afro-alpine plant *Lobelia telekii*. This plant has a rather large amount of a slightly viscous solution containing ice-nucleating agents in the inflorescence. During evening when the temperature falls below the melting point of the solution in the inflorescence, ice formation is initiated in the solution a few degrees below the melting point. This ensures that the latent heat of fusion is released to the plant tissues during the night and thus the tissues do not freeze. When temperatures are high during the day, the solution in the inflorescence is thawed and the process can be repeated the next night (Krog et al. 1979).

4.5.3 Freeze Concentration and the Redistribution of Water

Freeze tolerant organisms are faced with a range of problems not only related to the low temperature but also to the ice forming in their tissues. Water expands as it transitions from liquid to ice, and the resulting crystals are, but in a few cases, pure ice. This means that dissolved and suspended molecules are excluded from the ice crystals and will thus be found in small volumes of liquid with high solute concentrations in between the ice crystals "freeze concentration" (Knight et al. 1995; Ramlov et al. 1996). As temperature is decreasing, the vapour pressure of the (supercooled) liquid solution is all the time very slightly above that of the growing ice and thus more water molecules join the ice and the crystals grow. This can be summarized by the fact that the vapour pressure of supercooled aqueous solution is higher than the vapour pressure of ice at the same temperature. Nucleation does not happen readily over the cell membrane, at least at temperatures above −10 °C (Mazur 1977), so the cytosol remains liquid, resulting in an osmotic difference between the cytosol and the extracellular fluid—the extracellular fluid continually being hypertonic to the cell while the temperature is decreasing due to the freeze concentration. This results in water diffusing out of the cells. Most of the water leaves the cells via aquaporins (AQPs) and these molecules are therefore important in promoting freeze tolerance in, for example the goldenrod gall fly *Eurosta solidaginis*, where Philip et al. (2011) have shown that the aquaporin EsAQP1 increased 8 times in abundance from October to December in all tissues tested but was most abundant in the brain of the winter larvae. Some of the AQPs may also transport glycerol (aquaglyceroporins) or other small polar solutes (Finn and Cerda 2015; Hub and de Groot 2008) although this is not the case for EsAQP1, which only transports water. However, some cell membranes are highly permeable to water even without AQPs (Goto et al. 2011).

4.5.4 Freezing Rate

Freezing rate (ice crystallization rate) is of importance in the survival of freeze tolerant animals. Many insects can readily be cooled at 1 °C/min (Ramlov et al. 1992) whereas freeze tolerant vertebrates require a much slower rate of freezing and thus a much lower cooling rate. In experiments with the freeze tolerant salamander *S. keyserlingi* a cooling rate of 4.2 °C/h was used successfully (unpublished results, Hans Ramløv) and successful survival of freeze tolerant adult frogs is achieved at cooling rates in the range of 0.6 °C/h to 4 °C/h (Costanzo and Lee Jr. 1995). Rapid (>100 °C/min) cooling rates may result in intracellular freezing at least in single-cell systems (Mazur et al. 1972). At these cooling rates, the intracellular fluid may supercool and eventually nucleate before an osmotic equilibrium between the intracellular and extracellular fluid has established due to the high cooling rate. At thawing, the intracellular ice crystals recrystallize and this together with the initial ice formation kill the cells. This is obviously most pertinent at slow thawing rates (Mazur et al. 1972) where recrystallization has time to proceed over a longer time period. Slow cooling may lead to more damage than fast as fewer nucleation points will arise and because ice crystals may grow on already existing crystal faces; this, in turn, may lead to larger ice crystals with a larger propensity of recrystallization (Martino et al. 1998). At low cooling rates, the cells will for a relatively long time be exposed to the alterations in solute concentrations during the extra- and intracellular ice growth, which may be damaging due to solution effects such as dehydration, concentration of solutes, changes in pH and precipitation of solutes (Mazur et al. 1972). Summarizing the results of various experiments on freezing rate and survival results in an inverted U-shaped curve where the apex illustrates the optimal cooling rate for that particular animal or cell type. As the osmotic loss of intracellular water across the cell membranes is of importance for cell survival, the water conductance of the cell membrane must have an impact on the position of the apex with relation to the cooling rate (Mazur et al. 1972). The results referred here are based upon single-cell systems (Mazur et al. 1972) but leads to the assumption that similar properties are present in integrated tissues. However, that animals, consisting of a variation of tissues and cell types, are able to survive freezing, indicates that they have evolved adaptations to handle the above-mentioned problem. When freezing occurs in the extracellular fluids and the cells lose water from the cytosol, the volume of the cytosol decreases and the cells begin to shrink. Meryman (1970) suggested that there is a minimum tolerable cell volume below which the cells cannot survive. This may be called the critical minimum cell volume and in freeze tolerant animals this seems to be reached at a body ice content of about 65% of total body water in freeze tolerant vertebrates (Storey and Storey 1993). Although similar ice content seems to be the maximum for many insects (Zachariassen 1985) higher ice contents in these animals have been reported (Ramlov and Westh 1993).

4.5.5 Consequences of Freeze Dehydration

The inter-crystal pockets of liquid constituting the unfrozen fraction, whether that being (in the most cases) extracellular or intracellular, can have pH and ionic strength outside the physiological norm (Franks et al. 1990). This may lead to protein denaturation, cross linking, which causes aggregation and insolubilization (Franks 1985; Taylor 1987). Phase behaviour in membranes can be influenced by the pH of the solution surrounding the membrane. This is because the ionization state of the membrane lipids is, to a large extent, dependent on the pH. This further leads to the observation that the buffer capacity of the medium in which the membrane lipids are dissolved is also of importance for molecular packing of the lipids in the bilayer (Hazel et al. 1992). The volumetric expansion of freezing water may also lead to damage on cellular membranes and tissues (Lee Jr. 2010; Martino et al. 1998; Tursman and Duman 1995; Webb et al. 1994).

Increase in various ions other than H^+, for example cations such as Ca^{2+} may influence signalling. According to a model put forward by Teets et al. (2013), $Ca2^+$, when inside the cell, activates processes via, amongst otherwise unknown mechanisms, calcium/calmodulin-dependent protein kinase II (CaMKII). Those processes are carbohydrate, apoptosis and MAP kinase signalling. Calcium entrance into the cell is via leak canals and intracellular regulation of calcium is via temperature-dependent ATP-dependent calcium export mechanisms, which are inhibited by low temperature. Thus, when freezing occurs the extracellular calcium concentration increases and as the calcium export mechanisms are inhibited both events point towards the establishment of a higher intracellular calcium concentration. An increase in extracellular K^+ (hyperkalemia) disrupts muscle function in some insects due to the disruption of the ion balance (MacMillan et al. 2014). In freeze tolerant insects, this balance will be restored upon thawing (Kristiansen and Zachariassen 2001). During freeze concentration, oxidative damage may be caused when reactive oxygen species (ROS) are formed via Fe^{2+} and Cu^+ through the Fenton reaction although these ions most often are bound to blood proteins or are part of functional groups (Storey and Storey 2013).

In conclusion, it is apparently the dehydration caused by the ice formation leading to changes in hydration of membranes and proteins, changes in ionic strength and cell volume (cell shrinkage) and pH of the extracellular fluid and the cytoplasm that are the major causes of cryoinjury although the physical impact of growing ice crystals on the cells and tissues must also be taken into consideration.

4.5.6 Time Constraints in the Frozen State

Freeze tolerant organisms cannot endure the frozen state indefinitely and not only may the freezing of the body fluids be a challenge but also the thawing process will present challenges to the organisms. Although freeze tolerant organisms may occupy in hibernacula where the temperature rarely changes much some are exposed to highly variable temperatures during winter (Ramlov 1999) and at the end of hibernation temperatures may be rising very slowly or erratically. Such variations in temperature may give rise to recrystallization of the originally formed ice crystals. For organisms adapted to intracellular freezing, recrystallization is a double issue. Not only can the ionic strength and the pH change radically over time, but as the organism is also frozen, there is little it can actively do to prevent this. This presumably creates a strong selective force behind the evolution of AFP or Ice Structuring Proteins (ISPs) in freeze tolerant organisms although some freeze tolerant organisms do not show recrystallization inhibition (Ramlov et al. 1996; Raymond and Wharton 2016).

Freeze tolerant organisms have to be able to endure anoxia (Hoback and Stanley 2001) as a result of ischemia (Storey and Storey 1996) and because gasses do not easily diffuse through ice (Scholander et al. 1953). During anoxia, the frozen organisms accumulate various substances resulting from fermentation of endogenous fermentable fuels. These anaerobic end products are typically lactate, for example in hatchlings of the painted turtle *Chrysomyces picta* and the alpine weta *H. maori* (Hartley et al. 2000), succinate and alanine (Storey and Storey 1985), but also aspartate and urea has been reported to increase in the frozen state (Michaud et al. 2008). Toxopeus and Sinclair hypothesize that frozen insects are only partially hypoxic and that hypoxia may vary amongst various tissues (Toxopeus and Sinclair 2018). The reason for this suggestion is that large tracheae in certain insects do not collapse during freezing and thus exchange of CO_2 seems to take place (Sinclair et al. 2004, 2009). This is probably a special case for insects and may not have pertinence to other organisms where both vessels and respiratory structures are restricted by the ice. The accumulation of various end products of anaerobic metabolism may over time lead to toxic concentrations which will eventually acidify or in other ways be lethal to the frozen organism (Storey and Storey 1985). Such accumulation of anaerobic end products would also give rise to an increase in respiration after thawing when the oxidative breakdown of these substances takes place. This was observed in the alpine weta *H. maori*, where oxygen consumption was increased up to 24 h after thawing compared to the non-frozen controls (unpublished results, H. Ramløv). Considering that at least anaerobic respiration takes place in frozen freeze-tolerant animals leads to the assumption that energy reserves may be depleted during the frozen period and that this may be another limiting factor for the time in which freeze tolerant organisms can survive the frozen state (Storey and Storey 1985). Interestingly Irwin and Lee (2000) showed that overwintering in the frozen state increased survival and fecundity in the goldenrod gall fly larva *E. solidaginis* presumably because the energy expenditure was significantly suppressed in the frozen state compared to the unfrozen controls (Irwin and Lee 2000).

4.5.7 Reactive Oxygen Species and the Frozen State

Toxopeus and Sinclair suggest that reactive oxygen species (ROS) may accumulate in the frozen state (Toxopeus and Sinclair 2018). However, Joanisse and Storey showed that the freeze tolerant gall fly larva *Eurosta solidaginis* had decreased levels of antioxidant enzymes during winter (Joanisse and Storey 1996). This indicates that the oxidative stress is of minor importance in *E. solidaginis* in the frozen state as long as this is not interrupted by thawing events. Repeated thawing events have been shown to decrease survival to 30% compared to 82% survival in *E. solidaginis* larva exposed to a 20-day freeze (Doelling et al. 2014).

Presumably, it is not all organs, tissues or organs that are damaged equally in the frozen state. Marshall and Sinclair showed that repeated cycles of freezing in individuals of the arctiid woolly bear caterpillar *Pyrrharctia isabella* led to higher mortality and more damage to the malpighian tubules and to the haemocytes than in individuals that had not been frozen or had only experienced one freezing event. The fat body, on the other hand, did not show any sign of damage. The results indicate that damage is accumulated during freezing/thawing and that this damage was not sufficiently repaired during the time spent between the freezing events. Further, it also appeared that the damage to the animals was differential so that especially the malpighian tubules and the haemocytes were prone to damage (Marshall and Sinclair 2011). Similar results have been reported from *Eurosta solidaginis*, where Yi and Lee showed that malpighian tubules and haemocytes were most prone to freezing/thawing damage although significant damage had also occurred to the fat body cells (Yi and Lee Jr. 2003). Also in *B. antarctica,* it has been shown that repeated freezing/thawing events are more damaging than prolonged freezing at a constant temperature (Teets et al. 2011). The reasons for these damages are not clear. However, the decreased survival in *E. solidaginis* larvae exposed to repeated freezing/thawing events may be due to oxidative stress caused by a production of ROS that exceeds the capacity of the antioxidant defences (Doelling et al. 2014). Such a ROS production has been reported in several cases (reviewed in Doelling et al. (2014)). Doelling and co-workers showed that *E. solidaginis* larvae experiencing repeated cold exposures increased their aqueous antioxidant capacity by an average of 77%, whereas animals only exposed to a single freeze did not increase their antioxidant capacity. The animals exposed to repeated freeze–thaw events did not show an increase in oxidized proteins. These observations led the authors to conclude that *E. solidaginis* larvae have a robust inducible antioxidant defence (Doelling et al. 2014). The observations in the above-mentioned study are in good accordance with the situation of the hibernaculum of *E. solidaginis*, which is galls above the snow on the stem of goldenrod (Solidago) (Irwin and Lee 2003). These galls are presumably exposed to fluctuations in temperature crossing the freezing point of the larvae, at least during autumn and spring. Thus, time spent in the frozen state cannot last indefinitely and repeated freeze/thaw cycles may be lethal—if sufficient time for recovery is not available.

4.5.8 Control of Ice Formation and Localization

The control of ice formation, propagation and localization seem to be of crucial importance and adaptive in freeze tolerant organisms (Zachariassen and Hammel 1976).

Ice formation velocity and localization in freeze tolerant organisms are often controlled by INAs, which initiates extracellular ice formation at a high temperature below the melting point of the body fluids (Duman 1982; Loomis 1985; Sømme and Zachariassen 1981; Zachariassen and Hammel 1976; Zachariassen and Kristiansen 2000) see also previous remarks on the alpine cockroach *C. quinquemaculata*. Initiating ice formation at a high temperature promotes that the ice crystals grow relatively slowly and that large ice crystals are formed (Salt 1961). This combined with the extracellular situation of the INAs is thought to prevent intracellular freezing and to facilitate the osmotic dehydration of the cells (Zachariassen 1985) as well as extending the time available for the redistribution of cryoprotectants during freezing (Storey and Storey 1988). Comparison of cerambycid and chrysomelid beetles showed that the cerambycids were freeze avoiding with low transcuticular water permeability whereas the chrysomelids were freeze tolerant with high transcuticular water permeability (Zachariassen et al. 2008). Zachariassen et al. suggest that freezing tolerance is an adaptation in the chrysomelids to bring the body fluids into vapour pressure equilibrium with the external ice in the environment. Thereby preventing water loss across the relatively water-permeable cuticle. In this case, the initiation of ice formation at a relatively high temperature helps in minimizing the water loss from the freeze tolerant beetles (Zachariassen et al. 2008).

4.5.9 Low Molecular Weight Cryoprotectants

In combination with INAs ice formation velocity and the amount of ice formed is often modified by the synthesis of low molecular weight substances such as polyols, i.e. glycerol, sorbitol, ethylene glycol (Gehrken 1984; Kostál et al. 2004; Storey and Storey 1988, 1992, 1993, 2013; Zachariassen 1973), sugars, i.e. glucose (Storey and Storey 1993) and amino acids (Ramlov 1999; van der Laak 1982), which slow down ice formation rate (Lee Jr. and Lewis 1985) and reduce colligatively the amount of ice formed in the tissues (Zachariassen 1979), thereby modulating cell dehydration. As most of these cryoprotectants do not cross the cell membrane readily transport mechanisms must be involved if they are required to exert their activity also inside cells where they are not synthesized. Also, the synthesizing cells must be able to transport their products to the surrounding body fluid. The distribution of these cryoprotectants to the cytosol may for some organisms be regulated through the regulation of aquaporins (AQPs). Cope's gray tree frog, *Hyla chrysoscelis*, is freeze tolerant (Costanzo et al. 1992) and accumulates glycerol as a cryoprotectant (Layne Jr. and Jones 2001). In this frog, two AQPs (HC-1 and HC-2) as well as a glycerophorin (HC-3) which are homologs of the mammalian AQP1, AQP2 and AQP3 have been identified (Rojek et al. 2008; Zimmerman et al. 2007). Goldstein et al. hypothesized that in erythrocytes from the grey tree frog, glycerol is transported by the glycerophorin, HC-3, and that this transport would be enhanced in cold-acclimated animals. By measuring the uptake of glycerol and by evaluating the expression of HC-3 in the erythrocytes from warm- and cold-acclimated animals they showed that the expression of HC-3 was upregulated in cold-acclimated animals. However, glycerol permeability was high in warm-acclimated whereas glycerol permeability was lower in cold-acclimated erythrocytes (Goldstein et al. 2010). Thus, the hypothesized correlation was not supported by the study.

The freeze tolerant larvae of the Rice Stem Borer (*Chilo suppressalis*) produce high concentrations of glycerol as a cryoprotectant in both pre-diapause and diapause stages and an increase in freeze tolerance with a concomitant increase in glycerol was observed in diapausing larvae (Tsumuki 2000). Izumi et al. investigated the membrane transport of water and glycerol in the fat body tissues from the *C. suppressalis*. They observed that during freezing water leaves the tissues whereas glycerol enters the cells indicating that glycerol replaces water in the cells. Both the water and glycerol transport was disrupted by $HgCl_2$, a known water channel inhibitor, so that water in tissues did not decrease and neither did glycerol increase (Izumi et al. 2006). The inhibition of the water and glycerol transport by $HgCl_2$ indicates that a water channel is related to this transport. The human aquaporin 3 is transporting both water and glycerol so the authors assume that a similar glycerophorin may be involved in the water/glycerol transport in *C. suppressalis* and thereby avoidance of freezing injury (Izumi et al. 2007).

Other cryoprotectants such as trehalose, sorbitol and glucose must also be transported in and out of the cells via transporters but only a little information on these in relation to cold tolerance and freezing and thawing is available. Trehalose is the normal blood sugar of insects and a transporter of this sugar (TRET1) has been cloned from the cryptobiotic chironomid larva *Polypedilum vanderplanki* (Kikawada et al. 2007).

Glucose was very early shown as a cryoprotectant in freeze tolerant frogs (Storey and Storey 1988; Voituron et al. 2009). In vertebrates, glucose is transported across the membrane by a family of glucose transporters called GLUTs (Pessin and Bell 1992). Some of these are unidirectional (GLUT1, GLUT3, GLUT4) and transport glucose from the plasma into the cells. However, in the liver, a bidirectional (GLUT2) is found. This facilitates the uptake of glucose into the hepatocytes when plasma concentration is high and releases it again when plasma glucose concentration decrease. When comparing glucose transport in the freeze tolerant wood frog *Rana sylvatica* and the non-freeze tolerant leopard frog *Rana pipiens* (King et al. 1993), the level was eight times higher in liver tissue in *R. sylvatica* than in *R. pipiens* at 10 °C. Similarly, they found that at 22 °C the muscle glucose transport was eight times higher in *R. sylvatica* than in *R. pipiens*. The observed difference was in the liver membranes due to a c. five times higher number of glucose transporter sites whereas a similar difference in glucose transporters was not observed for muscle membranes. Later, this was shown that the number of transporters in the liver was increased 8.5 times in autumn collected *R. sylvatica* compared to summer collected specimens and that the transporters cross reacted with rat antibodies for the GLUT2 transporter. There were, however, no seasonal changes that could be detected in the number of transporters in the muscle tissue when summer and autumn collected specimens were compared (King et al. 1995). Unfortunately, no explanation is offered by the authors as to the observed high glucose transport in the *R. sylvatica* muscle tissue compared to the *R. pipiens* muscle tissue. However, some indication of the importance of the glucose transport in freeze tolerant frogs is indicated by a two- to threefold upregulation of *glut4* transcripts in heart of 24 h frozen frogs compared to the unfrozen control indicating a freeze responsive upregulation of the GLUT4 transporter when screening for differential gene expression (Storey 2004).

The low molecular weight cryoprotectants not only impact the rate of ice formation and the ice content. They may also have specific protectant properties through interaction with macromolecular structures in the organisms. Cryoprotectants belong to the group of substances collectively called compatible osmolytes that are accumulated as a response to stress in almost all taxa. These osmolytes must have no or negligible effect on macromolecular function, they may be relatively "cheap" energetically to synthesize and readily available so that their concentration can be quickly adjusted by turning on and off their synthesis and degradation (Hochachka and Somero 1984).

The low molecular weight cryoprotectants are efficient stabilizers of membrane integrity during cooling and dehydration (Yancey 2005). Proteins may be stabilized through the mechanism called preferential exclusion (preferential hydration) (Gekko and Timasheff 1981a). Protein stability is due to the influence of the carbohydrate or polyalcohol on the water structure and is concentration dependent (Avanti et al. 2014). This means that, for example glycerol as well as sugars are excluded from the domain of the protein, which leads to an increase in the chemical potential of the protein, which is thermodynamically unfavourable. Denaturation and unfolding of the protein lead to an increase in the surface of contact between the medium and the solvent and especially the exposure of hydrophobic residues to the solvent which then means that any reduction of the protein–solvent interface will render the system less favourable thermodynamically (Avanti et al. 2014; Gekko and Timasheff 1981a; Gekko and Timasheff 1981b). Some of the low molecular weight cryoprotectants such as trehalose and proline are stabilizing the plasma membranes during freezing and dehydration due to ice formation (Rudolph and Crowe 1985; Rudolph et al. 1986; Yancey 2005). There has been considerable debate as to the nature of this stabilization. Some authors argue that based on experiments both on intact cells and model systems like liposomes, sugars interact directly with the phospholipid interface (Crowe et al. 1984a, b; Luzardo et al. 2000) this is by Andersen et al. (2011) called the "interaction hypothesis" whereas others conclude that there is no specific interaction by, for example trehalose with the lipid head groups (Kent et al. 2014) but that the sugars are preferentially expelled from the hydration zone with the equivalent thermodynamic consequences as seen above for proteins. However, Andersen et al. (2011) conclude that both the hypothesizes may be correct. At low concentrations up to about 0.2 M they find that sugars accumulate at the membrane interface whereas a higher concentrations preferential exclusion of the sugar from the membrane interface takes over. Andersen et al. conclude that membrane–sugar interactions are consisting of two overlapping actions of an attraction and a repulsive component. The attraction component is governed by hydrogen bonding and saturates at intermediate sugar concentrations. The repulsive component is driven by the cosmotropic (water structuring) effect that expels the sugars from the aqueous interfaces. In the context of cryoprotectants, these considerations are of course of limited value, except if one is specifically interested in the mechanism of protection by these substances and their concentration dependence in the organisms (Andersen et al. 2011).

The strategies involved in freezing tolerance are generally focused on controlling ice formation, i.e. ice formation rate, location of the ice, as a minimum extracellular versus intracellular, but also the confinement of the ice to specific compartments as well as the ice fraction size (thereby the extent of organismal dehydration); recrystallization and supercooling capacity (supercooling point). However, also other properties are controlled, i.e. the three-dimensional structure of proteins and membranes; the distribution of water and low molecular weight cryoprotectants between the cytosol and the extracellular fluid via variations in transport capacity of these across the plasma membranes; the antioxidant defence as well as the energy consumption during the frozen period in some cases by the choice of hibernaculum.

4.6 Freeze Avoidance

The majority of cold tolerant organisms, or at least insects, survive the low temperatures in a supercooled state (Lee Jr. 2010). For solutions or a vapour to crystallize, nuclei onto which molecules can condense are required (Franks 1985). In aqueous solutions devoid of any impurities that can initiate heterogeneous nucleation, a large fluctuation in free energy is required to overcome the free energy difference per mole between microcrystals of 10–20 Å and the macroscopic crystalline phase in equilibrium with bulk water at 0 °C (Angell 1983; Franks 1985). As such a free energy fluctuation is a stochastic event, a supercooled metastable solution can crystallize at any time and the chance of it doing so is dependent of the extent of the degree of supercooling, the volume of the solution and the time at which it is supercooled (Bigg 1953). The considerations put forward by Bigg are based upon experiments with pure water, but they can be extended to aqueous solutions in general as long as their viscosity is within certain limits (Angell 1983; Franks 1985). As mentioned earlier, most biological system crystalizes as a consequence of heterogeneous nucleation (Wilson et al. 2003) although this viewpoint has been challenged by some authors (Zachariassen et al. 2004).

Freezing avoidance may, of course, be obtained by migration to other geographical locations or by adoption of hibernacula, which are not subjected to sub-freezing temperatures even during the coldest times of the year or by thick insulating coats and fat layers. However, ectothermic organisms, which do not have the possibility of migrating, rely on a number of behavioural or molecular adaptations to stay supercooled during cold periods. Many insects are rather small in size and has the capacity to supercool, as supercooling is reversely related to the volume (Angell 1982, 1983; Bigg 1953) many insects supercool just as a consequence of their small size when cooled below the melting point of their body fluids. In fact, small volumes of water (a few microlitres) may readily be cooled to −15 °C to −20 °C without freezing. However, the body fluids of supercooled organisms may crystallize at any time but the probability increases with decreasing temperature. Heterogeneous nucleation occurs because of the presence of various impurities in the solution and because of the relationship between the nucleation temperature and the size and the structure of the nucleator (Bigg 1953; Franks 1985; Vali 1995). As freeze avoiding insects survive the cold period in the supercooled metastable state, their supercooling capacity, i.e. the difference between the melting point of the body fluids and their supercooling point (the temperature at which ice crystallizes in the body fluids) often increase as a preparation for the cold period. Some insects survive at extremely low temperatures in the supercooled state, for example the larvae of the willow cone gall, *Rhabdophaga strobiloides*. In winter, these larvae supercool to −56 °C and their haemolymph melting point is −19 °C, thus a supercooling of 37 °C, during summer the supercooling point (SCP) is −26.3 °C which is 25 °C below the melting point (Miller and Werner 1987). Note that the melting point is changing considerably between summer and winter in these animals, which reflects high polyol concentrations in the haemolymph during winter, giving rise to the extremely low SCP. The low SCP seen in summer is reflected in the absence of both endogenous ice nucleators and ice-nucleating bacteria in the phloem constituting the larval food.

In order to survive in a supercooled state, freeze avoiding animals often, as a preparation for the cold period, cease feeding and empty their guts so that ice-nucleating bacteria, food and dust particles are removed (Cannon and Block 1988; Zachariassen 1985). However, this is not always the case as in the freeze avoiding larvae of the cerambycid beetle *Rhagium mordax* (Fig. 4.2) is overwintering in its hibernaculum with the gut filled with food (Alejevski 2012). Some substances, such as lipoproteins, which serve as lipid shuttles during summer are incidentally ice-nucleating agents and on a seasonal basis these Lipoprotein ice nucleators (LPINs) are removed in the winter by the stag beetle *Ceruchus piceus*, which this way achieve a SCP of −26 °C. Because of diapause this is apparently not a problem for the animal (Neven et al. 1986). Other insects may remove their endogenous INs during winter, but not all and hence antifreeze proteins are required to mask the IN activity (Duman 2002).

As is the case with freeze tolerant organisms, freeze avoiding organisms also synthesize low molecular weight cryoprotective substances as a response to cold or as a preparation for cold periods. Low molecular cryoprotectants exert their function primarily via two modes of action: (1) colligatively where they decrease the melting point and the supercooling point of the body fluids (Zachariassen 1985). The colligative melting point depression of water is 1,86 °C/molal. This means that in an 1 molal solution of a nonionizing solute has a melting point (equilibrium freezing

Fig. 4.2 Image of the freeze-avoiding longhorn beetle *Rhagium mordax*. The animal was found in its hibernaculum under the bark on a stump of a birch tree. Both larva and imago of the beetle synthesize very potent antifreeze proteins as a response to low temperatures in their environment (Photo: Thorbjørn Ramløv)

point) of −1.86 °C. The SCP is depressed one to three times the melting point depression by colligatively acting solutes (Duman et al. 1991; Mackenzie et al. 1977), thus stabilizing the supercooled state and (2) via interactions with proteins and membranes stabilizing these as explained above by direct interaction or preferential exclusion. These substances are primarily compatible osmolytes (Yancey et al. 1982) such as polyols, sugars and amino acids (Ramlov 2000) and in some species, the concentrations may reach considerable levels. For example, in the cerambycid beetle *Rhagium inquisitor* (both larvae and imago), the glycerol concentration may be 3 M (Zachariassen 1973) and in a larval wasp the glycerol concentration reaches 5 M (Salt 1961). In other species, the single cryoprotective substance does not reach such high concentrations but instead, a combination of several cryoprotectants are seen, the most common being glycerol and sorbitol (Duman 1980; Miller and Smith 1975; Morrissey and Baust 1976; Storey and Storey 1992) suggesting that the occurrence of a combination of cryoprotectants in certain species may have metabolic advantages as, for example the carbon in glycerol and sorbitol pools have different fates at the end of the cold period; sorbitol is reconverted into glycogen whereas glycerol is not. Another explanation may be that in a multicomponent cryoprotectant system, none of the involved substances reach poisonous levels even though a high osmolality and thereby a low melting point and supercooling point are achieved (Duman et al. 1991).

Some low molecular weight cryoprotectants may also enhance the activity of antifreeze proteins (Li et al. 1998). Duman showed that a combination of antifreeze proteins from the larva of the beetle *D. canadensis* and glycerol and citrate eliminated the activity of haemolymph ice nucleators as well as the ice-nucleating activity of *Pseudomonas syringae* ice-nucleating active bacteria and thereby lowering the supercooling points of solutions containing the bacteria or haemolymph ice nucleators (Duman 2002).

The freeze avoiding strategy is a dangerous one as nucleation, either exogenous or endogenous, or inoculation is a constant danger to supercooled organisms.

But if nucleation can be avoided and inoculation from the environment does not occur, many organisms are able to stay supercooled for long periods of time (Ramlov 2000).

Inoculation may be a problem in freeze-avoiding organisms when living in areas where they are exposed to ice in the environment (Eastman 1993; Sømme 1982). Such organisms rely to a certain extent on the avoidance of their body fluids coming into contact with ice. Antarctic fish, for example which lives in frigid waters where they are supercooled by approximately 1 °C for their whole life, have a very strong muscular sphincter at the termination of their urethra. This forms a blockage for potential dangerous inoculation of the urine via this avenue. The urine is void of AFPs and it is assumed that this is adaptive in the way that the energetic cost would be too high if the urine was protected with AFPs as these would be expelled and thus have to be synthesized "de novo" (Cheng and DeVries 1991). A study of the larvae of the beetle *D. canadensis* showed that AFPs are indeed present in the primary urine and midgut in this species. Thermal hysteresis in primary urine was lower than in midgut fluid and haemolymph. The authors speculate that this is because the primary urine is isolated within the hemocoel, lowering the risk of inoculative freezing. The haemolymph and gut fluid are also thought to contain more potent nucleators than the primary urine (Nickell et al. 2013).

In Antarctic fish most of the body fluids are fortified with AFPs except urine, endolymph and the ocular fluids, which only contain very low concentrations of AFPs, so as seen above these fluids must be protected from inoculation by other means (Cheng and DeVries 1991; Turner et al. 1985). Insects have a rather impermeable wax-covered cuticle and often their tracheae have muscular sphincters so that they can be closed, thus inoculation through the cuticle or via the trachea is unlikely. However, in both fish and insects nothing stops ice from entering the bodies via the mouth and in fish presumably also via the thin gill epithelium. In fish, the skin is evenly moist and as polar marine fish also drink ice-laden water most of the body fluids inevitably will get into contact with ice crystals (Præbel et al. 2009). The moist skin is also a plausible route for the inoculation of marine fish and consequently, skin antifreeze proteins have been described from a number of fishes (Low et al. 2002). These AFPs have no signal sequence and are therefore apparently not exported from their site of synthesis. Although their precise role still remains to be elucidated it can be speculated that they may fortify the skin against invading ice crystals when the fish come in contact with ice or from other exogenous ice nucleators also they may inhibit intracellular ice formation (Low et al. 2002).

In a large number of freeze-avoiding animals, nucleation events, as well as inoculation, seem to be counteracted by the presence of AFPs or ISPs (DeVries and Wohlschlag 1969; Duman 1977; Duman and Devries 1974). These are a diverse class of proteins where the coherence of the class is based more upon the function than on the structure although various structural similarities can be recognized throughout the class. However, what really coalesce these proteins into a class is mechanistic, i.e. the ability to recognize and bind to ice crystals thereby inhibiting their growth in a supercooled solution. Antifreeze proteins may neutralize the ability of nucleators to trigger freezing (Duman 2002; Olsen and Duman 1997) prevent inoculation by the environment whether this may occur via ingested ice crystals or across various thin epithelia (Gehrken 1992; Olsen et al. 1998) and may also counteract nucleation in highly supercooled organisms (Wilson and Leader 1995; Zachariassen and Husby 1982).

4.7 Antifreeze Proteins in Various Organisms

Since the discovery of antifreeze proteins in fish and insects, antifreeze proteins have been discovered in a large number of various organisms across many taxa which include bacteria (Gilbert et al. 2005; Yamashita et al. 2002), diatoms and other algae (Raymond and Fritsen 2001; Raymond et al. 2009), yeast (Lee et al. 2010), lichens (Sidebottom et al. 1999), fungi, plants (Duman and Olsen 1993; Hoshino et al. 2003; Newsted et al. 1994), Nematodes (Wharton et al. 2005), insects (Duman 1977; Duman et al. 2004), collembola (Meier and Zettel 1997) and polar fish (Brown and Sönnichsen 2002; Cheng and DeVries 2002; DeVries and Wohlschlag 1969).

Although the common feature of all these proteins is that they bind to ice crystals and inhibit the growth of these in supercooled solutions, the primary structures and the antifreeze activity (the thermal hysteresis) vary from taxon to taxon. In polar fish, the antifreeze activity, i.e. the separation of the melting point and the hysteresis freezing point are at maximum of about 1.2 °C (Cheng and DeVries 1991). In insects, which have the highest antifreeze activity, it is up to about 8 °C (Kristiansen et al. 2012; Wilkens and Ramlov 2008; Zachariassen et al. 2008) but activities of 13 °C under certain circumstances have been mentioned by Bennett et al. (2005). However, in plants, fungi, etc. the antifreeze activity is significantly lower often not more than a few tenths of a centigrade (Bredow and Walker 2017; Griffith et al. 1992). From the differences in structures and antifreeze activities, it may be inferred that AFPs have evolved independently multiple times. This has been shown amongst marine teleost fishes (Cheng and DeVries 2002) and in insects (Kristiansen et al. 2011) as well as chapters in this book. The large differences across taxa probably reflect the various environmental challenges, which the organisms have to cope with. The sea temperature can never fall below approximately −1.9 °C because of the salt content of the seawater, thus an antifreeze activity of about 1.5 °C ensures that the fish living in ice-laden seawater will never freeze because the low molecular weight substances (NaCl, etc.) together give a colligative freezing point depression of approximately 1.02 °C to 1.16 °C (Eastman 1993; O'Grady and DeVries 1982). Cold tolerant insects, on the other hand, may encounter temperatures significantly lower than what is encountered by fish. Reports on freeze-avoiding insects surviving below −60 °C in nature with supercooling points below −80 °C have been published (Sformo et al. 2010). These insects vitrify at a temperature of approximately −76 °C and survive exposure to −100 °C. It is assumed that their AFPs have a thermal hysteresis of 13 °C but this is partly due to the highly concentrated haemolymph (Bennett et al. 2005). Obviously, a thermal hysteresis of even 13 °C is not enough to prevent ice growth when the animal is exposed to temperatures below −50 °C. However, due to the synthesis of low molecular weight cryoprotectants as well as high ion concentrations due to dehydration the thermal hysteresis is presumably much higher in the intact animal than in vitro (Duman et al. 2004; Olsen et al. 1998; Sformo et al. 2010; Zachariassen and Husby 1982). In contrast to fish and insect AFPs, plant AFPs do not show strong thermal hysteresis, for example winter rye AFP shows a thermal hysteresis of 0.33 °C when this is measured using a Clifton nanolitre osmometer (Griffith et al. 1992). However, the plant AFPs are produced as a response to cold and in freeze tolerant plants and it is speculated that they inhibit the propagation of ice through the cell wall, slow its growth outside the cells as well as inhibit ice recrystallization and ice nucleation (Griffith and Yaish 2004). A similar mechanism of inhibiting recrystallization in animals, which survive intracellular freezing, has been proposed by Wharton et al. (2005).

4.8 Conclusions

Exothermic organisms adapted to tolerate temperatures below their normal temperature of activity or even temperatures below the melting point of their body fluids are faced with a number of problems on both the organismal and the cellular level; changes in Q_{10}, anoxia, formation of reactive oxygen species, ischaemia, imbalance in ion gradients, changes in pH, changes in membrane fluidity, phase changes in membranes, cold denaturation of enzymes and finally ice formation in their tissues extracellular or even intracellular (as well as ice in the environment) and thereby dehydration of the cells as well as expansion of the growing ice and recrystallization. Some animals may tolerate a certain degree of chilling, but will not survive temperatures below the melting point of their body fluids. A large number of organisms do survive temperatures below the melting point of their body fluids either in a frozen, supercooled or dehydrated state. The adaptations to this survival maybe centred around the control of the ice formation and location the ice itself or maybe the control of membrane fluidity, maintaining protein structure, pH and the resurrection of ion gradients upon warming. The adaptations include the synthesis of low molecular weight substances, changes in phospholipid complement in membranes, changes in aquaporin concentration and transporters in membranes, the synthesis of recrystallization inhibiting substances and antifreeze proteins. Many of these adaptations work in concert and one has to realize that there is never ONE key to solve a specific problem faced by an organism but rather a range of more or less subtle adaptations working in concert is making survival in an environment possible. However, should one point to a single adaptation that has evolved several times to survive cold environments by ectothermic organisms it must be the evolution of antifreeze proteins.

References

Alejevski F (2012) Seasonal transcription of antifreeze protein genes in larvae of *Rhagium mordax* and its expression in *Drosophila melanogaster*. M.Sc. Thesis. Department of Science and Environment, Roskilde University

Andersen HD, Wang C, Arleth L, Peters GH, Westh P (2011) Reconciliation of opposing views on membrane-sugar interactions. Proc Natl Acad Sci U S A 108:1874–1878

Angell CA (1982) Supercooled water. In: Franks F (ed) Water and aqueous solutions at subzero temperatures. Springer, Boston, MA, pp 1–81

Angell CA (1983) Supercooled water. Annu Rev Phys Chem 34:593–630

Anishkin A, Loukin SH, Teng J, Kung C (2014) Feeling the hidden mechanical forces in lipid bilayer is an original sense. Proc Natl Acad Sci U S A 111:7898–7905

Ansart A, Vernon P (2003) Cold hardiness in molluscs. Acta Oecol 24:95–102

Avanti C, Saluja V, van Streun EL, Frijlink HW, Hinrichs WL (2014) Stability of lysozyme in aqueous extremolyte solutions during heat shock and accelerated thermal conditions. PLoS One 9:e86244

Bahrndorff S, Tunnacliffe A, Wise MJ, McGee B, Holmstrup M, Loeschcke V (2009) Bioinformatics and protein expression analyses implicate LEA proteins in the drought response of Collembola. J Insect Physiol 55:210–217

Bayley JS, Winther CB, Andersen MK, Gronkjaer C, Nielsen OB, Pedersen TH, Overgaard J (2018) Cold exposure causes cell death by depolarization-mediated Ca(2+) overload in a chill-susceptible insect. Proc Natl Acad Sci U S A 115:E9737–E9744

Ben-Naim A (2013) Theory of cold denaturation of proteins. Adv Biol Chem 3:11

Bennett VA, Sformo T, Walters K, Toien O, Jeannet K, Hochstrasser R, Pan Q, Serianni AS, Barnes BM, Duman JG (2005) Comparative overwintering physiology of Alaska and Indiana populations of the beetle *Cucujus clavipes* (Fabricius): roles of antifreeze proteins, polyols, dehydration and diapause. J Exp Biol 208:4467–4477

Berman DI, Meshcheryakova EN, Bulakhova NA (2016) Extreme negative temperatures and body mass loss in the Siberian salamander (*Salamandrella keyserlingii*, amphibia, hynobiidae). Dokl Biol Sci 468:137–141

Bigg EK (1953) The supercooling of water. Proc Phys Soc Sect B 66:688–694

Bredow M, Walker VK (2017) Ice-binding proteins in plants. Front Plant Sci 8:2153–2153

Brown DJ, Sönnichsen FD (2002) The structure of fish antifreeze proteins. In: Fish antifreeze proteins, vol 1. World Scientific, River Edge, NJ, pp 109–138

Calderon S, Holmstrup M, Westh P, Overgaard J (2009) Dual roles of glucose in the freeze-tolerant earthworm *Dendrobaena octaedra*: cryoprotection and fuel for metabolism. J Exp Biol 212:859–866

Cannon RJC, Block W (1988) Cold tolerance of microarthropods. Biol Rev 63:23–77

Cheng CC, DeVries AL (1991) The role of antifreeze glycopeptides and peptides in the freezing avoidance of cold-water fish. In: di Prisco G (ed) Life under extreme conditions. Springer, Berlin, pp 1–14

Cheng CCM, DeVries AL (2002) Origins and evolution of fish antifreeze proteins. In: Fish antifreeze proteins, vol 1. World Scientific, River Edge, NJ, pp 83–107

Clerc SG, Thompson TE (1995) Permeability of dimyristoyl phosphatidylcholine/dipalmitoyl phosphatidylcholine bilayer membranes with coexisting gel and liquid-crystalline phases. Biophys J 68:2333–2341

Cossins AR, Murray PA, Gracey AY, Logue J, Polley S, Caddick M, Brooks S, Postle T, Maclean N (2002) The role of desaturases in cold-induced lipid restructuring. Biochem Soc Trans 30:1082–1086

Costanzo JP, Claussen DL (1990) Natural freeze tolerance in the terrestrial turtle, *Terrapene carolina*. J Exp Zool 254:228–232

Costanzo JP, Lee RE Jr (1995) Supercooling and ice nucleation in vertebrate ectotherms. APS Press, St. Paul, pp 221–237

Costanzo JP, Lee RE Jr (2013) Avoidance and tolerance of freezing in ectothermic vertebrates. J Exp Biol 216:1961–1967

Costanzo JP, Wright MF, Lee RE (1992) Freeze tolerance as an overwintering adaptation in Cope's grey treefrog (*Hyla chrysoscelis*). J Exp Zool 283(3):221–225

Crowe JH, Crowe LM, Chapman D (1984a) Infrared spectroscopic studies on interactions of water and carbohydrates with a biological membrane. Arch Biochem Biophys 232:400–407

Crowe JH, Whittam MA, Chapman D, Crowe LM (1984b) Interactions of phospholipid monolayers with carbohydrates. Biochim Biophys Acta 769:151–159

DeVries AL, Wohlschlag DE (1969) Freezing resistance in some Antarctic fishes. Science 163:1073–1075

Doelling AR, Griffis N, Williams JB (2014) Repeated freezing induces oxidative stress and reduces survival in the freeze-tolerant goldenrod gall fly, *Eurosta solidaginis*. J Insect Physiol 67:20–27

Duman JG (1977) The role of macromolecular antifreeze in the darkling beetle, *Meracantha contracta*. J Comp Physiol 115:279–286

Duman JG (1980) Factors involved in overwintering survival of the freeze tolerant beetle, *Dendroides canadensis*. J Comp Physiol 136:52–59

Duman JG (1982) Insect antifreezes and ice-nucleating agents. Cryobiology 19:613–627

Duman JG (2002) The inhibition of ice nucleators by insect antifreeze proteins is enhanced by glycerol and citrate. J Comp Physiol B 172:163–168

Duman JG, Devries AL (1974) Freezing resistance in winter flounder *Pseudopleuronectes americanus*. Nature 247:237–238

Duman J, Horwath K (1983) The role of hemolymph proteins in the cold tolerance of insects. Annu Rev Physiol 45:261–270

Duman JG, Olsen MT (1993) Thermal hysteresis protein activity in bacteria, fungi, and phylogenetically diverse plants. Cryobiology 30:322–328

Duman JG, Patterson JL, Kozak JJ, DeVries AL (1980) Isopiestic determination of water binding by fish antifreeze glycoproteins. Biochim Biophys Acta 626:332–336

Duman JG, Neven LG, Beals JM, Olson KR, Castellino FJ (1985) Freeze-tolerance adaptations, including haemolymph protein and lipoprotein nucleators, in the larvae of the cranefly *Tipula trivittata*. J Insect Physiol 31:1–8

Duman JG, Wu DW, Xu L, Tursman D, Olsen MT (1991) Adaptations of insects to subzero temperatures. Q Rev Biol 66:387–410

Duman JG, Olsen TM, Yeung KL, Jerva F (1995) The roles of ice nucleators in cold tolerant invertebrates. APS Press, St. Paul, pp 201–219

Duman JG, Bennett V, Sformo T, Hochstrasser R, Barnes BM (2004) Antifreeze proteins in Alaskan insects and spiders. J Insect Physiol 50:259–266

Eastman JT (1993) 11 - Antifreeze Glycopeptides. In: Eastman JT (ed) Antarctic fish biology. Academic Press, San Diego, pp 178–201

Elnitsky MA, Lee RE (2009) The rapid cold-hardening response in insects: ecological significance and physiological mechanisms. J Exp Biol 216:3937–3945

Finn RN, Cerda J (2015) Evolution and functional diversity of aquaporins. Biol Bull 229:6–23

Fisker KV, Overgaard J, Sorensen JG, Slotsbo S, Holmstrup M (2014) Roles of carbohydrate reserves for local adaptation to low temperatures in the freeze tolerant oligochaete *Enchytraeus albidus*. J Comp Physiol B 184:167–177

Franks F (1985) Biophysics and biochemistry at low temperatures. Cambridge University Press, Cambridge [Cambridgeshire]

Franks F, Mathias SF, Hatley RH (1990) Water, temperature and life. Philos Trans R Soc Lond Ser B Biol Sci 326:517–531. discussion 531–533

Garlick KM, Robertson RM (2007) Cytoskeletal stability and heat shock-mediated thermoprotection of central pattern generation in *Locusta migratoria*. Comp Biochem Physiol A Mol Integr Physiol 147:344–348

Gehrken U (1984) Winter survival of an adult bark beetle *Ips acuminatus* Gyll. J Insect Physiol 30:421–429

Gehrken U (1992) Inoculative freezing and thermal hysteresis in the adult beetles *Ips acuminatus* and *Rhagium inquisitor*. J Insect Physiol 38:519–524

Gehrken U, Strømme A, Lundheim R, Zachariassen KE (1991) Inoculative freezing in overwintering tenebrionid beetle, *Bolitophagus reticulatus* Panz. J Insect Physiol 37:683–687

Gekko K, Timasheff SN (1981a) Mechanism of protein stabilization by glycerol: preferential hydration in glycerol-water mixtures. Biochemistry 20:4667–4676

Gekko K, Timasheff SN (1981b) Thermodynamic and kinetic examination of protein stabilization by glycerol. Biochemistry 20:4677–4686

Gilbert JA, Davies PL, Laybourn-Parry J (2005) A hyperactive, Ca2+-dependent antifreeze protein in an Antarctic bacterium. FEMS Microbiol Lett 245:67–72

Goldstein DL, Frisbie J, Diller A, Pandey RN, Krane CM (2010) Glycerol uptake by erythrocytes from warm- and cold-acclimated Cope's gray treefrogs. J Comp Physiol B Biochem Syst Environ Physiol 180:1257–1265

Goto SG, Philip BN, Teets NM, Kawarasaki Y, Lee RE Jr, Denlinger DL (2011) Functional characterization of an aquaporin in the Antarctic midge *Belgica antarctica*. J Insect Physiol 57:1106–1114

Griffith M, Yaish MW (2004) Antifreeze proteins in overwintering plants: a tale of two activities. Trends Plant Sci 9:399–405

Griffith M, Ala P, Yang DS, Hon WC, Moffatt BA (1992) Antifreeze protein produced endogenously in winter rye leaves. Plant Physiol 100:593–596

Hartley LM, Packard MJ, Packard GC (2000) Accumulation of lactate by supercooled hatchlings of the painted turtle (*Chrysemys picta*): implications for overwinter survival. J Comp Physiol B 170:45–50

Hazel JR (1995) Thermal adaptation in biological membranes: is homeoviscous adaptation the explanation? Annu Rev Physiol 57:19–42

Hazel JR, Williams EE (1990) The role of alterations in membrane lipid composition in enabling physiological adaptation of organisms to their physical environment. Prog Lipid Res 29:167–227

Hazel JR, McKinley SJ, Williams EE (1992) Thermal adaptation in biological membranes: interacting effects of temperature and pH. J Comp Physiol B 162:593–601

Hoback WW, Stanley DW (2001) Insects in hypoxia. J Insect Physiol 47:533–542

Hochachka PW, Somero GN (1984) Biochemical adaptation. Princeton University Press, New Jersey

Hochachka PW, Somero GN (2002) Biochemical adaptation. Oxford University Press, Oxford, New York

Holmstrup M (2014) The ins and outs of water dynamics in cold tolerant soil invertebrates. J Therm Biol 45:117–123

Holmstrup M, Westh P (1994) Dehydration of earthworm cocoons exposed to cold - a novel cold-hardiness mechanism. J Comp Physiol B 164:312–315

Holmstrup M, Costanzo JP, Lee RE (1999) Cryoprotective and osmotic responses to cold acclimation and freezing in freeze-tolerant and freeze-intolerant earthworms. J Comp Physiol B 169:207–214

Holmstrup M, Bayley M, Ramlov H (2002) Supercool or dehydrate? An experimental analysis of overwintering strategies in small permeable arctic invertebrates. Proc Natl Acad Sci U S A 99:5716–5720

Hoshino T, Kiriaki M, Ohgiya S, Fujiwara M, Kondo H, Nishimiya Y, Yumoto I, Tsuda S (2003) Antifreeze proteins from snow mold fungi. Can J Bot 81:1175–1181

Hub JS, de Groot BL (2008) Mechanism of selectivity in aquaporins and aquaglyceroporins. Proc Natl Acad Sci U S A 105:1198–1203

Irwin JT, Lee RE (2000) Mild winter temperatures reduce survival and potential fecundity of the goldenrod gall fly, *Eurosta solidaginis* (Diptera: Tephritidae). J Insect Physiol 46:655–661

Irwin JT, Lee JRE (2003) Cold winter microenvironments conserve energy and improve overwintering survival and potential fecundity of the goldenrod gall fly, *Eurosta solidaginis*. Oikos 100:71–78

Izumi Y, Sonoda S, Yoshida H, Danks HV, Tsumuki H (2006) Role of membrane transport of water and glycerol in the freeze tolerance of the rice stem borer, *Chilo suppressalis* Walker (Lepidoptera: Pyralidae). J Insect Physiol 52:215–220

Izumi Y, Sonoda S, Tsumuki H (2007) Effects of diapause and cold-acclimation on the avoidance of freezing injury in fat body tissue of the rice stem borer, *Chilo suppressalis* Walker. J Insect Physiol 53:685–690

Joanisse DR, Storey KB (1996) Oxidative damage and antioxidants in *Rana sylvatica*, the freeze-tolerant wood frog. Am J Phys 271:R545–R553

Kawarasaki Y, Teets NM, Denlinger DL, Lee RE Jr (2013) The protective effect of rapid cold-hardening develops more quickly in frozen versus supercooled larvae of the Antarctic midge, *Belgica antarctica*. J Exp Biol 216:3937–3945

Kent B, Hunt T, Darwish TA, Hauss T, Garvey CJ, Bryant G (2014) Localization of trehalose in partially hydrated DOPC bilayers: insights into cryoprotective mechanisms. J R Soc Interface 11:20140069

Kikawada T, Saito A, Kanamori Y, Nakahara Y, Iwata K, Tanaka D, Watanabe M, Okuda T (2007) Trehalose transporter 1, a facilitated and high-capacity trehalose transporter, allows exogenous trehalose uptake into cells. Proc Natl Acad Sci U S A 104:11585–11590

Kim M, Robich RM, Rinehart JP, Denlinger DL (2006) Upregulation of two actin genes and redistribution of actin during diapause and cold stress in the northern house mosquito, *Culex pipiens*. J Insect Physiol 52:1226–1233

King PA, Rosholt MN, Storey KB (1993) Adaptations of plasma membrane glucose transport facilitate cryoprotectant distribution in freeze-tolerant frogs. Am J Phys 265:R1036–R1042

King PA, Rosholt MN, Storey KB (1995) Seasonal changes in plasma membrane glucose transporters enhance cryoprotectant distribution in the freeze-tolerant wood frog. Can J Zool 73:1–9

Knight CA, Wen D, Laursen RA (1995) Nonequilibrium antifreeze peptides and the recrystallization of ice. Cryobiology 32:23–34

Kostál V, Tamura M, Borovanska M, Zahradníčková H (2004) Enzymatic capacity for accumulation of polyol cryoprotectants changes during diapause development in the adult red firebug, *Pyrrhocoris apterus*. Physiol Entomol 29(4):344–355

Kristiansen E, Zachariassen KE (2001) Effect of freezing on the transmembrane distribution of ions in freeze-tolerant larvae of the wood fly *Xylophagus cinctus* (Diptera, Xylophagidae). J Insect Physiol 47:585–592

Kristiansen E, Ramlov H, Hojrup P, Pedersen SA, Hagen L, Zachariassen KE (2011) Structural characteristics of a novel antifreeze protein from the longhorn beetle *Rhagium inquisitor*. Insect Biochem Mol Biol 41:109–117

Kristiansen E, Wilkens C, Vincents B, Friis D, Lorentzen AB, Jenssen H, Lobner-Olesen A, Ramlov H (2012) Hyperactive antifreeze proteins from longhorn beetles: some structural insights. J Insect Physiol 58:1502–1510

Krog JO, Zachariassen KE, Larsen B, Smidsrød O (1979) Thermal buffering in afro-alpine plants due to nucleating agent-induced water freezing. Nature 282:300–301

Lange R, Staaland H, Mostad A (1972) The effect of salinity and temperature on solubility of oxygen and respiratory rate in oxygen-dependent marine invertebrates. J Exp Mar Biol Ecol 9:217–229

Layne JR Jr, Jones AL (2001) Freeze tolerance in the gray treefrog: cryoprotectant mobilization and organ dehydration. J Exp Zool 290:1–5

Layne JR Jr, Lee RE Jr (1987) Freeze tolerance and the dynamics of ice formation in wood frogs (*Rana sylvatica*) from southern Ohio. Can J Zool 65:2062–2065

Layne JR Jr, Lee RE Jr (1989) Seasonal variation in freeze tolerance and ice content of the tree frog *Hyla versicolor*. J Exp Zool 249:133–137

Lee AG (2004) How lipids affect the activities of integral membrane proteins. Biochim Biophys Acta 1666:62–87

Lee RE, Denlinger DL (2010) Rapid cold-hardening: ecological significance and underpinning mechanisms. In: Denlinger DL, Lee JRE (eds) Low temperature biology of insects. Cambridge University Press, Cambridge, pp 35–58

Lee RE, Chen C-P, Denlinger DL (1987) A rapid cold-hardening process in insects. Science 238:1415

Lee RE, McGrath JJ, Todd Morason R, Taddeo RM (1993) Survival of intracellular freezing, lipid coalescence and osmotic fragility in fat body cells of the freeze-tolerant gall fly *Eurosta solidaginis*. J Insect Physiol 39:445–450

Lee MR, Lee RE Jr, Strong-Gunderson JM, Minges SR (1995) Isolation of ice-nucleating active bacteria from the freeze-tolerant frog, *Rana sylvatica*. Cryobiology 32:358–365

Lee JK, Park KS, Park S, Park H, Song YH, Kang S-H, Kim HJ (2010) An extracellular ice-binding glycoprotein from an Arctic psychrophilic yeast. Cryobiology 60:222–228

Lee RE Jr (2010) A primer on insect cold-tolerance. Cambridge University Press, Cambridge

Lee RE Jr, Costanzo JP (1998) Biological ice nucleation and ice distribution in cold-hardy ectothermic animals. Annu Rev Physiol 60:55–72

Lee RE Jr, Lewis EA (1985) Effect of temperature and duration of exposure on tissue ice formation in the gall fly *Eurosta solidaginis* diptera tephritidae. Cryo Letters 6:25–34

Lee RE Jr, Elnitsky MA, Rinehart JP, Hayward SAL, Sandro LH, Denlinger DL (2006a) Rapid cold-hardening increases the freezing tolerance of the Antarctic midge *Belgica antarctica*. J Exp Biol 209:399

Lee RE Jr, Damodaran K, Yi SX, Lorigan GA (2006b) Rapid cold-hardening increases membrane fluidity and cold tolerance of insect cells. Cryobiology 52:459–463

Li N, Andorfer CA, Duman JG (1998) Enhancement of insect antifreeze protein activity by solutes of low molecular mass. J Exp Biol 201:2243–2251

Loomis SH (1985) Seasonal changes in the freezing tolerance of the intertidal pulmonate gastropod *Melampus bidentatus* say. Can J Zool 63:2021–2025

Low W-K, Lin Q, Ewart KV, Fletcher GL, Hew CL (2002) The skin-type antifreeze polypeptides: a new class of type I AFPs. In: Fish antifreeze proteins, vol 1. World Scientific, River Edge, NJ, pp 161–186

Luzardo MC, Amalfa F, Nunez AM, Diaz S, Biondi De Lopez AC, Disalvo EA (2000) Effect of trehalose and sucrose on the hydration and dipole potential of lipid bilayers. Biophys J 78:2452–2458

Lytvyak E, Olstad DL, Schopflocher DP, Plotnikoff RC, Storey KE, Nykiforuk CI, Raine KD (2016) Impact of a 3-year multi-centre community-based intervention on risk factors for chronic disease and obesity among free-living adults: the healthy Alberta communities study. BMC Public Health 16:344

Mackenzie AP, Derbyshire W, Reid DS, Richards Rex E, Franks F (1977) Non-equilibrium freezing behaviour of aqueous systems. Philos Trans R Soc Lond B Biol Sci 278:167–189

MacMillan HA, Findsen A, Pedersen TH, Overgaard J (2014) Cold-induced depolarization of insect muscle: differing roles of extracellular K+ during acute and chronic chilling. J Exp Biol 217:2930–2938

Marshall KE, Sinclair BJ (2011) The sub-lethal effects of repeated freezing in the woolly bear caterpillar *Pyrrharctia isabella*. J Exp Biol 214:1205–1212

Martino M, Otero L, Sanz P, Zaritzky N (1998) Size and location of ice crystals in pork frozen by high-pressure-assisted freezing as compared to classical methods. Meat Sci 50(3):303–313

Mazur P (1977) The role of intracellular freezing in the death of cells cooled at supraoptimal rates. Cryobiology 14:251–272

Mazur P, Leibo SP, Chu EH (1972) A two-factor hypothesis of freezing injury. Evidence from Chinese hamster tissue-culture cells. Exp Cell Res 71:345–355

Meier P, Zettel J (1997) Cold hardiness in *Entomobrya nivalis* (Collembola, Entomobryidae): annual cycle of polyols and antifreeze proteins, and antifreeze triggering by temperature and photoperiod. J Comp Physiol B 167:297–304

Mellanby K, Gardiner JS (1939) Low temperature and insect activity. Proc R Soc L Ser B 127:473–487

Meryman HT (1970) The exceeding of a minimum tolerable cell volume in hypertonic suspension as a cause of freezing injury. In: Ciba foundation symposium - the frozen cell. Ciba, Churchill, pp 51–67

Mesa ML, Vacchi M (2001) Age and growth of high Antarctic notothenioid fish. Antarct Sci 13:227–235

Michaud MR, Benoit JB, Lopez-Martinez G, Elnitsky MA, Lee RE, Denlinger DL (2008) Metabolomics reveals unique and shared metabolic changes in response to heat shock, freezing and desiccation in the Antarctic midge, *Belgica antarctica*. J Insect Physiol 54:645–655

Miller LK, Smith JS (1975) Production of threitol and sorbitol by an adult insect: association with freezing tolerance. Nature 258:519–520

Miller LK, Werner R (1987) Extreme supercooling as an overwintering strategy in three species of willow gall insects from interior Alaska USA. Oikos 49:253–260

Morris GJ, Clarke A (1987) Cells at low temperatures. In: Grout BWW, Morris GJ (eds) The effects of low temperatures on biological systems. Edward Arnold, London, pp 72–119

Morrissey RE, Baust JG (1976) The ontogeny of cold tolerance in the gall fly, *Eurosta solidagensis*. J Insect Physiol 22:431–437

Mugnano J, Lee R, Taylor R (1996) Fat body cells and calcium phosphate spherules induce ice nucleation in the freeze-tolerant larvae of the gall fly *Eurosta solidaginis* (Diptera, Tephritidae). J Exp Biol 199:465–471

Neven LG, Duman JG, Beals JM, Castellino FJ (1986) Overwintering adaptations of the stag beetle, *Ceruchus piceus*: removal of ice nucleators in the winter to promote supercooling. J Comp Physiol B 156:707–716

Newsted JW, Polvi S, Papish B, Kendall E, Saleem M, Koch M, Hussain A, Cutler AJ, Georges F (1994) A low molecular weight peptide from snow mold with epitopic homology to the winter flounder antifreeze protein. Biochem Cell Biol 72:152–156

Nickell PK, Sass S, Verleye D, Blumenthal EM, Duman JG (2013) Antifreeze proteins in the primary urine of larvae of the beetle *Dendroides canadensis*. J Exp Biol 216:1695–1703

O'Grady SM, DeVries AL (1982) Osmotic and ionic regulation in polar fishes. J Exp Mar Biol Ecol 57:219–228

Olsen TM, Duman JG (1997) Maintenance of the supercooled state in the gut fluid of overwintering pyrochroid beetle larvae, *Dendroides canadensis* : role of ice nucleators and antifreeze proteins. J Comp Physiol B 167:114–122

Olsen T, Sass S, Li N, Duman J (1998) Factors contributing to seasonal increases in inoculative freezing resistance in overwintering fire-colored beetle larvae *Dendroides canadensis*. J Exp Biol 201(Pt 10):1585–1594

Overgaard J, MacMillan HA (2017) The integrative physiology of insect chill tolerance. Annu Rev Physiol 79:187–208

Pessin JE, Bell GI (1992) Mammalian facilitative glucose transporter family: structure and molecular regulation. Annu Rev Physiol 54:911–930

Philip BN, Kiss AJ, Lee RE Jr (2011) The protective role of aquaporins in the freeze-tolerant insect *Eurosta solidaginis*: functional characterization and tissue abundance of EsAQP1. J Exp Biol 214:848–857

Præbel K, Hunt B, Hunt LH, DeVries AL (2009) The presence and quantification of splenic ice in the McMurdo Sound Notothenioid fish, *Pagothenia borchgrevinki* (Boulenger, 1902). Comp Biochem Physiol A Mol Integr Physiol 154:564–569

Ramlov H (1999) Microclimate and variations in haemolymph composition in the freezing-tolerant New Zealand alpine weta *Hemideina maori* Hutton (Orthoptera : Stenopelmatidae). J Comp Physiol B 169:224–235

Ramlov H (2000) Aspects of natural cold tolerance in ectothermic animals. Hum Reprod 15(Suppl 5):26–46

Ramlov H, Westh P (1993) Ice formation in the freeze-tolerant alpine weta *Hemideina maori* Hutton (Orthoptera, Stenopelmatidae). Cryo-Letters 14:169–176

Ramlov H, Bedford J, Leader J (1992) Freezing tolerance of the New-Zealand Alpine weta, *Hemideina maori* Hutton [Orthoptera, Stenopelmatidae]. J Therm Biol 17:51–54

Ramlov H, Wharton DA, Wilson PW (1996) Recrystallization in a freezing tolerant antarctic nematode, *Panagrolaimus davidi*, and an alpine weta, *Hemideina maori* (Orthoptera: Stenopelmatidae). Cryobiology 33:607–613

Raymond JA, Fritsen CH (2001) Semipurification and ice recrystallization inhibition activity of ice-active substances associated with Antarctic photosynthetic organisms. Cryobiology 43:63–70

Raymond MR, Wharton DA (2016) The ability to survive intracellular freezing in nematodes is related to the pattern and distribution of ice formed. J Exp Biol 219:2060–2065

Raymond JA, Janech MG, Fritsen CH (2009) Novel ice-binding proteins from a psychrophilic Antarctic alga (Chlamydomonadaceae, chlorophyceae)1. J Phycol 45:130–136

Rojek A, Praetorius J, Frokiaer J, Nielsen S, Fenton RA (2008) A current view of the mammalian aquaglyceroporins. Annu Rev Physiol 70:301–327

Rudolph AS, Crowe JH (1985) Membrane stabilization during freezing: the role of two natural cryoprotectants, trehalose and proline. Cryobiology 22:367–377

Rudolph AS, Crowe JH, Crowe LM (1986) Effects of three stabilizing agents--proline, betaine, and trehalose--on membrane phospholipids. Arch Biochem Biophys 245:134–143

Salt RW (1961) Principles of insect cold-hardiness. Annu Rev Entomol 6:55–74

Scholander PF, Flagg W, Hock RJ, Irving L (1953) Studies on the physiology of frozen plants and animals in the Arctic. J Cell Physiol Suppl 42:1–56

Semper K (1883) Natural conditions of existence as they affect animal life. Kegan Paul, Trench & CO., London

Sformo T, Walters K, Jeannet K, Wowk B, Fahy GM, Barnes BM, Duman JG (2010) Deep supercooling, vitrification and limited survival to −100{degrees}C in the Alaskan beetle *Cucujus clavipes puniceus* (Coleoptera: Cucujidae) larvae. J Exp Biol 213:502–509

Shimada K, Riihimaa A (1988) Cold acclimation, inoculative freezing and slow cooling: essential factors contributing to the freezing-tolerance in diapausing larvae of *Chymomyza costata*. Drosophilidae, Diptera

Sidebottom CM, Smallwood MF, Byass LJ (1999) Frozen product, vol. WO99/37673

Sinclair BJ, Klok CJ, Scott MB, Terblanche JS, Chown SL (2003) Diurnal variation in supercooling points of three species of Collembola from Cape Hallett, Antarctica. J Insect Physiol 49:1049–1061

Sinclair BJ, Klok CJ, Chown SL (2004) Metabolism of the sub-Antarctic caterpillar *Pringleophaga marioni* during cooling, freezing and thawing. J Exp Biol 207:1287–1294

Sinclair BJ, Gibbs AG, Lee WK, Rajamohan A, Roberts SP, Socha JJ (2009) Synchrotron x-ray visualisation of ice formation in insects during lethal and non-lethal freezing. PLoS One 4(12): e8259

Sømme L (1982) Supercooling and winter survival in terrestrial arthropods. Comp Biochem Physiol A Mol Integr Physiol 73:519–543

Sømme L, Zachariassen KE (1981) Adaptations to low temperature in high altitude insects from Mount Kenya. Ecol Entomol 6:199–204

Storey KB (2004) Strategies for exploration of freeze responsive gene expression: advances in vertebrate freeze tolerance. Cryobiology 48:134–145
Storey JM, Storey KB (1985) Freezing and cellular metabolism in the gall fly larva, *Eurosta solidaginis*. J Comp Physiol B 155:333–337
Storey KB, Storey JM (1988) Freeze tolerance in animals. Physiol Rev 68:27–84
Storey KB, Storey JM (1992) Natural freeze tolerance in ectothermic vertebrates. Annu Rev Physiol 54:619–637
Storey KB, Storey JM (1993) Cellular adaptations for freezing survival of amphibians and reptiles, vol 2. JAI Press, London, pp 101–129
Storey KB, Storey JM (1996) Natural freezing survival in animals. Annu Rev Ecol Syst 27:365–386
Storey KB, Storey JM (2013) Molecular biology of freezing tolerance. Compr Physiol 3:1283–1308
Tantos A, Friedrich P, Tompa P (2009) Cold stability of intrinsically disordered proteins. FEBS Lett 583:465–469
Taylor MJ (1987) Physico-chemical principles in low temperature biology. In: Grout BWW, Morris GJ (eds) The effects of low temperatures on biological systems. Edmond Arnold, London, pp 3–71
Teets NM, Denlinger DL (2013) Physiological mechanisms of seasonal and rapid cold-hardening in insects. Physiol Entomol 38:105–116
Teets NM, Kawarasaki Y, Lee RE Jr, Denlinger DL (2011) Survival and energetic costs of repeated cold exposure in the Antarctic midge, *Belgica antarctica*: a comparison between frozen and supercooled larvae. J Exp Biol 214:806–814
Teets NM, Yi SX, Lee RE Jr, Denlinger DL (2013) Calcium signaling mediates cold sensing in insect tissues. Proc Natl Acad Sci U S A 110:9154–9159
Toxopeus J, Sinclair BJ (2018) Mechanisms underlying insect freeze tolerance. Biol Rev Camb Philos Soc 93:1891–1914
Tsumuki H (2000) Review of low temperature tolerance and ice nuclei in insects, with special emphasis on larvae of the rice stem borer, *Chilo suppressalis* Walker. Jpn J Appl Entomol Zool 44(3):149–154
Tsumuki H, Konno H (1991) Tissue distribution of the ice-nucleating agents in larvae of the rice stem borer, *Chilo suppressalis* Walker (Lepidoptera: Pyralidae). Cryobiology 28(4):376–381
Turner JD, Schrag JD, Devries AL (1985) Ocular freezing avoidance in antarctic fish. J Exp Biol 118:121
Tursman D, Duman JG (1995) Cryoprotective effects of thermal hysteresis protein on survivorship of frozen gut cells from the freeze-tolerant centipede *Lithobius forficatus*. J Exp Zool 272 (4):249–257
Vali G (1995) Principles of ice nucleation. APS Press, St. Paul, pp 1–28
van der Laak S (1982) Physiological adaptations to low temperature in freezing-tolerant *Phyllodecta laticollis* beetles. Comp Biochem Physiol A Physiol 73:613–620
Voituron Y, Paaschburg L, Holmstrup M, Barre H, Ramlov H (2009) Survival and metabolism of *Rana arvalis* during freezing. J Comp Physiol B 179:223–230
Waagner D, Bouvrais H, Ipsen JH, Holmstrup M (2013) Linking membrane physical properties and low temperature tolerance in arthropods. Cryobiology 67:383–385
Webb MS, Uemura M, Steponkus PL (1994) A comparison of freezing injury in oat and rye: two cereals at the extremes of freezing tolerance. Plant Physiol 104:467
Wharton DA, Ferns DJ (1995) Survival of intracellular freezing by the Antarctic nematode *Panagrolaimus davidi*. J Exp Biol 198:1381–1387
Wharton DA, Barrett J, Goodall G, Marshall CJ, Ramlov H (2005) Ice-active proteins from the Antarctic nematode *Panagrolaimus davidi*. Cryobiology 51:198–207
Wharton DA, Pow B, Kristensen M, Ramlov H, Marshall CJ (2009) ICE-active proteins and cryoprotectants from the New Zealand alpine cockroach, *Celatoblatta quinquemaculata*. J Insect Physiol 55:27–31

Wilkens C, Ramlov H (2008) Seasonal variations in antifreeze protein activity and haemolymph osmolality in larvae of the beetle *Rhagium mordax* (Coleoptera : Cerambycidae). CryoLetters 29:293–300
Wilson PW, Leader JP (1995) Stabilization of supercooled fluids by thermal hysteresis proteins. Biophys J 68:2098–2107
Wilson P, Ramlov H (1995) Hemolymph ice nucleating proteins from the New-Zealand alpine Weta *Hemideina maori* (Orthoptera, Stenopelmatidae). Comp Biochem Physiol B 112:535–542
Wilson PW, Heneghan AF, Haymet AD (2003) Ice nucleation in nature: supercooling point (SCP) measurements and the role of heterogeneous nucleation. Cryobiology 46:88–98
Worland MR, Convey P (2001) Rapid cold hardening in Antarctic microarthropods. Funct Ecol 15:515–524
Worland MR, Block WI, Grubor-Lajsic G (2000) Survival of *Heleomyza borealis* (Diptera, Heleomyzidae) larvae down to − 60°C. Physiol Entomol 25:1–5
Worland MR, Wharton DA, Byars SG (2004) Intracellular freezing and survival in the freeze tolerant alpine cockroach *Celatoblatta quinquemaculata*. J Insect Physiol 50:225–232
Yamashita Y, Nakamura N, Omiya K, Nishikawa J, Kawahara H, Obata H (2002) Identification of an antifreeze lipoprotein from *Moraxella sp.* of Antarctic origin. Biosci Biotechnol Biochem 66:239–247
Yancey PH (2005) Organic osmolytes as compatible, metabolic and counteracting cytoprotectants in high osmolarity and other stresses. J Exp Biol 208:2819–2830
Yancey PH, Clark ME, Hand SC, Bowlus RD, Somero GN (1982) Living with water stress: evolution of osmolyte systems. Science 217:1214–1222
Yi SX, Lee RE Jr (2003) Detecting freeze injury and seasonal cold-hardening of cells and tissues in the gall fly larvae, *Eurosta solidaginis* (Diptera: Tephritidae) using fluorescent vital dyes. J Insect Physiol 49:999–1004
Zachariassen KE (1973) Seasonal variation in hemolymph osmolality and osmotic contribution of glycerol in adult *Rhagium inquisitor* L. (Col., Cerambycidae). Cryo Letters 29(4):293–300
Zachariassen KE (1979) The mechanism of the cryoprotective effect of glycerol in beetles tolerant to freezing. J Insect Physiol 25:29–32
Zachariassen KE (1985) Physiology of cold tolerance in insects. Physiol Rev 65:799–832
Zachariassen KE, Hammel HT (1976) Nucleating agents in the haemolymph of insects tolerant to freezing. Nature 262:285–287
Zachariassen KE, Husby JA (1982) Antifreeze effect of thermal hysteresis agents protects highly supercooled insects. Nature 298:865–867
Zachariassen KE, Kristiansen E (2000) Ice nucleation and antinucleation in nature. Cryobiology 41:257–279
Zachariassen KE, Kristiansen E, Pedersen SA, Hammel HT (2004) Ice nucleation in solutions and freeze-avoiding insects-homogeneous or heterogeneous? Cryobiology 48:309–321
Zachariassen KE, Li NG, Laugsand AE, Kristiansen E, Pedersen SA (2008) Is the strategy for cold hardiness in insects determined by their water balance? A study on two closely related families of beetles: Cerambycidae and Chrysomelidae. J Comp Physiol B 178:977–984
Zimmerman SL, Frisbie J, Goldstein DL, West J, Rivera K, Krane CM (2007) Excretion and conservation of glycerol, and expression of aquaporins and glyceroporins, during cold acclimation in Cope's gray tree frog *Hyla chrysoscelis*. Am J Physiol Regul Integr Comp Physiol 292:R544–R555

Part II
The Biology of Antifreeze Proteins

Chapter 5
Fish Antifreeze Proteins

Arthur L. DeVries

5.1 Introduction

Temperature is one of the key factors governing the distributions of living organisms. For ectothermic animals unable to conserve heat derived from their metabolism by means of thermoregulation, the low end of the thermal range is represented by the polar regions. High-latitude polar oceans are perennially frigid and even though water and ice lose heat to the frigid air, the polar seawater temperatures seldom drop below the freezing point (fp) of −1.9 °C because of the insulating effects of the ice cover. Despite the freezing seawater, polar marine environments are found to sustain life in most of the major animal taxa, and some of them in large numbers.

It is well known that low temperatures depress the rate of chemical reactions involved in energy-producing metabolic pathways and other biochemical and physiological processes. In the polar and northern cold-temperate oceans, one of the major physiological/biochemical challenges for ectothermic animals is to generate sufficient energy for activity, growth, and reproduction at near zero or subzero temperatures. The presence of abundant active ectothermic invertebrate and vertebrate faunas in the polar seas indicate that they have adapted to the low temperatures over evolutionary time. The largest marine vertebrate taxon is teleost fish, and polar species have provided many examples of biochemical and physiological adaptations to subzero temperatures (Somero et al. 2017). While metabolic and biochemical adaptations are essential for life in the cold, the most obvious environmental threat to survival is freezing death. Marine teleost fishes have body fluids that are hyposmotic to seawater, which means they have a higher freezing point than seawater, and would freeze at −1.9 °C in ice-laden marine waters. Unlike some other ectothermic animals

A. L. DeVries (✉)
University of Illinois, Urbana, IL, USA
e-mail: adevries@illinois.edu

H. Ramløv, D. S. Friis (eds.), *Antifreeze Proteins Volume 1*,
https://doi.org/10.1007/978-3-030-41929-5_5

that are freeze tolerant, freezing is always lethal in teleost fish (Scholander et al. 1957) and thus the ability to avoid freezing is of the utmost importance. To avoid freezing, many polar fishes have evolved antifreeze proteins (APs) that lower the freezing point of their body fluids below the freezing point of seawater (Devries 1971). Since the discovery of the first fish antifreeze in the Antarctic fish in 1968 (Devries and Wohlschlag 1969; Devries 1970), significant advances have been made in the identification and characterization of the different types of APs in fish, insects, and microorganisms all with a considerable compositional and structural variation. The structure function of these various APs is an active area of research (see Chap. 4 in Vol. II). Technological advances in protein chemistry have enabled very detailed descriptions of AP structures, identification of ice-binding faces of the APs and hypotheses of how they bind to ice crystals and inhibit ice growth, leading to many reports in the literature in these aspects (see reviews Harding et al. 1999, 2003; Davies et al. 2002). In contrast, understanding the in vivo role of antifreeze proteins in freeze avoidance in their natural environments has lagged behind. Ultimately, the biological importance of fish APs are their contributions to the avoidance of freezing and thus maintaining teleost diversity in otherwise uninhabitable freezing seas, and decidedly worth our thorough understanding. This chapter reviews the role of APs in freezing avoidance of polar and north temperate fishes, with particular attention to the integumental resistance to ice entry, the presence of endogenous ice, and its fate, how much blood AP is needed to survive in the various polar sea habitats and how APs interface with the anatomical and physiological systems of the fish.

5.2 The Freezing Challenge for Hyposmotic Fish

Oceanic seawater has a salinity of around 34 parts per a thousand water molucules (34 o/oo), equivalent to an osmotic concentration of about 1000 mmol/kg and a fp of approximately −1.9 °C. The body fluids of marine teleost fishes are strongly hyposmotic to seawater with their primary osmolyte being NaCl. The osmotic concentration expressed as osmolality of most marine teleost blood is around 350 mmol/kg, which is equivalent to an fp of −0.65 °C (Black 1951). Fishes living in the polar waters generally have a higher blood salt content than temperate fishes, which is reflected in their relatively high blood osmolality (~450–550 mmol/kg) and it accounts for the majority of the depression of the equilibrium fp or melting point (mp) of −1.0 to −1.1 °C (Enevoldsen et al. 2003; Devries 1982; Denstad et al. 1987). However, these equilibrium fps are still significantly higher than the fp of seawater (−1.9 °C), and in the high latitude ice-covered polar seas, the fishes are therefore undercooled by as much as 0.9 °C relative to seawater (Devries and Wohlschlag 1969). Although this undercooling appears small, in the presence of ice it cannot exist and freezing occurs. In ice-laden polar waters and shallow winter temperate waters, fishes frequently come into contact with ice and in polar regions may acquire ice crystals on their skin, gill surfaces, and in their intestinal fluids.

Fishes drink seawater and partially desalinate it in the intestine in order to maintain their water balance resulting in intestinal fluid, that is, isosmotic to the blood but again hyposmotic to seawater. Ingestion of food and drinking seawater may result in ingested ice, which provides the necessary template for ice growth of the undercooled intestinal fluid and if not protected would lead to freezing and death. It only takes one nucleation source or entry of a single ice crystal to initiate freezing in an undercooled fish even if the amount of undercooling is small.

Separation of environmental ice from undercooled body fluids by impermeable epithelial tissues may preserve the undercooled state in some cases, and the intact skin of fish has been shown to be a physical barrier to ice propagation (Valerio et al. 1992b; Turner et al. 1985). However, the skin can occasionally be breached by physical abrasion and the thin single-cell delicate epithelium of the gill filaments as well as the intestinal epithelium remains vulnerable to injury and ice entry. Thus the undercooled blood requires some form of protection from freezing in the presence of ice.

To avoid freezing, few fishes have synthesized and accumulated colligative solutes in the blood but this sort of protection is rare. There is only one known example of teleost fish in ice-laden environments; the rainbow smelt *Osmerus mordax*, which raises its serum osmolyte concentration until isosmotic with seawater. It achieves this by the synthesis of large amounts of glycerol, which it secretes into the blood where it is continually lost via the gills (Raymond 1992). Polar fishes and most north temperate fishes in danger of freezing have evolved and rely on APs, which bind to and inhibit the growth of ice crystals that enter their blood or other body fluids at temperatures below their equilibrium fp (mp). The APs substantially lower the freezing temperature (Devries 1971) and may slightly alter the mp so that it is higher than what is expected based on colligative relationships (Cziko et al. 2014). The separation between the fp and the mp first described in 1971 is the hallmark of fish APs and is referred to as a "thermal hysteresis" in the freezing–melting behavior (Devries 1971). The hysteresis value is often equated with antifreeze activity and gives reasonable estimates of the amount of APs in the blood when comparing fish species having the same type of APs.

5.3 Types of Antifreeze Proteins

The antifreeze proteins were first discovered in 1967 and were called that because their function was clearly one that allowed fish to avoid freezing in ice-laden seawater (Devries and Wohlschlag 1969; Devries 1970). With the discovery of APs in insects and microbes, it has been suggested that they be called ice structuring proteins (Clarke et al. 2002) but recently they have generally been referred to as ice-binding proteins (Davies 2014) although they do alter the ice growth morphology in their presence. Although they all bind to ice not all effectively, inhibit ice growth as they do in fishes and in some insects. With fishes, their function is clearly one of ensuring freeze avoidance by binding to ice crystals that enter the fish, inhibiting their growth, and therefore avoiding freezing to death. Thus it is reasonable for this class of ice-binding proteins to be called "antifreeze" proteins because that clearly is their function. With many of the microbial and plant ice-binding proteins, there is no significant ice growth inhibition and thus their function is not entirely clear although they do have moderate inhibition of recrystallization activities. In one microbe, they appear to be involved in attachment to the ice cover in frozen lakes to ensure that they remain in the photic zone (Bar et al. 2016).

Fish APs were first isolated from the Antarctic fishes of the suborder Notothenioidei, which includes five endemic Antarctic families, and were identified as a series of different sized glycoproteins of similar composition (Devries and Wohlschlag 1969). Notothenioid fishes refer to the species within the five Antarctic families most of which possess AFGPs. The first peptide or small protein antifreezes were identified in the winter flounder, *Pseudopleuronectes americanus* (Duman and Devries 1974b, 1976). Three additional types of peptide antifreeze were subsequently discovered in northern and southern hemisphere fishes, and characterized. These different types of APs differ in amino acid composition, sequence, and secondary structure (Fig. 5.1) (Fletcher et al. 2001; Cheng and Detrich III 2007). In some cases, near-identical APs have evolved in unrelated taxa, while different types are present in closely related taxa, epitomizing the processes of independent and convergent evolution (Cheng 1998).

5.3.1 Antifreeze Glycoproteins (AFGPs)

The Antarctic notothenioid fishes and several northern and Arctic cods, and gadids, synthesize alanine-rich glycopeptide antifreeze molecules (AFGPs) (Van Voorhies et al. 1978; Raymond et al. 1975; Fletcher et al. 1981; Devries 1970; Praebel and Ramløv 2005). The AFGPs are composed of tripeptide repeats consisting of a peptide backbone of Ala/Pro-Ala-Thr, that is, *O*-glycosylated at the threonine side chain with the disaccharide β-D-galactosyl-(1 $\rightarrow$ 3)-α-*N*-acetyl-D-galactosamine (Shier and Devries 1975). The different sizes are made up of differing numbers of this basic glycotripeptide repeat and vary from 4 to 55 repeats (Cheng 1996). Gel electrophoresis initially identified eight distinct sizes based on their mobilities and were designated AFGPs 1–8, with AFGP 1 being the largest size (~33 kDa) and the AFGP 8 the smallest (2.6 kDa) (Devries et al. 1970). The two smallest sizes, AFGPs 7 and 8, are the most abundant isoforms, comprising about two-thirds of circulating AFGPs in most notothenioid species but with a few having equal amounts of the large and small sizes. These eight sizes have been subcategorized into two groups, with the large molecular mass ones (AFGPs 1–5) having a peptide backbone of only Ala-Ala-Thr repeats while the smaller ones, AFGPs 6–8, have a Pro occasionally replacing the first Ala of some of the internal tripeptide repeats. Better gel resolution and protein visualization techniques subsequently revealed several more bands within each of these two groups (Fig. 5.2), but the historical numbering system persists. Besides size heterogeneity, the small AFGPs though appearing as single

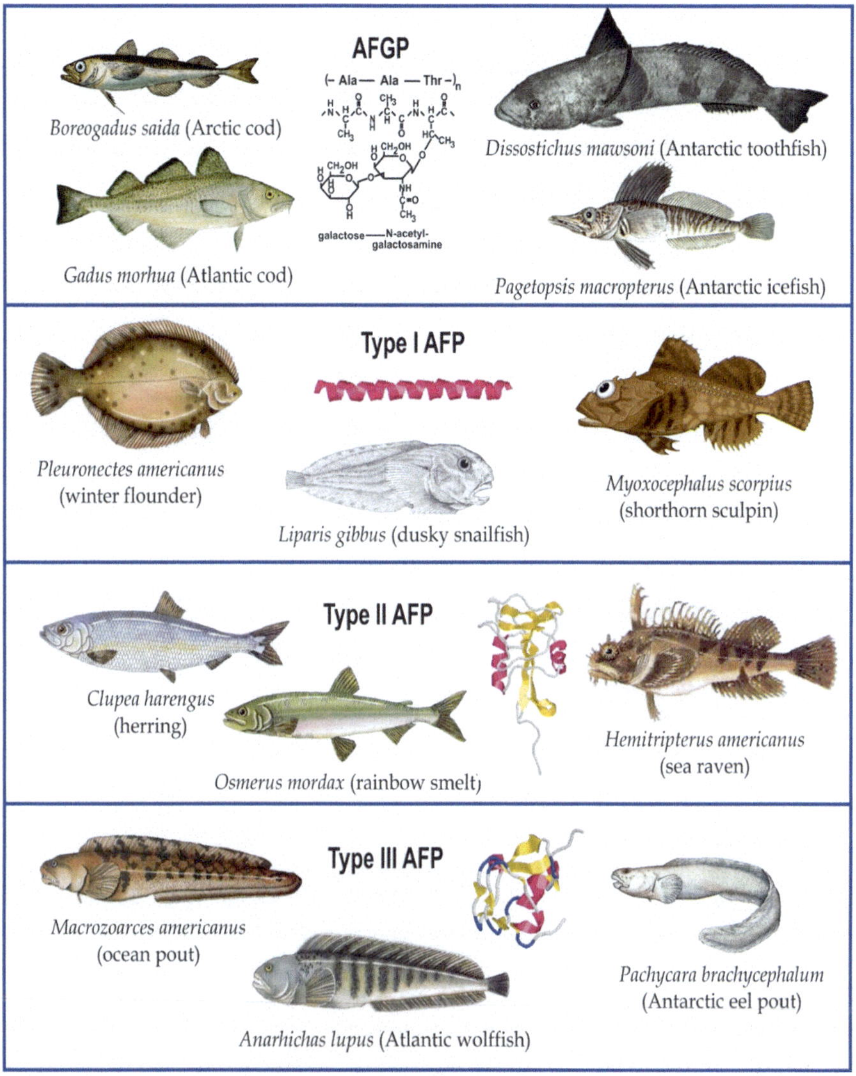

Fig. 5.1 Antifreeze protein structures are shown along with the corresponding fish species. The structure of the gadid and notothenioid AFGP is from structural modeling. The structures of type I, II, and III AFPs are experimentally solved by X-ray crystallography and/or NMR. Modified and reprinted from *The Physiology of Polar Fishes, Fish Physiology* Vol. 22, Elsevier Academic Press

bands on electrophoretic gels were found to be heterogeneous in composition with the Pro for Ala replacements occurring at one or more Ala positions in the internal tripeptide repeats (Morris et al. 1978). Recent AFGP gene sequences show that the large size AFGPs also contain a few Pro for Ala substitutions, and the coding sequence for a large size isoform with as many as 88 tripeptide repeats (equivalent

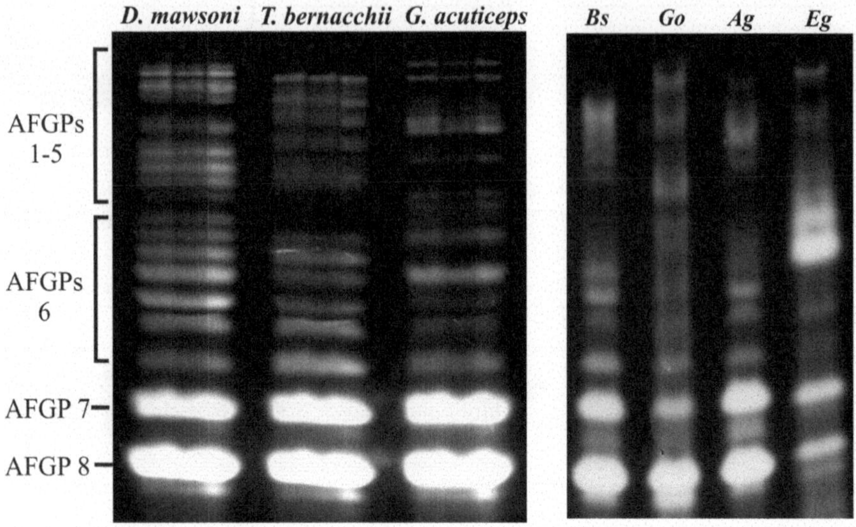

Fig. 5.2 AFGPs from Antarctic notothenioid fish (left) and northern codfishes (right) on gradient polyacrylamide gel electrophoresis showing the size heterogeneity. The AFGPs were labeled with a fluorescent tag (fluorescamine) and the fluorescent intensity of each isoform is representative of physiological concentration. AFGPs 6 and above resolved into multiple bands. The notothenioid AFGPs were run in triplicates. The cod species are, Bs—*Boreogadus saida*, Ag—*Arctogadus glacialis*, Go—*Gadus ogac*, and Eg—*Eleginus gracilis*. Reprinted from *The Physiology of Polar Fishes, Fish Physiology* Vol. 22, Elsevier Academic Press

to ~54 kDa AFGP) was found (Chen et al. 1997a) but not resolved with gel electrophoresis.

Although not as well characterized in the Arctic gadids, *Boreogadus*, *Arctogadus*, *Gadus*, *Microgadus*, and *Eleginus*, their AFGPs also occur as a heterogeneous family of size isoforms (Fig. 5.2) (Praebel and Ramløv 2005; Chen et al. 1997b), but in some of the north temperate gadids the largest AFGP sizes are smaller than their Antarctic counterparts. The saffron cod, *Eleginus gracilis* has primarily two very prominent isoforms in the AFGP 6 size range (Raymond et al. 1975) (Fig. 5.2). The cod AFGPs are nearly identical in primary structure, including Pro for Ala substitutions, to the Antarctic version with the minor difference being some of the Thr residues in the tripeptide repeats are replaced by an Arg residue (O'grady et al. 1982c; Chen et al. 1997b).

Both the Antarctic notothenioid and northern cod AFGPs are encoded by large polyprotein genes, where each gene encodes a series of AFGP molecules linked in tandem by small cleavable spacer amino acid residues (O'grady et al. 1982c; Hsiao et al. 1990; Chen et al. 1997a, b). Detailed analyses of the AFGP genes from these two unrelated groups of fish show that they have evolved their respective AFGP from different genomic origins. The Antarctic notothenioid AFGP gene was derived from a trypsinogen-like protease gene, and its evolutionary origin was estimated to be about 7–15 million years ago, which correlates with the time estimate of sea-level glaciation and freezing of the Antarctic waters (Cheng and Detrich III 2007; Chen et al. 1997b). The evolutionary ancestry of the cod AFGP gene has been investigated and it is clear that it was not derived from a trypsinogen-like protease gene (Chen et al. 1997a; Cheng 1998) but appears to have arisen de novo (Zuang et al. 2019). Thus the AFGPs of Antarctic notothenioid fish and the northern cod arose by convergent evolution, and represent a rare case of protein sequence convergence (see Chap. 9 for a detailed discussion of their origin).

5.3.2 Antifreeze Peptides (AFP)

A number of Arctic and north temperate fishes and two species of Antarctic fish synthesize one of three distinct types of AFP (Fig. 5.1). Flat fishes, sculpins, and snail fish have helical alanine-rich peptides (type I) with isoforms of molecular weights of approximately 3 and 4 kDa in the flat fishes and sculpin while the snail fish has a larger type I of about 9 kDa. Another Ala-rich antifreeze in winter flounder has recently been reported that exists in very low concentration in the blood but is larger ~16.7 kDa and has much greater activity (Marshall et al. 2004), but whether it plays a significant role in freezing avoidance remains to be determined because its blood concentration is so low. Type II AFPs are cysteine-rich and β-structured peptides of about 14–17 kDa and are found in sea raven, smelt, and herring (Slaughter et al. 1981; Ng and Hew 1992; Ewart and Fletcher 1990) and are related to the carbohydrate-binding domain of Ca^{++}-dependent and -independent (C-type) lectins with which they share a ~30% protein sequence identity (Ewart et al. 1992). Type III AFPs are small globular proteins found in Atlantic and Arctic ocean pouts (Hew et al. 1984), wolf fish (Desjardins et al. 2012), Antarctic eel pouts (Schrag et al. 1987; Cheng and Devries 1989), and a viviparous eel pout in Danish coastal waters (Sørensen and Ramløv 2002). Their sizes range from 6.5 to 7 kDa except for a 14-kDa two-AFP domain variant that is present in the Antarctic eel pout, *Lycodichthys dearborni* (Wang et al. 1995a). Type III AFPs arose from modification of the C-terminus half of sialic acid synthase based on sequence similarity (Deng et al. 2010; Baardsnes and Davies 2001). Besides the high levels of AFGP present in the notothenioid fishes, a carbohydrate-free AFP has recently been isolated from their blood. With a mass of 15 kDa, it has about the same antifreeze activity on a molar basis as most of the other AFPs but is present at only 2 mg/ml of blood. Combined with AFGPs 1–5, there is a synergistic effect in that the hysteresis fp is almost double that of the sum of the individual AFGP and AFP fps (Jin 2003). It is suggested that it potentiates the activity of the large AFGPs and therefore was termed an antifreeze potentiating protein (AFPP). An AFP has also been isolated from concentrated blood serum from the longhorn sculpin, *Myoxocephalus octodecemspinosus*, has a molecular mass of 12 kDa and shares about 20% sequence identity with members of the exchangeable apolipoprotein superfamily (Deng and Laursen 1998; Deng et al. 1997) and is designated a type IV AFP. The role of type IV AFP in longhorn sculpin with respect to preventing organismal freezing is insignificant because at its very low in vivo concentrations, no hysteresis is observed and furthermore it is also present in fishes that live only in tropical waters (Gauthier et al. 2007). Also, this sculpin avoids the freezing shallow water during the winter by moving into deep warmer water (Leim and Scott 1966). When concentrated in vitro it shows appreciable hysteresis and therefore can be considered as an AFP on a mechanistic basis. The significance of this observation is that it suggests a possible pathway as to how evolution could recruit a protein with some antifreeze activity and with positive selection become a physiologically functional AFP.

The AFPs, as a rule, have far fewer isoforms than the AFGPs. Type III AFP of the Antarctic eel pout *Pachycara brachycephalum* is at one extreme of the spectrum as its AFP exists as one predominant isoform (Cheng and Devries 1989) while the Atlantic and notched-fin eel pouts have several isoforms some of which have little antifreeze activity (Takamichi et al. 2009; Nishimiya et al. 2005). The distinct evolutionary origins of the AFPs are reflected in their distinct protein sequences and higher-order structures, and their common ice-binding and ice growth inhibition ability exemplifies functional convergence (Cheng 1998).

5.4 Non-colligative Lowering of the Freezing Point by Antifreeze Protein

Ideal solutes lower the freezing point of the water linearly in a concentration-dependent manner by 1.858 °C per mole of dissolved solute particles per kg of water. In physiological studies, the osmotic effect is usually expressed as osmolality with the unit mmol/kg H_2O and accounts for the number of solute molecules contributing to the osmotic pressure. Antifreeze proteins occur in substantial physiological concentrations (mass/volume), but these translate into very small molar concentrations due to their macromolecular nature. For example, the ~6.8 kDa type III AFP of the Antarctic eel pout *P. brachycephalum* circulates at about 20 mg/ml in the blood, which is equivalent to only 2.9 mM that would lower the freezing point by only 0.0056 °C based on its colligative effect. Although, a seed ice crystal in the same AFP III solution will start to melt close to this colligative freezing point of −0.0056 °C, it will not grow until the temperature is lowered to about −1.3 °C much lower than the colligative fp (or mp). This freezing temperature or "freezing point" is a non-colligative or non-equilibrium fp, signifying a non-colligative mechanism as the basis of the freezing point depression. The thermal hysteresis refers to this temperature difference between the mp and the non-colligative fp in the presence of AP and is a measure of antifreeze activity (Devries 1971).

5.4.1 Freezing Behavior of Antifreeze Protein Solutions

The unusual lowering of the fp, but not the colligative fp (mp), first observed in AFGP-bearing Antarctic notothenioid fish *Trematomus borchgrevinki* (common cryopelagic fish belonging to the family Nototheniidae) blood serum, lead to detailed studies of the freezing behavior of solutions of purified APs and evaluations of the techniques to achieve accurate and reproducible freezing points (Raymond and Devries 1977). Freezing point osmometers were first used for measuring blood fps and gave high fps in most fishes exposed to freezing seawater (Pearcy 1961). It underestimated fp values of blood fortified with AFGPs (Fletcher et al. 1982a, 1985a; Devries and Wohlschlag 1969). In fact, purified AFGPs 7 and 8, the most abundant AFGPs in most notothenioid fishes, showed little fp depression when determined with a freezing point osmometer. Later, it was observed that for some of the APs the freezing point depends strongly on the rate at which the sample was cooled to its fp (Schrag and Devries 1982; Raymond and Devries 1972), and the high fps obtained by freezing point osmometery were due to rapid freezing of the sample when nucleated at approximately 5 °C below its equilibrium fp. Later studies utilized solutions of APs in 10 μl capillary tubes seeded with small (~25 μm diameter) polycrystalline ice crystals and visual observation of the freezing and melting process with a slow rate of cooling or warming (Schrag and Devries 1982; Devries 1971, 1986), and AFGPs 7 and 8 were found to have substantial antifreeze activity by this method (Schrag et al. 1982). Most recent studies make use of the Clifton Nanoliter Osmometer/Cryoscope, which allows for visual observation of the melting and freezing processes and controlled melting of the polycrystalline seed ice to a single crystal. The temperature-dependent growth of the single seed ice crystal can be accurately accessed, and the effect of APs on ice crystal growth morphology can be visualized and captured with digital images (Kao et al. 1985). Most freezing–melting points are now determined with this device or a similar platform (Braslavsky and Drori 2013) and the values are reproducible if the same cooling rates are used.

The freezing behavior of purified AFGPs, type I winter flounder AFP and eel pout AFP III have been most thoroughly studied. These three APs show similarities in their antifreeze activity but differences in their crystal growth habit at similar mM concentrations. At low mM concentrations, almost all fish APs have identical or similar antifreeze activities with the exception of the small AFGPs 7 and 8 (Fig. 5.3). However, the ice growth morphology differs upon freezing. Using a very small seed ice crystal (10 μm) in a 20 mg/ml solution of AFGPs 1–5 no detectable microscopic growth can be observed until the temperature is lowered to the nonequilibrium fp of about −1.2 °C. Freezing involves the propagation of a cluster of fine spicules of diameters between 10 and 50 μm, with their long axis parallel to the *c*-axis, the non-preferred axis of growth for ice in pure water and common salt solutions (Raymond and Devries 1977; Devries 1982, 1986). With the same concentration of winter flounder type I AFP, microscopic growth occurs within the hysteresis gap as the discoid seed crystal grows into a hexagonal bipyramid that lengthens with decreasing temperature until the fp is reached (Chakrabartty et al. 1989). At the fp,

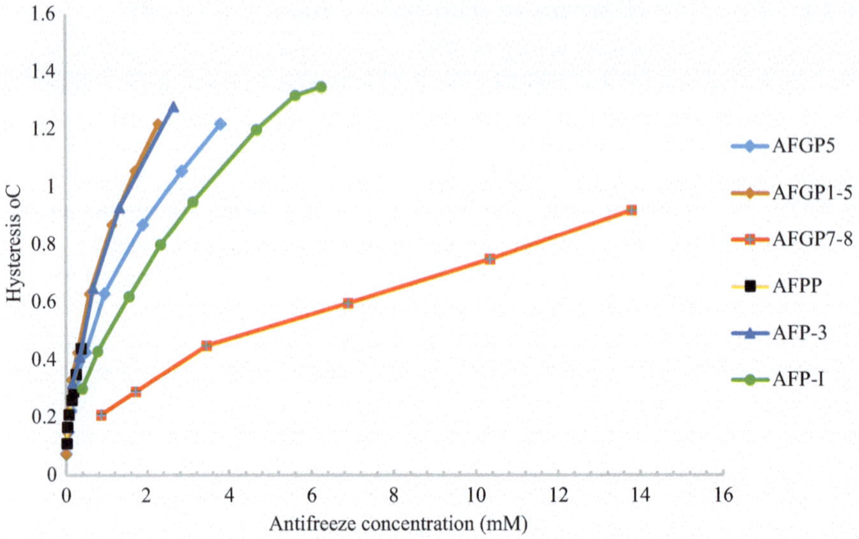

Fig. 5.3 Freezing temperatures and melting points of aqueous solutions of antifreeze glycoproteins and selected antifreeze proteins (Type III) as a function of molar concentration. The freezing point of AFGPs 7 and 8 is only about two thirds those of the large AFGPs, 1–5. All melting points are very similar. At temperatures below the hysteresis gap rapid thin spicular ice growth occurs with AFGPs 1–5 while with most of the others growth is in the form of thick spears

ice spicules rapidly grow initially from the tips of the bipyramid but they tend to be thicker than those observed with the large AFGPs. Likewise, with the small AFGPs 7 and 8, hexagonal bipyramids form in the hysteresis gap, lengthen in the gap as the fp is approached and at the fp coarse spicules propagate from the tips. In general, most of the pure AFPs solutions show growth in the hysteresis gap in the form of hexagonal bipyramids at least at lower concentrations (2–10 mg/ml) (Raymond et al. 1989; Chao et al. 1995). Very large single crystals (1 cm diam) in the presence of AFGPs 1–5 show a small amount of stepped growth on the prism planes and a small amount on the basal plane resulting in hexagonal pitting of its surface in the hysteresis gap (Raymond et al. 1989), which cannot be microscopically observed with small single ice crystals (10–20 μm). However, differential scanning calorimeter measurements indicate a small amount of submicroscopic growth most likely corresponding to those growth sites observed with large single crystals (Ramlov et al. 2005). With large single crystals when the basal plane is completely covered with hexagonal pits no further growth in the hysteresis gap is observed (Raymond et al. 1989). The walls of the pits are visibly stepped and their associated fluorescence in the presence of labeled AFGPs implies binding at the steps. Although not thoroughly documented, the growth of hexagonal bipyramids has been observed in the native serum of some of the AFGP-bearing cods and AFP-bearing fishes where blood fps are higher than −1.5 °C. With the AFGP-bearing Antarctic notothenioid fishes and high Arctic gadids, although the small AFGPs 7 and 8 in pure solution yield hexagonal bipyramids over a wide range of concentrations, there is no indication of the pyramidal ice growth morphology in the native serum regardless of the magnitude of the hysteresis. There is only spicular growth from the small seed crystals at the fp. Apparently, the strong ice growth inhibition effect of the large AFGPs and the presence of a newly discovered AFPP in the notothenioid fishes are sufficient to prevent any visible hexagonal bipyramidal growth within the hysteresis gap. Electron micrographs of replicas of the spicular ice needles/spears formed in the presence of both AFGPs and AFPs reveal that they are hexagonal in cross section and some have blunt ends while others have pyramidal faces at their terminus (Wilson et al. 2002) consistent with their adsorption patterns on single ice crystal hemisphere etching experiments (Knight et al. 1991, 1993).

5.5 Adsorption Inhibition Mechanism of Non-colligative Antifreeze Activity

Even with the diversity in composition, structure, and size, all fish APs function by binding to ice and arresting growth in aqueous solutions including most of the body fluids of fishes. Recently it has been shown that with fish and insect AFPs, the binding is irreversible (Meister et al. 2018; Celik et al. 2013) and no doubt this will eventually be demonstrated with all the APs. The mechanism of antifreeze activity most generally accepted is that of adsorption of AP molecules to ice resulting in inhibition of ice growth (Raymond et al. 1989; Raymond and Devries 1977; Knight et al. 1991). Initially, it was postulated that APs bind to the ice by means of hydrogen bonding, and involves a lattice match between regularly spaced hydrogen bonding moieties in the APs and the water molecules in the ice crystal lattice (Knight et al. 1991; Devries and Lin 1977). From extensive mutational studies, it was shown that the majority of the residues necessary for antifreeze activity were hydrophobic in different AFPs (Wen and Laursen 1992; Haymet et al. 1999; Chao et al. 1997). From the consideration of the necessary binding energies of the adsorption process, it was concluded that the number and strength of the hydrogen bonds of the putative hydrogen bonding residues involved were insufficient to explain adsorption to ice. Hydrophobic interactions are now thought to be the primary driving force involved in binding to ice (Schauperl et al. 2017), but still with a limited role for hydrogen bonding (see Chap. 4 in Vol. II). With the AFGPs, there is no direct evidence for hydrophobic interactions and hydrogen bonding between the hydroxyls of the saccharide moieties appears to be a plausible hypothesis. A brief conceptual description of the adsorption–inhibition mechanism is given here.

In vitro ice-binding studies have revealed that different APs recognize and preferentially adsorb to specific crystallographic planes of hexagonal ice (Knight et al. 1991, 1993), consistent with a different docking requirement by the structural binding sites specific to each AP, and the crystallographic orientation in the ice crystal lattice that best fits this requirement. The ice crystallography of antifreeze adsorption was examined experimentally by growing a single ice crystal in a dilute solution of APs (0.1% w/v) into an ice hemisphere, followed by ice surface sublimation to determine the location and orientation of adsorbed antifreeze molecules (Knight et al. 1991). Ice growth is not arrested at these low concentrations, and AP molecules bound to specific crystallographic planes of the growing ice surface become incorporated into the ice as ice steps grow over them. Sublimation of the final ice hemisphere removes the surface layers of water molecules, and the regions where antifreeze molecules were bound to appear as rough, opaque surfaces, while the rest of the hemisphere surface is clear and smooth. In general, the diverse APs bind to low index planes in ice, i.e. primary, secondary prism planes, and pyramidal planes originating on prism planes, and the alignment direction for adsorbed molecules could be determined with reasonable confidence for APs that have an elongated structure—the notothenioid AFGPs and type I AFP of the flat fishes and sculpin.

The AFGPs were originally thought to exist in solution as disordered structures (Raymond and Devries 1977; Brown and Sonnichsen 2002) but other CD studies indicated a substantial presence of left-handed polyproline type II helix in the structure (Bush and Feeney 1986). In the latter conformation, the disaccharides would be positioned on one side of the helix, and this orientation would position the disaccharide hydroxyl groups so they could potentially hydrogen bond to the oxygens in the ice lattice. The small AFGPs 7 and 8 were found to adsorb to the primary prism planes $\{10\bar{1}0\}$ of ice and because of the asymmetry of the adsorption signature, it was deduced to align parallel to the *a*-axes. The spacing of the tripeptide repeats (9.31 Å) in a type II helix configuration is about twice the periodicity of the *a*-axis (4.519 Å). Thus hydrogen bonding presumably could involve the hydroxyls of the Thr-linked disaccharides of each repeat with alternate water molecules along the *a*-axis (Knight et al. 1993). Model building indicates that one hydroxyl group in each saccharide would match the spacing requirement and if buried in the ice each could participate in three hydrogen bonds consistent with the hydrogen bonding hypothesis for the AFGPs (Knight et al. 1993). Recently it has been suggested, based on spectroscopic evidence, that the AFGPs assume an alpha-helical configuration when adsorbed to ice (Furukawa et al. 2017). The disaccharide moieties would, in this case, be separated by 4.5 Å perpendicular to the *c*-axis, a repeat spacing along the *a*-axis of hexagonal ice. However, the disaccharides would no longer align on one side of the helical molecule. Chemical modifications of the disaccharide hydroxyls or blocking some of them results in loss of thermal hysteresis, which supports the involvement of the disaccharide hydroxyls in ice binding (Shier et al. 1972; Devries 1971). Organic synthesis of AFGP analogues has shown that the N-acetyl group on the galactosamine and the alpha 1–3 linkage joining the disaccharide to the threonine are also necessary for activity (Tachibana et al. 2004). The details of AFGP ice binding and their orientation with respect to their binding plane present a challenge because they cannot be crystallized and thus details of the mechanism of binding remain to be elucidated. Recent molecular simulations (Mochizuki and Molinero 2018) suggest that the hydrophobic side of the AFGP in a polyproline II helical conformation, with its abundance of alanine residues with their methyl groups, is the ice-binding face; however, so far there is no experimental evidence consistent with this simulated model.

Type I AFP of winter flounder and Alaskan plaice are near perfect alpha-helices. They are found to adsorb on the pyramidal planes $\{20\bar{2}1\}$ of ice aligned in the $\{01\bar{1}2\}$ direction, which has a periodicity that closely matches the lengths of the 11-residue repeats (16.5 Å) of these AFPs, and binding presumably involves the four repeated polar Thr residues (Knight et al. 1991). The helical type I AFP is amphipathic, which would position the putative ice-binding moieties linearly on one side of the molecule that could align with the planar surface of the ice. However, conservative replacement of Thr with Ser, preserving the putative ice-binding hydroxyl, lead to substantial loss of antifreeze activity, while replacement of Thr by the nonpolar Val also lead to a decrease in activity, but not as much as with Ser replacements, which suggests that hydrogen bonding may not be the predominant force for binding (Haymet et al. 1999; Chao et al. 1995). It was also observed that the nonpolar Ala residues on each side of the Thr are perfectly conserved in all type I AFP isoforms, and their replacement strongly reduces activity (Baardsnes et al. 1999, 2001) indicating their important role in AFP activity. In addition, replacing Ala residues on the hydrophobic side of the helical type I AFP of sculpin antifreeze with charged Lys residues results in loss of activity (Baardsnes et al. 1999). It has been suggested that these results and structural studies of a microbial AFP indicate that the ice-like water associated with its ice-binding face results from hydrophobic interactions with hydrogen bonding residues anchoring the "ice" to the AFP (Garnham et al. 2011). The associated "ice" also has repeat spacings that match those on the basal and prism planes of hexagonal ice.

Type II and type III AFPs are globular proteins that do not contain repetitive sequences, thus the ice-lattice match model does not appear to apply. The ice-binding residues of type II AFP have not been definitively identified, but are inferred by site-directed mutagenesis to be the same as the homologs of the Ca^{++}-binding sites in herring AFP (Baardsnes et al. 1999) or by computational analysis to be a surface patch of 19 residues (Cheng et al. 2002). The 3-dimensional structure of type III AFP has been well resolved and the putative ice-binding residues have been identified and mapped in detail through extensive site-directed mutagenesis of selected residues (Yang et al. 1998; Deluca et al. 1998; Brown and Sonnichsen 2002). Some of these studies indicate that the arrangement of the ice-binding residues constitute a flat surface that matches with the planar surface of ice for binding to occur through hydrogen bonding (Yang et al. 1998). Others contend that the putative binding surface of type III is relatively hydrophobic and the tight packing of the side chains on this surface precludes effective hydrogen bonding to ice (Sonnichsen et al. 1996). The ice crystallography of type III AFP adsorption is not entirely clear, showing binding primarily to the prism and pyramidal faces involving the compound ice-binding site of the molecule (Garnham et al. 2010) but also at orientations between prism and basal faces (Antson et al. 2001), which could not be correlated to any particular binding motif in the protein. Thus despite extensive research on the structure–function relationship of type III AFP and ice

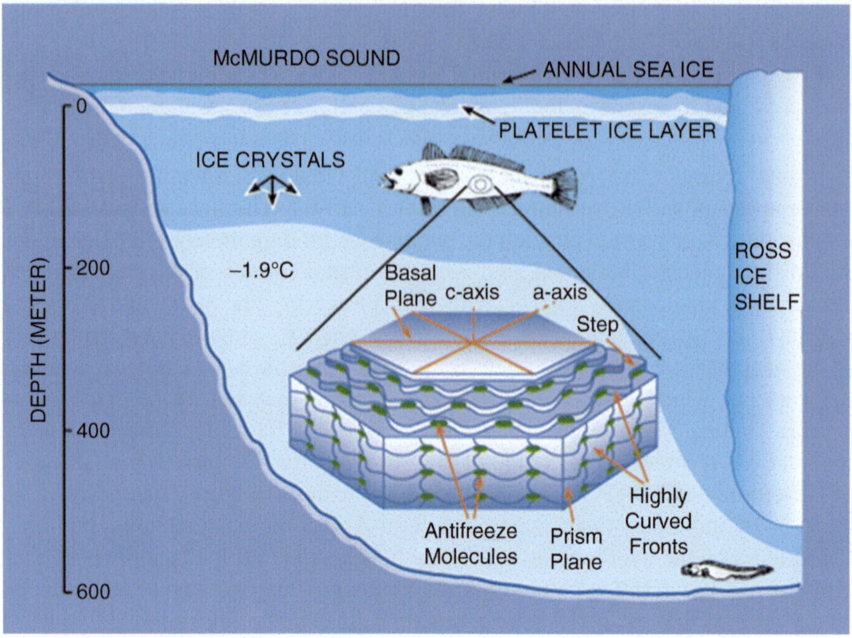

Fig. 5.4 A schematic of a notothenioid fish in McMurdo Sound, Antarctica swimming in ice-laden freezing seawater near the Ross Ice Shelf. An ice crystal is shown within the fish with AFGP molecules adsorbed onto its prism planes, altering the crystal planes surfaces and preventing growth of the crystal. Reprinted from *The Physiology of Polar Fishes, Fish Physiology* Vol. 22, Elsevier Academic Press

hemisphere etch studies, the precise nature of where and how it binds to ice is not definitive. Recent research indicates that these AFP structures, water molecules in a configuration similar to ice on its putative ice-binding face at physiological temperatures and even at room temperature. The presence of this preordered water may be important for docking and adhering to the ice face (Meister et al. 2014). As well, crystallography of a fusion protein–AFP III complex indicates the presence of water molecules associated with the putative ice-binding site, which is consistent with the above observation. It is postulated that the "structured water" may insert into the quasi-liquid layer leading to its solidification with the underlain ice lattice. However, it is unclear how such a mechanism could result in binding to a specific crystal face.

The essential element of the inhibition mechanism is that APs adsorb to specific crystal planes and incompletely cover them (Raymond and Devries 1977). Ice growth, therefore, can only occur between the adsorbed AP molecules, resulting in growth fronts that are highly curved (Fig. 5.4). Curved growth fronts become less stable relative to planar fronts and the local equilibrium fp is lowered. Growth of these fronts will cease once their local fp is equivalent to the undercooling of the solution. Further cooling will lead to a further increase in curvature but no visible macroscopic growth until the nonequilibrium fp is reached. Conceptually, this process can be visualized as water molecules in highly curved fronts having fewer neighbors to hydrogen bond to than those on a planar front (Devries 1984). At a given temperature within the hysteresis gap, the water molecules will spend more time in the liquid phase than in the solid phase. In order for them to remain in a highly curved front, energy must be taken out of the system and lowering the temperature is equivalent to lowering the fp of the solution. The generally accepted inhibition mechanism is explained on the basis of the Kelvin effect (Raymond and Devries 1977). Several reviews give an in-depth discussion of the mechanism (Wilson 1993; Brown and Sonnichsen 2002) as well as Chap. 4 in Vol. II. From a biological perspective, the adsorption inhibition mechanism does imply that ice must be present. Indeed ice does enter the bodies of some polar fish and its presence has been clearly demonstrated (Praebel et al. 2009; Cziko et al. 2014).

5.6 Integrative View of Freeze Avoidance in Fishes Is Needed

Much of the recent research on the APs and IBPs has been directed toward identification, isolation, structural characterization, and efforts to determine how they bind to ice and inhibit its growth. Most of the adsorption/inhibition mechanism studies have been done on fish and insect AP single isoforms generated from cDNA clones and expressed in microbes (Fletcher et al. 2001) while others have been chemically synthesized (Wen and Laursen 1992). The reductionist approach has been taken to the extreme as evidenced by the many recent publications involving molecular simulations of the published AF structures attempting to match putative IBP faces to ice crystal faces and hypotheses as to the nature of the driving forces resulting in irreversible adsorption (Schauperl et al. 2017; Mochizuki and Molinero 2018). Such simulations lack a physiological context in that they are simulated in water whereas in nature APs function in a complex biological solution of salts and blood proteins which contribute to the body fluid equilibrium fp and some AF proteins by themselves lack antifreeze activity but combined with active isoform contribute to the total AP activity. In most fishes, the circulating AP complement is composed of different size isoforms as well as compositional variation with the exception of one Antarctic eel pout where it appears the AFP is largely a single isoform (Cheng and Devries 1989). The variation in size reaches extremes in the Antarctic notothenioid and Arctic gadid fishes where there may be as many as 16 different size isoforms with only minor compositional variation. In the flounders and sculpins, there are usually 2 or 3 different sizes with minor compositional variation. In general, eel pout (zoarcid) AFPs have a uniform size but many of their isoforms have substantial variation in those residues not involved in the putative ice-binding face. Some even have variants that lack antifreeze activity by themselves but if combined with an active isoform they then show significant antifreeze activity (Nishimiya et al. 2005).

Thus with most fishes, the blood hysteresis is a result of contributions from several isoforms, some with less antifreeze activity than others (Fig. 5.3). Furthermore, in the notothenioid fishes, there are two very different molecules (an AFGP and an AFPP) that contribute to the AP-based freeze avoidance system. Also, the AP generated hysteresis can be slightly enhanced by the presence of salts such as NaCl (Fletcher et al. 2001). In order for an in-depth understanding of freeze avoidance in fishes, one would like to have a complete picture of the contributions of all the solutes (salts, APs, and to be able to ascertain whether there are interactions among them as well as with other blood proteins) that contribute to the low blood fps. Also, one would like to accurately know the tissue and fluid distribution of the APs and whether body fluids lacking APs can avoid freezing simply by isolation from nucleating ice crystals by the surrounding tissue fluids fortified with APs. Before considering how the APs promote freeze avoidance in fishes, it is informative to describe non-physicochemical mechanisms employed to avoid freezing. These include behavioral avoidance of freezing in ice-laden seawater and survival in near freezing seawater in an undercooled state in the absence of ice.

5.7 Migration to Ice-Free Habitats

Fishes that live in habitats that freeze during the winter can avoid freezing by migrating to ice-free environments during the winter. Migration to warmer offshore waters, which is a behavioral means utilized by some northern fishes, or occupying deep waters that are ice-free due to pressure-dependent depression of the in situ fp (in situ fp is the fp at any given depth with the effect of pressure and other physical properties of seawater considered) (Millero 1978). One example of offshore migration is the north Atlantic longhorn sculpin, *M. octodecemspinosus* that moves from the coastal northeast Atlantic shallow water into warmer deep water (+4 °C) during the winter (Leim and Scott 1966). In the high Arctic, Arctic char, *Salvelinus alpinus*, leave their resident streams and lakes in the summer and feed in the nearby marine waters but return to fresh water before freeze-up where the lowest water temperature possible is 0 °C, which is well above the freezing point of their body fluids (Dempson and Kristofferson 1987). Some white fish species (another salmonid) in the Bering Sea exhibit the same freeze avoidance strategy. None of these species has been shown to possess protective APs.

5.8 Undercooling in the Absence of Ice

Water can exist in the liquid state at temperatures well below its equilibrium fp and is referred to as supercooled or more correctly as undercooled water. In biological systems or large volumes of water, homogeneous nucleation (spontaneous freezing at −38.5 °C) is unimportant because most aqueous solutions nucleate at much higher temperatures usually 15–30 degrees below the equilibrium fp and is termed heterogeneous nucleation. The instability of the metastable state is largely due to the presence of nucleators, which may be particulates in the solution, scratches on a container wall, biological nucleators, or physical disturbances.

In non-AP bearing polar marine fishes whose equilibrium fp is solely based on blood salts and other small osmolytes, the fp is −0.7 °C and the amount of undercooling needed to avoid freezing is at most 1.2 °C, since the fp of even the most saline ocean water is at about −1.9 °C. In the absence of ice, fishes can be undercooled even more, as long as the bathing seawater is free of nucleators and the fish is free of endogenous nucleation sources. Ice-free high latitude Antarctic fishes can be experimentally undercooled by 5 °C in an ice-free glycerol–seawater mixture for at least an hour, and they survive normally when returned to ambient seawater (Praebel et al. 2009). Such extensive undercooling of ice-free specimens indicates the absence of both exogenous and endogenous non-ice nucleators, and the stability of the undercooled state in fish. Similar observations have been made in temperate fish such as killifish and cunner where they were undercooled to −3 °C in ice-free seawater (Scholander et al. 1957), and thus Arctic fishes very likely are also capable of extensive undercooling although there are no reports in the literature. Because of the stability of this slight undercooling in the absence of ice, some fish can complete their life cycle at −1.9 °C without the danger of freezing despite lacking protective APs. In the deep fjords of the Arctic region, undercooled fishes inhabiting −1.8 °C water have been caught that quickly freeze when exposed to surface ice-laden water (Scholander et al. 1957). These undercooled fishes are in no danger of freezing as long they remain at depth where no ice formation occurs. Freezing point–melting point analyses of the blood of some deepwater fishes in the Greenland fjords indicate the absence of a hysteresis indicating they lack APs (Enevoldsen et al. 2003), indicative of this freeze-avoidance strategy.

In contrast to the Arctic, ice can form even in the coastal waters as deep as several hundred meters off the Antarctic continent because of the presence of massive thick ice shelves. Those in the Arctic are relatively small and thin and it is unlikely that they would generate significant ice at depth. A substantial part of the Antarctic continental margin is associated with ice shelves, which are floating extensions of glaciers that drain the immense inland ice sheet. The largest ice shelves, the Ross Ice Shelf bordering the Ross Sea and the Filchner–Ronne Ice Shelf bordering the Weddell Sea, have a thickness of 1 km and 600 m at their continental origin and at their oceanic termination, respectively. The water in contact with the shelf base near the grounding line may become as cold as −3 °C because of the depth-dependent effect of pressure on the fp. Oceanic currents drive water under the shelves where it is cooled and tidal currents can entrain cold ice shelf water seaward and upward near the edges of the shelves (Hunt et al. 2003) and as it rises it becomes undercooled with respect to the in situ fp which in turn sometimes results in ice crystal formation deep in the water column (Foldvik and Kvinge 1974). In McMurdo Sound, which is adjacent to the Ross Ice Shelf, water exits slightly undercooled and as it is advected toward the surface minute ice crystals can be observed in the water column. It is unclear whether the ice crystals originate from the frozen underside of the shelf or if they spontaneously form in situ in the undercooled water. The small crystals are thought to be involved in the formation of the abundant sub-ice platelet layer on the underside of the annual sea ice cover, as well as the platelets in the anchor ice mats on shallow bottoms (40 m) during the winter in that they attach to substrates and grow as the supercooled water flows over them (Cziko et al. 2014). In some ice shelf locations near McMurdo Sound, hydrographic measurements indicate that during austral winters freezing seawater can be found as deep as 175 m at times near the termination of the Ross Ice Shelf (DeVries unpublished) and in the Weddell Sea large ice crystals or platelets have been collected in a self-closing net at depths of 250 m indicating ice formation at that depth (Dieckmann et al. 1986) near the Filchner Ice Shelf. Conductivity, salinity, and depth profiles indicate that in some locations in the Weddell Sea near the Filchner Ice Shelf freezing isotherms (depth above which ice formation can occur) could be as deep as 500 m (Foldvik and Kvinge 1974).

Despite ice formation in the deep Antarctic water, some undercooled fishes are found at depths that may not be encroached upon by ice derived from the cold ice shelf water. The snail fish *Paraliparis devriesi* lives at 600–700 m on the bottom of McMurdo Sound, Antarctica, where the water temperature is about −1.93 °C year round. Its blood contains small amounts of AFP, but it only lowers the blood fp to −1.3 °C (Jung et al. 1995) and thus it is undercooled by 0.6 °C with respect to the surface water. It freezes when brought up through ice-laden surface waters indicating inadequate antifreeze protection to survive in the icy surface condition. At its habitat depth of 700 m though, it would have an in situ fp of −1.73 °C due to the effect of pressure, but is still undercooled by two-tenths of a degree. The presence of ice in the liparid's habitat is unlikely because the terminal edge of the Ross Ice Shelf, at the transition to annual sea ice in McMurdo Sound, is less than 100 m thick and would not cause ice formation below that depth. The two species of benthic Antarctic zoarcid fish in the Sound, *L. dearborni* and *P. brachycephalum* live at about 500–600 m, and at the surface have fps equal to that of the seawater (−1.9 °C) or slightly above. When brought to the surface many will begin to freeze but at depth their fps would be well below the in situ temperature. In contrast, the predominant Antarctic fish group, the notothenioid fishes, caught throughout the water column of the McMurdo Sound have fps slightly below in situ seawater fps (Devries 1988) and do not freeze when brought through ice-laden surface water with the exception of the deep dwelling *Trematomus loennbergii* that occasionally freezes at the surface. Thus the Antarctic zoarcid and liparid fishes are the two non-notothenioid species that lack sufficient AP to avoid freezing in high Antarctic shallow water, and exist in a slightly undercooled state at the surface but not in their deepwater habitats where because of the pressure their fps would be lower than the temperature of the water at that depth. The low level of AP in these two species may be related to the fact that they are recent immigrants to the Antarctic continental shelf, and their life history is completed in ice-free deep water.

The undercooling observed in various Arctic and Antarctic fishes appears to be metastable for the life of the fish as long as they remain at depths where ice formation is absent. Fishes that survive in the undercooled state avoid the energetic costs of synthesis and conservation of APs which would be substantial.

5.9 Blood and Organismal Freezing Point Correlations

For temperate fishes lacking APs, blood fps have been equated with organismal fps (Devries and Cheng 2005) and there is a close correspondence between the two regardless of how blood fps were determined. With polar fishes that are fortified with APs, this relationship is not as strong and depends on the method used for determining the freezing point of the fish's blood. Especially in the Antarctic notothenioid fishes, the measured blood fps strongly depend upon the technique used. Early studies on polar and north temperate fishes used the freezing point osmometer that gave high fps and no hysteresis was reported because mps were not determined. When it was discovered that the AF activity of some APs was cooling rate dependent, fps were determined visually with a polycrystalline ice seed in capillary tubes viewed in a refrigerated viewing chamber that could be slowly cooled or warmed (Devries 1986). With this technique, lower blood fps were observed and corresponded closely with organismal fps (Devries 1988). Those determined with the Clifton Cryoscope using a very small (10 μm diameter) single ice crystal using a slow cooling rate gave values a few tenths of a degree lower than those obtained with the capillary tube method for Antarctic notothenioid fishes. Thus antifreeze activity of notothenioid blood is not only strongly dependent on the cooling rate in the presence of ice (Schrag and Devries 1982) but is also dependent on size of the test seed ice crystal. This is also observed in some north temperate fish blood fortified with AFP (Takamichi et al. 2007). In the notothenioids, this size dependence is associated with the recently discovered AFPP. Cooling rate-dependent fps have also been observed with some insect AFPs as well as with their hemolymph (Duman 2001) and with other fish AFPs such as the notched-fin eel pout (Takamichi et al. 2007). The lower blood fp values obtained with the Clifton Cryoscope suggest that freezing of the intact fish in the presence of ice involves other factors such as variability in their resistance to ice entry.

Of utmost importance for teleost survival in ice-laden freezing environments is that the organismal freezing point must be lower than the ambient water temperature. Systematic whole fish freezing point determinations have not been carried out for Arctic polar and north temperate fishes because of logistical difficulties such as capturing them and maintaining them at their environmental temperature so that any associated ice will not melt or lead to freezing of the fish. Organismal fps have been systematically determined for a number of Antarctic notothenioid fishes from McMurdo Sound, which is possible because they can be captured through holes in the ice covered with heated huts and transferred to ice-free water at their habitat temperature. The whole fish fp determinations involved lowering the temperature of ice-free seawater to below -1.9 °C, which can be achieved by adding an appropriate amount of glycerol to lower the seawater solution fp. Glycerol does not alter the

Table 5.1 Environmental temperature, presence of ice, organismal freezing points, blood melting/freezing points, and hysteresis for some Antarctic, Arctic, and north temperate fishes

Species	Environmental temperature (°C)	Ice present	Organismal freezing point (°C)	Blood freezing point (°C)	Blood melting point (°C)	Hysteresis (°C)
High latitude						
Trematomus borchgrevinki	−1.9	Yes	−2.3	−2.75	−1.1	1.64
Trematomus bernacchii	−1.9	Yes	−2.2	−2.5	−1.14	1.36
Lycodichthys dearborni	−1.9	No	−1.9	−1.9	−0.9	1.0
Boreogadus saida	−1.9	Yes	−2.0[a]	−2.2	−1.12	1.06
Arctogadus glacialis	−1.9	Yes	−2.0[a]	−2.09	−0.95	1.14
Sub-Arctic						
Eleginus gracilis	−1.8	Yes	−1.9[a]	−2.1	−1.1	1.0
Myoxocephalus verrucosus	−1.8	Yes	−2.01[a]	−2.3	−0.9	1.4
North temperate						
Zoarces viviparus	−1.1	Yes	−1.9[a]	−2.1	−0.94	1.16

[a]Organismal freezing point estimated as 0.2–0.3 °C higher than blood serum freezing point

blood osmolality or the intrinsic fp of fish for short periods of exposure (1–2 h) (Praebel et al. 2009). During slow cooling when the specimen was periodically touched with a piece of ice, rapid freezing resulted when its fp was reached. This undercooling nucleation assay was also used to determine whether wild caught specimens carried endogenous or exogenous ice derived from the environment at various times of the year. Regardless of whether the ice was internal or external freezing occurred when the organismal fp was reached. With this technique, most shallow-water Antarctic notothenioid fishes of McMurdo Sound were found to have organismal fps that were a few tenths of a degree below the ambient temperature of −1.9 °C (Table 5.1). The same organismal fp values were obtained with environmental fishes that naturally harbor endogenous ice and associated exogenous ice, which would be present on the integument as well as in the intestinal fluid and spleen.

5.10 Source of Exogenous and Endogenous Ice in Fish

The adsorption–inhibition mechanism presupposes the presence of ice somewhere in the fishes' body fluids, such that the APs would be functionally and physiologically pertinent (Devries 1988). APs can only exert their antifreeze effect if ice crystals are present to which they can bind resulting in ice growth inhibition, which in turn prevents organismal freezing. The source of ice is the perennial freezing marine environments of both polar oceans, and the shallow coastal waters of the north temperate oceans that freeze in the winter months. Exogenous ice expectedly is acquired via ice contact with surface tissues (integuments and gills), and ingestion through diet and drinking seawater. The gastrointestinal tract lumen can be considered as a continuation of the external surface because of its exposure to ingested ice-laden seawater.

5.10.1 Endogenous Ice in Antarctic Fishes

In the waters adjacent to ice shelves in Antarctica, notothenioid fishes are exposed to minute ice crystals in the water column for much of the year. Tests for the presence of ice indicate that it was associated with the skin, gills, and intestinal fluid (Table 5.2). Most surprisingly, the spleen also contained ice while no other internal

Table 5.2 Presence or absence of ice in organs, tissues, and fluids in the shallow-water Antarctic, *Trematomus borchgrevinki*

Tissue or fluid	Ice present or absent
Organs	
Spleen	+
Gills	+
Stomach	+
Skin	+
Muscle	–
Liver	–
Kidney	–
Heart	–
Brain	–
Fluids	
Intestinal fluid	+
Blood	–
Ocular fluid	–
Urine	–
Bile	–

Presence or absence of ice was determined by nucleation of undercooled (−4.5 °C) physiological saline using approximately 200 mg of environmental fish tissue or fluid (modified from Tien 1995)

tissues or fluids were ice positive (Tien 1995; Praebel et al. 2009). The presence of ice in the spleen, which is a deep-seated internal organ, indicates that ice must somehow enter across the body surface into the blood circulation, and become sequestered in the spleen. In teleost fishes, some of the circulating blood passes through the spleen and a portion of both red blood and white cells, which include macrophages are stored there (Franklen et al. 1993; Fänge and Nilsson 1985). Recently using AFGP and AFPP decorated small fluorescent silica beads as a proxy for ice crystals, it was shown that several hours after injection into the circulation of a notothenioid fish, fluorescence could be detected in the spleen and specifically associated with the macrophages (Evans et al. 2011). Presumably, the silica beads are taken up by systemic macrophages, then migrate to the spleen and are sequestered there. Based on this study it is proposed that both AFGP and AFPP adsorb to ice crystals that enter the fish and systemic macrophages recognize them as foreign and phagocytose them. Assuming ice is transported to the spleen via macrophages, paradoxically ice has not been detected in blood samples from fish that test positive for splenic ice. However, the absence of ice crystals in the blood is not all that surprising given the small number (<100) in the spleen (Praebel et al. 2009) and the fact that the blood volume of a fish is ~3% of its body weight. Considering the time needed for accumulation of the low number of splenic ice crystals, the chance of finding an ice crystal in a small blood sample is very small.

During the winter the cryopelagic, notothenioid fish, *T. borchgrevinki* that inhabits the sub-ice platelet layer where the water is freezing and ice most abundant, all test positive for both exogenous and splenic ice. Tests for ice in serial dilutions of splenic cell suspensions showed their spleens contained between 5 and 88 ice crystals between October and January (Praebel et al. 2009). The presence of relatively few splenic ice crystals indicates that ice entry must be a relatively rare event even though 100% of the specimens tested positive for splenic ice. There also appears to be a species difference in the spleen "iciness" among notothenioid fishes. The McMurdo shallow-water benthic species *T. bernacchii* and *T. hansoni* often rest on mats of anchor ice at 20 m or shallower, and would have similar ice and low-temperature exposure as *T. borchgrevinki*, but only ~60% of individuals of these benthic species that tested positive for exogenous ice were positive for splenic ice. The reason for this difference from *T. borchgrevinki* is unknown, but it is possible that their gill and gut epithelia are more resistant to ice entry than those of *T. borchgrevinki*. Quantification of the number of splenic ice crystals in a population of benthic species may provide indirect evidence as to whether they are more resistant to ice entry than *T. borchgrevinki* when sampled at the same time of the year if they have on the average fewer splenic ice crystals than *T. borchgrevinki*.

The rate of ice acquisition in the environment has been examined by returning ice-free (1 h in 4 °C seawater) notothenioid fishes to their respective ice-laden habitats. Within a few hours, ice was present on the skin and gills and within 1–2 days it was also present in their intestinal fluid. However, it took several days for the appearance of ice in the spleen (Praebel et al. 2009). Furthermore, the appearance of splenic ice is not a linear accumulation with time but sporadic suggesting that the environmental conditions, such as temperature and nature of ice in the water column, resulting in ice entry are unique and probably are not present on a daily basis. During the winter, artificial freezing conditions in aquaria failed to result in endogenous ice even after the addition of very small ice crystals over extended periods of time at −1.95 °C (Tien 1995). Thus the environmental conditions that lead to ice accumulation in the spleen are at present not completely understood. By obtaining real-time temperatures in the shallow-water column, one could determine when plumes of undercooled water with suspended small ice crystals arrive from under the ice shelf and their duration. Placing ice-free fish in a plume of undercooled water might result in a linear accumulation of splenic ice with time. With such information, it may be possible to precisely describe the environmental conditions that lead to ice entry and its rate of accumulation in the spleen.

5.10.2 Fate of Endogenous Ice

Presumably, ice enters the notothenioid fish either through the gills, skin, or intestinal epithelial lining resulting from injuries. Its growth in the extracellular fluids would be arrested by adsorption of the AFGPs and AFPP. Regardless of how ice eventually reaches the spleen, a steady accumulation of growth-arrested ice crystals in the spleen throughout the lifetime of the fish, which may span 10–15 years, would eventually create physiological problems for the fish without a mechanism to remove it. Until recently, the thermal melting of endogenous ice was thought impossible because the water of McMurdo Sound was believed to be at or near its fp of −1.9 °C throughout the year (Littlepage 1965). Indeed there are habitats in McMurdo Sound where the water temperature remains below −1.7 °C regardless of the time of the year and thus it would be impossible to melt endogenous ice because the equilibrium fp (mp) is approximately −1 °C (Cziko et al. 2014). However, recent high-resolution multiyear temperature records of several shallow-water habitats (9–30 m) in the eastern McMurdo Sound have revealed temperature fluctuations and warming during the austral summer months (January–February) reaching peak water temperatures of +0.5 °C depending upon the year. The total time of temperature excursions above −1.1 °C, the temperature at which endogenous ice should melt, could be in the order of days (Hunt et al. 2003; Cziko et al. 2014). Thus thermal melting is a mechanism to eliminate endogenous ice in the local fish during those warming episodes is a distinct possibility. However, there is another complication in explaining the fate of endogenous ice, which is that even if temperatures rise above the equilibrium fp (−1 °C), it still may not be warm enough to melt the ice because the adsorbed APs cause a time-dependent inhibition of the melting of the ice crystals. Even at temperatures as high as 0 °C ice may remain in the spleen for several hours (Cziko et al. 2014). Thus in order to melt the ice, water temperatures would have to remain well above the equilibrium fp for hours to days and this may not happen during some summer seasons. That ice may persist throughout the year in some McMurdo Sound fishes is indicated by the presence of splenic ice in *T. borchgrevinki* in mid-February near the ice shelf well past the summer temperature highs (Praebel et al. 2009). In addition, the temperatures of deeper waters of the McMurdo Sound (>100 m) do not rise above −1.5 °C regardless of the time of year (Littlepage 1965), which means that ice acquired in shallow-water fish in their early life stages will not melt when they move into deeper water to complete their life cycle. Since there is evidence for uptake of AFGP-adsorbed ice crystals by macrophages in notothenioid fish, it is possible that some sort of disposal mechanism involving macrophages may exist to counteract ice build up to unacceptable levels. Macrophages are known to generate free radicals as well as acidify their lysosomes by proton pumping accompanied by chloride entry (Sonawane et al. 2002), and thus there is a possibility that endocytosed ice may be melted in endosomes by means of accumulation of ions depressing the local mp. The fate of endogenous ice in these fish is an interesting problem that awaits further investigation albeit a technically difficult one.

Whether north polar fishes acquire endogenous ice like the Antarctic fish has not been investigated. It is expected that the high latitude fishes will at least acquire exogenous ice during winter months from contact with the sea ice cover and ice crystals in the water column generated by wave action during freeze up. Although the Arctic Ocean is surrounded by landmasses, it is in open communication and influenced by southern oceanic currents such as the Gulf–Atlantic and therefore undergo large seasonal variation in water temperatures. Summer surface water temperatures can reach 8 °C even in high latitudes (Enevoldsen et al. 2003; Devries and Steffensen 2005) and thus thermal melting could readily eliminate endogenous ice if acquired by fish during the winter months.

5.11 The Integument as a Physical Barrier to Ice Propagation

Since the stability of the undercooled state in fish has been adequately established and the absence of nucleators, it is apparent that the presence of internal splenic ice has to be acquired from the environment. It follows that ice must enter through the body surface, which includes the intestinal tract when ice-free fishes are exposed to the ice-laden waters. What is uncertain is which tissue is the "weak link." It is well documented that the cell membrane constitutes a physical barrier to ice propagation. For some cells, an undercooling of several degrees is required for ice to pass through pores in the cell membrane (Mazur 1966). Therefore, for polar fishes where the undercooling is small, the driving force for propagation through cell membranes is small. Ice entry most likely occurs at cell junctions or through damaged cells. Propagation of ice through the corneal epithelial tissue (the clear head skin over the eye and corneal epithelium underneath) into the 0.5 °C undercooled ocular fluid does not occur in Antarctic notothenioid fishes at its environmental temperature of −1.9 °C (Turner et al. 1985) indicating tight junctions. The relatively large external surface, the skin, which consists of multiple cellular layers overlain with scales in some species and by mucous in all, would be expected to constitute a physical barrier resistant to ice entry. With winter flounder 1.0 °C of undercooling relative to the blood fp (−1.5 °C) is required for in vitro trans-skin ice propagation for skin derived from AFP-fortified winter specimens (Valerio et al. 1992b). In summer when AFP is absent only 0.5 °C undercooling is required, and the addition of AFP to the serosal side of the skin of summer flounder resulted in greater undercooling needed for trans-skin ice propagation. The presence of APs in the interstitial spaces of the skin appears to be an important component of the resistance to ice propagation across the skin.

Further evidence for the skin as a strong barrier to ice entry comes from experiments with larval Antarctic notothenioid fishes. Immediately upon hatching in the austral Winter (late August to early September), the larvae of the naked dragon fish, *Gymnodraco acuticeps* swim from the rocks where they hatch directly to the sub-ice platelet layer at the underside of the solid ice cover, where ice crystals are abundant and the water temperature is at its coldest for the year. Underwater observations showed that these larvae do not freeze, and laboratory experiments indicate that they are resistant to freezing at −3.6 °C in the presence of ice. Micro sampling of the blood of these larvae showed they have very low hysteresis levels, with a blood fp of only −1.3 °C and thus are undercooled by 0.6 °C, and theoretically should freeze. Even though they can tolerate temperatures below −2 °C, in the presence of ice, if their skin is breached, they quickly freeze at a temperature corresponding to their blood fp upon contact with ice (Cziko et al. 2006). Apparently, the integument is an adequate barrier to ice entry as long as it remains intact. In adults, the gills are probably the most vulnerable to damage and ice entry, and in some early larva species gills are poorly developed and consist primarily of gill arches with lamella developing later in the juvenile stage. Intestinal fluid hysteresis of dragon fish larvae is initially twice that of the blood indicating perhaps that the gut lining is also a potential site of ice entry, since they begin feeding as soon as they hatch, and thus intestinal fluid freezing by dietary ice must be prevented.

Although the pre-hatch dragon larvae have already begun to synthesize AFGP their blood levels do not reach adult levels until 22 weeks post-hatch. The only conclusion that can be drawn is that their integumental surfaces including gill arches and their nascent filaments are adequate barriers to ice entry during the larval stage (Cziko et al. 2006). At 22 weeks post-hatch, the blood hysteresis levels are similar to those of the adults, suggesting that when the gill lamella has developed they become prone to ice entry and require a full complement of AFGP for protection.

The Antarctic silver fish *Pleuragramma antarctica* spawns immediately below the ice cover in the Antarctic coastal waters and their buoyant eggs float to the underside of the ice where they hatch in the sub-ice platelet layer (Evans and Devries 2017). They spend their larval stage in the upper part of the water column. The eggs are resistant to freezing because of the tough chorion that blocks ice entry. The hatchlings emerge with very little AFGP (body fluid fp of −0.82 °C) yet freeze only below −2.7 °C and some survive to −5 °C (Cziko et al. 2006) in the presence of ice. In contrast to the dragon fish and silver fish, *T. borchgrevinki* larvae hatch with an AFGP compliment that is only slightly below that of the adults and therefore along with an intact integument should be well protected from ice entry regardless of the developmental stage of the gills.

Similar observations have been made with Atlantic cod larvae of the northern cold stock that have an fp of −0.88 °C (determined from a larval homogenate) and no hysteresis, but can resist inoculative freezing when touched by ice down to −1.35 °C, indicating integumental resistance to ice entry (Valerio et al. 1992a). Gill lamellae in cod larvae only appear 5 weeks post-hatch (Hunt Von Herbing et al. 1996) and therefore the more vulnerable single-cell gill epithelium is absent for that period of time. In summary, it appears that the intact integument is an excellent barrier to ice propagation in the larval stages where gill filaments are absent or rudimentary. The intact integument in adults is also most likely an excellent barrier to ice entry, except for breaches that are possible at the gill, intestinal tract, and even in the multilayer skin. Again it should be emphasized in the absence of APs, the entry of one ice crystal is sufficient to initiate freezing of the entire larva or adult in an undercooled state.

A number of AFP-bearing fishes have been found to produce a "skin-type" AFP that has been suggested to prevent ice entry. Winter flounder and shorthorn sculpin, skin-type AFP is type I in the Ala-rich feature, but distinct from the circulatory AFP synthesized by the liver in these species (Low et al. 1998, 2001; Gong et al. 1996). The absence of secretory signal sequences in the coding sequences indicates that they are intracellular peptides and it has been suggested that they function intracellularly, which is inconsistent with the known function of APs in protecting extracellular fluids from freezing (Low et al. 2001; Fletcher et al. 2001). In the case of the winter flounder skin AFP, though without a secretory signal, surprisingly it also appears to be found outside the cells (Murray et al. 2003). However, it is not clear in this study how immunohistochemical detection could differentiate between interstitial fluid AFP derived from the circulation (liver type) versus the skin type, which are both Ala-rich and have similar sequence stretches that would constitute similar epitopes, resulting in cross-reacting antibodies. Various suggestions have been raised as to what the biological significance of the skin-type AFP maybe, but its actual role as a freezing prevention protein is far from certain and requires more definitive evidence. There has been no estimation of their physiological concentrations probably because it is difficult to make such a measurement. In all the reported studies, the activity of skin-type AFP was measured by concentrating AFP isolated from the skin or recombinant proteins to a similar concentration in the blood, which may be far above the physiological concentration in the skin. If the in vivo concentrations are insufficient to inhibit ice growth, then their role as antifreeze in fish is dubious. Furthermore, skin extracts with a thermal hysteresis indicative of the presence of an AFP was detected in the north temperate cunner, *Tautogolabrus adspersus* (Valerio et al. 1990), which has no blood antifreeze, the later indicating no requirement for the freeze-avoidance function in this fish, consistent with the known behavioral means employed by the adults and juveniles to avoid freezing (Olla et al. 1975). A survey of teleost fishes would be instructive, and if putative skin-type AFP is found to be prevalent in both AP and non-AP-bearing fishes, it is possible that its presence has a primary non-antifreeze function, and that its ice-binding activity is incidental and its activity resides in its structural characteristics which would still be interesting. Skin antimicrobial agents are short peptides prevalent in fish (Cole et al. 1997) and other vertebrate ectotherms and it is certainly worth investigating whether the putative skin AFP has antimicrobial properties.

Regardless of the role of the skin-type AFPs, undamaged skin in itself is an effective barrier to ice entry. The question remains as to how ice enters the shallow-water Antarctic notothenioid fishes. Given that *T. borchgrevinki* spleens contain less than 100 ice crystals, ice entry must be infrequent. The site of ice entry may be the gill epithelium, or the intestinal epithelium, where a single layer of cells on the apical side separates the ice-laden water from the blood, and the ice in the intestinal fluid from the interstitial fluid, respectively. Fragments of chitin from chitinous prey if exiting through the gill arches or as it transits along the intestine could cause damage. Another is the presence of epi-parasites on the gills and intestinal parasites that create lesions during attachment to the epithelium. The gills and skin of notothenioid fishes are commonly infested by parasitic copepods and bloodsucking leeches (Bielecki et al. 2008). Their digestive tracts are often infested by cestode and nematode worms (Zdzitowiecke 1998). Such lesions are possible sites of ice entry and definitive studies should be carried out to verify the validity of the supposition.

5.12 Synthesis and Distribution of Antifreeze Proteins

In northern fishes, the liver is the site of AF synthesis where upon it is secreted into blood, and from the circulation it diffuses down its concentration gradient through leaky endothelial junctions or pores in the capillaries into other fluid compartments including the cerebral, extradural, peritoneal, and pericardial fluids (Ahlgren and Devries 1984). Strong AFP mRNA signals were found in the liver RNA of fishes that express their respective type of AFP (Wang et al. 1995b; Lin and Long 1980; Lin 1979; Hew and Yip 1976; Hew et al. 1988; Ewart and Fletcher 1993), consistent with the liver being the major synthetic and secretory source of circulatory AFP. Northern cod fishes also have AFGP mRNA expression in their liver, like the AFP-bearing fishes (Cheng et al. 2006). Although it has been reported that the liver synthesizes the AFGPs in the Antarctic notothenioid fishes (O'grady et al. 1982a; Hudson et al. 1979; Haschemeyer and Mathews 1980), later studies clearly showed the absence of AFGP mRNA in the liver of the Antarctic notothenioid fishes (Chen et al. 1997a), and thus it became unclear whether liver was in fact involved at all in the synthesis of AFGPs in the notothenioid fishes.

Since marine fishes must drink seawater to maintain water balance and therefore ingest ice, there is a danger of ice growth when the seawater becomes isosmotic to the blood in the small intestine. In the small intestine, where the bulk of digestion occurs, large amounts of fluids including pancreatic bicarbonate and enzymes, bile, and intestinal mucosal secretions are added, decreasing the salt concentration. In addition, osmoregulation involves active transport of NaCl from the ingested sea-water into the blood across the intestinal epithelium, and ultimately out into the seawater through chloride cells in the gill (Evans 1993). Absorption of NaCl from the intestinal fluid leaves behind a fluid isosmotic to the blood but hyposmotic (mp about −1.0 °C) to seawater and would likely freeze as it contains ice crystals were it not for the presence of AFGP in the notothenioid and gadid intestinal fluids (O'grady et al. 1982b; Praebel and Ramløv 2005). The source of intestinal AFGPs in notothenioid fishes is the pancreas, where there is very high AFGP mRNA expression, and a full complement of AFGPs are present in the pancreatic fluid that can be sampled in sufficient quantity from the large notothenioid fish, *Dissostichus mawsoni* for analysis (Cheng et al. 2006). Pancreatic expression of APs has also been verified in the pancreatic tissue of AFGP-bearing northern cods, and in various AFP-bearing species, and in cases where enough intestinal fluid could be sampled, the AF proteins are present. The AFGPs are not digested as they transit the intestinal track (Cheng et al. 2006) and their concentration in the anterior portion of the intestine are low, but become concentrated as water and nutrients are absorbed along the tract toward the rectum where the fluid fp is around −2.2 °C, well below that of ambient seawater temperature (O'grady et al. 1982b). Thus the phylogenetically diverse AP-bearing fishes have converged on a common physiological solution, pancreatic synthesis, and secretion into the intestinal fluid where it is concentrated, thus preventing freezing of the fluid.

The explanation for the appearance of liver AFGP synthesis in the early studies of the notothenioid fishes can most likely be explained because whole livers were used and the teleost pancreas is not a discrete organ, but a diffuse tissue scattered within mesenteries associated with all the components of the gastrointestinal (GI) tract and non-GI organs in the abdominal cavity (Eastman and Devries 1997). In some teleosts, it infiltrates the liver forming hepatopancreas islets (Kurokawa and Suzuki 1995; Hinton et al. 1972). Thus the observed liver synthesis of AFGPs in the early studies may have its origin in the liver tissue associated with pancreatic tissue surrounding the bile duct branches. The question remains as to what is the synthetic site of circulatory AFGPs in the notothenioid fishes, if liver, which is the major source of secretory proteins in vertebrates, does not express AFGP mRNA. There is no histological evidence that the pancreatic tissue secretes AFGP into the circulation (Evans et al. 2012). Except for the pancreatic tissue, thus far no other tissues or organs examined in both adult and juvenile notothenioid fishes have been shown to express AFGP mRNA.

5.13 Rectal Absorption of Intestinal Antifreeze Glycoproteins

The AFGPs are not digested as indicated by AF activity measurements and gel electrophoresis as they move along the digestive tract. Histological studies indicate some reabsorption of AFGPs in the rectum in one of the notothenioid fish species and thus far this is the only mechanism identified for transferring AFGPs into the blood (Evans et al. 2012). However, only a small amount appears to be absorbed and this may explain the long period required for dragon fish larva to acquire adult levels of circulating AFGPs (Cziko et al. 2006). Rectal absorption of the intestinal fluid AFGPs seems like an inefficient mechanism for transferring it to the circulation if in fact, it is the only source of blood AFGPs. It is possible that some other tissue may also be an additional source, however, northern blots of RNA from the various tissues have so far not identified any. In any case even with some rectal absorption a large amount of AFGP is lost with the feces, an energetically costly situation. Rectal absorption has not been extensively investigated in northern fishes but the assumption is that a significant amount of AP is also lost with evacuation of their feces (Praebel and Ramløv 2005).

5.14 Stability of Undercooled Fish Fluids Antifreeze Proteins

Some polar fishes have body fluids that lack or are nearly devoid of APs and are undercooled between 0.5 °C and 1 °C (Turner et al. 1985; Dobbs and Devries 1975). Only trace amounts of the small AFGPs can be found in the ocular fluids and endolymph of the Antarctic notothenioid fish, *T. borchgrevinki* (Ahlgren et al. 1988). AFGPs are essentially absent in the urine of all high-latitude Antarctic notothenioid fishes because their kidneys lack glomeruli and are composed only of blind tubules (Dobbs and Devries 1975). Thus urine formation involves only secretion (Eastman and Devries 1986; Dobbs et al. 1974). Interestingly a few low latitude notothenioid fishes that lack AFGPs appear to have a few glomeruli that may reflect the ancestral condition rather than a functional necessity of these structures (Eastman and Devries 1986). The bladder urine of the high latitude Antarctic notothenioids is undercooled by about 1 °C. It is separated from seawater by a thick body wall and scale-covered integument and the interstitial spaces are well fortified with AFGPs (Ahlgren et al. 1988). The only opening to the exterior is through the urethra but it is closed with a muscular sphincter and the opening is also occluded by mucous except during micturition. There are no reported incidences of captured fishes with frozen urinary bladders. Only an occasional aquarium specimen whose bladder was catheterized for experimental purposes froze when exposed to small ice crystals in the flowing seawater (DeVries unpublished). Thus the body wall appears to constitute an excellent barrier and the modest urine undercooling can persist over the life span of the fish. The AFP-bearing Antarctic eel pouts *L. dearborni* and *P. brachycephalum*, however, they have kidney glomeruli, but no filtration occurs because the filtration barrier is thick, and the neck connecting Bowman's capsule to the tubule does not appear to have a patent connection. They appear to be functionally aglomerular with urine formation resulting only from tubular secretion, and thus should not lose their AFP into the formative urine (Eastman et al. 1979).

In the case of north temperate and Arctic fishes, few comprehensive studies have been done, but most AP-bearing fishes have glomeruli. Histological examination again indicates they are functionally aglomerular (Eastman et al. 1987) and thus APs should not be present in the urine. An interesting observation, that is, an exception to functional aglomularism is that the saffron cod (*E. gracilis*) glomeruli are smaller and hypertrophied and lack red blood cells in their capillaries during the winter while in the summer they are larger with red blood cells present in their capillaries (Kitagawa et al. 1990) correlated with the presence of AFGPs in the winter and their absence in the summer.

Another exception to this functional aglomerularism is the winter flounder that has functional glomeruli and urine formation involves filtration, secretion and reabsorption of nutrients, selected ions, and water from the formative urine (Petzel and Devries 1980). Its type I AFPs are small (3.2 and 4.2 kDa) and would be expected to be readily filtered into the formative urine in Bowman's space. However, flounder AFP molecules are anionic, and are largely prevented from passing through the anionic basement membrane of the glomerular capillaries because of a charge–charge repulsion mechanism, resulting in their conservation in the circulation (Petzel and Devries 1980; Boyd and Devries 1983). Other studies have indicated that there are substantial AFPs in the urine of the winter flounder, sea raven, cod, and ocean pout (Fletcher et al. 1989) even though histological studies indicate some are functionally aglomerular (Eastman et al. 1987). It is possible that the renal conservation mechanisms are not 100% effective and that small amounts of AP are lost because of incomplete blockage of filtration at the glomerulus or leaks through cellular junctions of the kidney tubules and bladder wall of aglomerular species. The teleost bladder wall is known to actively reabsorb both salt and water (Demarest 1984) and the high concentration of AFPs observed in the bladder urine of these fishes may be the consequence of concentrating small amounts of AFP from a large volume of ureteral urine and represent only a small amount of loss despite its high concentration in the bladder urine. Even considering the relatively low glomerular filtration rates in most marine fishes, if the APs were not conserved at the filtration barrier, the fish would be unable to maintain the necessary protective blood levels of AFPs as the synthesis rates would be slow at these low temperatures. In the case of the winter flounder where the filtration rate is 0.5 ml/kg/h (Petzel and Devries 1979) and if AFPs were filtered and not reabsorbed, all their circulating AFP would be lost in 48 h. More definitive physiological research is needed to document whether kidney filtration is completely absent in northern AFP bearing fishes where histological examination indicates the presence of glomeruli that lack glomerular function. It would also be informative to determine whether the relatively high AP levels reported in bladder urine are largely the result of the bladder concentration process of ureteral urine that originally contained only trace amounts of AP that leaked across cell junctions of the tubules. Quantification of the loss could be determined by analyzing the AP content in ureteral urine as a function of time. More definitive research is needed to document AFP concentrations in ureteral and bladder urine, as well as amount of urine, voided per unit time. This would give an estimate of the loss of AFP which has to be small given that its rate of synthesis would be slow at such low temperatures.

The endolymph of the Antarctic notothenioid fishes lacks hysteresis even when the most sensitive techniques are employed and this finding is consistent with fluid secretion by active transport of ions followed by osmotic influx of water. The undercooled endolymph in the semicircular canals within the skull is surrounded by tissues fortified with AFGPs, and thus safe from nucleation. The ocular fluids, on the other hand, are separated from the environment by a relatively thin clear head skin and cornea. The vitreous and aqueous fluids do contain small amounts of the small AFGPs as determined by gel electrophoresis and hysteresis measurements (Turner et al. 1985) but not enough to prevent freezing. However, the ocular fluid undercooling is small (0.5 °C), and together, the cornea and the overlying transparent head skin constitute an effective barrier to the inward propagation of ice at −1.9 °C (Turner et al. 1985). Although the tissues lack visible vascularization, there is a small capillary blood supply and thus their interstitial fluid would be fortified with AFGPs. Collectively, these factors prevent ice nucleation of the undercooled ocular fluids. Undercooling of several of the internal fluid compartments is an essential component of the freezing avoidance strategy of polar fishes.

5.15 Blood Serum Freezing Points, Hysteresis Levels, and Environmental Severity

There is a strong correlation between organismal and blood fps and environmental severity (presence of ice and freezing seawater). For example, populations of *T. borchgrevinki* in McMurdo Sound are exposed to supercooled water, laden with small ice crystals and have organismal and blood fps of −2.3 °C and −2.7 °C, respectively (Table 5.1). In the deep warmer waters off the Antarctic Peninsula some notothenioid fish blood fps are around −1.5 °C (Table 5.3). Because the AP contribution to the low blood fps is greater than 50% and the mps are relatively constant (−0.9 to −1.1 °C) the hysteresis values also reflect this correlation with a few caveats. In general, the hysteresis values increase with the severity of the environment (Table 5.3). Although fps are cooling rate and crystal size-dependent and as well a small hysteresis associated with the melting point, the hysteresis value is a useful indicator of the level of resistance to freezing (Fields and Devries 2015) as long as a constant slow cooling rate and small seed ice crystal (10–20 μm) is used. To obtain a more robust mps and estimates of the hysteresis, the relatively small mp hysteresis (−0.1 to −0.3 °C) can be removed from the hysteresis value by measuring the serum osmolality with a Wescor Vapor Pressure Osmometer and converting the value to mp (Cziko et al. 2014). In both Arctic and Antarctic fishes, there is a reasonable correlation between decreasing environmental temperature, ice abundance, and the level of the serum thermal hysteresis (Fields and Devries 2015; Devries 1982) (Table 5.3). Comparison of freezing avoidance in the southern and

Table 5.3 Comparison of depth, environmental temperature, presence or absence of ice, blood-freezing/melting points and hysteresis, and blood AFGP concentrations in polar fish

							AFGP (mg/ml)			
Species	Depth (m)	Water temp. (°C)	Ice present	Blood FP (°C)	Blood MP (°C)	Hysteresis (°C)	1–5	6	7–8	Total
High Antarctic—McMurdo Sound										
Trematomus borchgrevinki	3–30	−1.93	+	−2.7	−1.1	1.6	7.4	2.5	25.3	35.2
Trematomus bernacchii	10–250	−1.93	+	−2.5	−1.1	1.4	9	1.8	23.5	34.3
Trematomus loennbergii	500	−1.93	–	−2.2	−1.1	1.1	13.6	2	12.6	28.2
Lycodichthys dearborni	500	−1.93	–	−1.9	−1	1	–	–	–	–
Antarctic Peninsula										
Notothenia coriiceps	2–300	−1.8	+, –	−2.2	−1	1.2	6.1	1.8	22	29.9
Chaenocephalus aceratus	30–300	−1	–	−1.5	−0.9	0.6	0.3	Trace	4.7	5
Chamsocephalus gunnari	30–300	−1	–	−1.44	−0.87	0.57	Trace	Trace	10	10
High Arctic										
Boreogadus saida	2	−1.8	+	−2.2	−1.1	1.1	6	1.5	20	27.5
Sub-Arctic										
Eleginus gracilis	5	−1.8	+	−2.1	−1	1.1	1	10	8	19
Myoxocephalus verrucosus	3–50	−1.8	+, –	−2.3	−0.9	1.4	–	–	–	–

northern hemisphere polar fishes indicates hysteresis in the Arctic and north temperate fish fauna is less than Antarctica, which is not surprising. The year-round water temperature is lower and ice cover greater in the Antarctic with small seasonal changes in temperature in contrast to the large seasonal changes in the Arctic region (Devries and Steffensen 2005).

Antarctica is a lone south polar continent surrounded by the vast Southern Ocean. Although it interfaces with other oceans in the Southern Hemisphere, it is in effect isolated from them by the massive clockwise flow of the Antarctic Circumpolar Current (ACC), which is 200–1000 km wide and reaches the sea floor (Foster 1984). The ACC is both a thermal barrier, decoupling the warm subtropical gyres from the cold Antarctic waters, as well as a physical barrier to migration of fish in either direction. The opening of the Drake Passage and the unrestricted flow of the ACC presumably began around 25 Mya, leading to the thermal isolation and subsequent continental glaciation and freezing of the surrounding seas, reaching present day conditions around mid-Miocene (10–15 Mya) (Kennett 1977, 1982). Thus the Antarctic waters today, though spanning a vast latitudinal range (about 55°S–78°S) are the coldest in the world, extremely cold stable at high latitudes, and experience only minor variations at the lower latitudes, with peak austral summer temperature at about +0.5 °C and 2 °C, respectively (Cziko et al. 2014; Hunt et al. 2003; Clarke 1987). Correlated with these high latitude frigid conditions is the constitutive expression of high levels of AFGP in the endemic notothenioid fishes. Although expression is also constitutive in fishes inhabiting the warmer west Antarctic Peninsula water, in some species the AFGP levels are slightly less and in some cases only half of those in the high latitude species (Table 5.3) (Jin and Devries 2006; Devries 1970; Ahlgren and Devries 1984; Bilyk and DeVries 2010; Fields and Devries 2015). Relatively small variations in environmental temperatures do exist, however, in the high latitude waters, and are associated with the presence of nearby ice shelves, and depth of the fish habitat, which affect the extent of iciness in the water column that is already near its fp. These minor environmental variations are reflected in the slightly different hysteresis fps, with fish living in the coldest iciest shallow waters having the lowest fps while those in deeper ice-free water have slightly higher fps (Fields and Devries 2015). Comparison of populations of the same species inhabiting the cold high latitude waters with the ~2 °C warmer water of the western Antarctic Peninsula have similar hysteresis fps which is consistent with constitutive expression (Fields and Devries 2015). The relatively modest changes in hysteresis in *T. borchgrevinki* following laboratory warm acclimation (+4 °C) for 4 months suggests that AFGP expression levels must be genetically fixed and adaptation to their extreme freezing environment must have occurred over evolutionary time (Jin and Devries 2006).

The water temperature spanning the water column of McMurdo Sound varies by only a few hundredths of a degree during the winter that provides an opportunity to examine the effects of depth and presence of environmental ice on the serum hysteresis in closely related fish at the same location. The cryopelagic fish, *T. borchgrevinki* that is associated with the sub-ice platelet layer have the greatest serum hysteresis (1.64 °C) as this water column habitat is the coldest, and the sub-ice platelet layer is continually growing during most of the winter, making it also the iciest habitat. Benthic species such as *T. bernacchii* that spend time foraging on mats of anchor ice in shallow water have slightly less hysteresis (1.35 °C), while the deepwater species like *T. loennbergii* exposed to similar temperatures have substantially less serum hysteresis (1.04 °C), consistent with the absence of ice at depth. When freezing conditions exist in the surface waters, *T. loennbergii* brought to the surface will sometimes freeze consistent with their lower serum hysteresis levels. Likewise, the McMurdo Sound benthic eel pout *L. dearborni*, which lives at 500–600 m, have a similar depth related low serum hysteresis (1.0 °C), and suffer freezing death in the icy water at the surface. As pointed out previously, these species are in no danger of freezing in their deepwater habitat because of the absence of ice and their lowered freezing point due to the effect of hydrostatic pressure.

In contrast to McMurdo Sound, the western Antarctic Peninsula marine environment near Anvers Island (64°45′S; 64°03′W) is comparatively mild due to its lower latitudinal position, and to encroachment of warm circumpolar deep water (+1 °C) on the continental shelf. Additionally, there is no ice formation in the water column below the surface due to the absence of nearby ice shelves. With these less extreme conditions, the common local shallow water notothenioid fish, *Notothenia coriiceps* has a hysteresis of 1 °C (Devries 1988), which is less than their McMurdo Sound shallow water ecological counterparts even though it is exposed to ice when in the shallow water (1–2 m) (Table 5.3) during the winter (Ahlgren et al. 1988). A few of the red blooded and channichthyiid fishes (family Channichthyidae) have half the hysteresis as the high latitude fishes and their lack of change in hysteresis levels with seasonal warming are again consistent with constitutive expression, the levels that must have been genetically fixed during their evolutionary histories. Although some low latitude fishes have only half the AFGP level that high latitude fish species have and are in no danger of freezing in their present habitats, it is unclear whether in the past they faced more severe environmental conditions and had higher levels of AFGPs but are now in the process of losing their AFGP genes in their present warmer environment. However, some channichthyiid fishes in the Peninsula waters have blood fps of −2 °C indicating they can avoid freezing regardless of their water column habitat.

In contrast to isolated Antarctic, the Arctic Ocean is surrounded by northern landmasses with few connections to the Pacific and Atlantic Oceans, and it is covered by dynamic multilayer pack ice. The Arctic Ocean and sub-Arctic seas are open to temperate oceanic influences such as the Gulf–Atlantic and Japanese currents, and experience greater annual temperature changes than the Southern Ocean (Devries and Steffensen 2005). Thus some northern fish ranges include the cold Arctic Ocean as well as the warmer sub-Arctic seas such as some gadid and cottid fishes (Love et al. 2016) and some make migrations across large geographic distances and thus temperature clines (Kurlansky 1997; Howe 1991). Correlated with these variables are latitudinal and seasonal variations in antifreeze levels in different AP-bearing northern fishes. In contrast to the Antarctic, it is difficult to make generalizations as to the responses of Arctic and north temperate fishes to environmental severity, in part because of spatial and seasonal temperature variability in the various Arctic regions, and the response of fishes that belong to several unrelated taxa. However, a few generalizations can be made where the water temperature is near freezing throughout the year.

Both of the high Arctic gadids, the polar cod, *Boreogadus saida* and the ice cod, *Arctogadus glacialis* are present in the Arctic basin (Cohen et al. 1990) and as well in the freezing fjords of east and west Greenland (70°N; 51°W) (Praebel and Ramløv 2005) where they are associated with the ice cover and freezing seawater through most of the year. During the summer, the polar cod generally avoid the warm surface water and remain in the colder deep water (Love et al. 2016). The ice cod have also been documented to avoid warm surface water and inhabit a cold layer of water (−1 °C) at 20–50 m even in July in a fjord near Uummannaq, west Greenland (Jordan et al. 2001) and they are also common in some east Greenland deep Fjords at −1.8 °C in late summer. In part, the polar and ice cod can be considered the north polar counterparts of the high latitude Antarctic notothenioids in being stenothermal and in the extent of exposure to environmental severity. A few winter studies of these two Arctic gadids have been done, and *B. saida* caught in mid-January through the ice in Ny-Ålesund, Spitsbergen in January had a blood hysteresis of 1.06 °C while specimens from the Ice Fjord in the same area kept in 0.6 °C aquarium water had a mean serum hysteresis of 1.15 °C as well as those sampled during the summer in the deep cold fjords (−1.8 °C) of east Greenland and Spitzbergen. The ice cods from the same fjords have a hysteresis of 1 °C in September the warmest month of the year. The fjord cod, *Gadus ogac* from Labrador has a winter hysteresis of 1.17 °C while those from Disco Bay, west Greenland caught in the ~+6 °C surface water during the summer have a hysteresis of 0.8 °C (Enevoldsen et al. 2003) indicating that they lose their AFGPs during the summer. For the most part, the high latitude gadid species retain their high blood APs levels throughout the year although there may be slight losses if they move into warmer waters for extended periods during the summer months (Enevoldsen et al. 2003; Denstad et al. 1987). The high latitude Alaskan Arctic sculpin *Myoxocephalus verrucous* (synonymous with *M. scorpius*), a permanent shallow water inhabitant of the Bering Sea has a serum hysteresis of 1.4 °C in early summer (Raymond et al. 1975) (Table 5.3) as well as those in late summer in Grise Fjord (76 N), Ellesmere Isld. Fletcher et al. (1982a) consistent with the absence of seasonal cycling of their AFP. However, populations in Newfoundland and Nova Scotia waters do lose their AFPs in the summer (Fletcher et al. 1982a; Duman and Devries 1974a). Whether the Arctic populations would retain their AFPs if exposed to +6 °C water for an extended period of time is unknown but if little change then one could conclude that its expression is genetically fixed. It may be that this species, regardless of location, synthesizes AFP throughout the year but in the southerly populations' protein degradation outpaces synthesis during the warm summer temperatures. If there are genetic differences between populations, it could be easily verified by comparing the amounts of liver AFP mRNA present during the summer and winter seasons.

Some species such as the north temperate ocean pout, *Macrozoarces americanus* have high winter levels of AFP which during the summer are reduced to half the winter levels (Fletcher et al. 1985b). Also, winter population differences between

Nova Scotia and New Brunswick can be explained by a higher gene dosage correlated with a more severe environment in the Newfoundland population (Hew et al. 1988). In contrast to latitudinal temperature responses of this species, another north temperate eel pout *Zoarces viviparus* inhabiting the coastal waters of Denmark is well fortified with AFPs in the winter but lose all of it during the summer (Sørensen and Ramløv 2001).

Another species that lose their AFPs in the summer is the winter flounder (Petzel et al. 1980; Fletcher et al. 1985a) which begins in early summer when the water has warmed from freezing to 0 °C. Two North Atlantic gadids that show seasonal cycling are the tomcod, *Microgadus tomcod* from Shinnecock Bay, NY, and Nova Scotia (Reisman et al. 1984; Fletcher et al. 1982a; Duman and Devries 1975) and the Newfoundland Atlantic cod, *Gadus morhua* (Fletcher et al. 1982b). The saffron cod, *E. gracilis* from the Bering Sea also shows seasonal cycling (Raymond et al. 1975; Burcham et al. 1984) with a wintertime hysteresis of 1.1 °C (Table 5.3) while they all lose their AFGPs in the summer. With these fishes, the seasonal cycling is correlated with the onset of warm summer and freezing winter temperatures.

Little is known about the control of seasonal AFP cycling except in the winter flounder. Although AFP levels are correlated with decreasing temperature (Fletcher et al. 2001; Duman and Devries 1974a) it does not directly initiate synthesis. It appears that short days cause the pituitary to stop producing growth hormone (GH), which regulates the synthesis of IGF-1. The lack of GH in the winters inhibit the synthesis of IGF-1 and in its absence, AFP gene transcription and translation proceeds and AFP is secreted into the blood (Fletcher et al. 2001). The exact timing of the rise, peak, and decline phases of APs differ between species as well as between different geographic populations of the same species correlated with the warming of the water but partly due to an endogenous rhythm apparently genetically fixed (Fletcher et al. 1985a, b).

Hysteresis values have been used as a proxy for antifreeze activity in the blood of polar fishes (Fields and Devries 2015). However, there is not a strong correlation between the blood AP concentrations and antifreeze activity because isoforms of the APs in many fishes have different activities (Schrag and Devries 1982; Marshall et al. 2004; Kao et al. 1985). Furthermore, in AFP-bearing fishes, the multiple purification steps result in low yields of pure AFP. Only in the case of the AFGP-bearing fishes has quantification been successful because the AFGPs are completely soluble in 5% trichloroacetic acid (TCA) while all other proteins are precipitated. The AFGPs in the supernatant can be applied directly to high-resolution HPLC analytical sizing columns and the large and small isoforms cleanly separated and quantified (Jin and Devries 2006; Fields and Devries 2015). Quantification of the blood AFGPs in the notothenioid fishes does not fully account for the blood hysteresis because the hysteresis contribution of the AFPP is not included as it is precipitated by the TCA. There is a correlation between the total AFGP content and the blood hysteresis, but it is not as strong as it would be in the polar and northern gadids, which lack the AFPP contribution to the hysteresis. So far only a few gadid species have been analyzed (Table 5.3) and a survey of all of the AFGP-bearing species would be worthwhile.

5.16 Conclusions

Freeze avoidance is associated with the presence of APs in fishes that live in ice-laden seawater. They are present in most body fluids of polar fishes where they bind to ice crystals that enter the fish's circulation and inhibit their growth. In the Antarctic notothenioid fishes and Arctic cods, the APs are AFGPs while in other unrelated northern fishes they are small carbohydrate-free proteins (AFPs) of different compositions and structures. The mechanism of antifreeze action is adsorption inhibition and the AP activity for most is similar on a molar basis despite the different sizes and structures. The APs are present in the blood, interstitial and intestinal fluids and are only present in trace amounts in the ocular fluid, urine, and none in the semicircular canal fluid. The latter fluids are undercooled by 1 °C and this small undercooling is stable. Nucleation is prevented because they are surrounded by tissues fortified with APs. The site of AP synthesis in most fishes is the liver with the exception of the Antarctic notothenioids; so far the only identified source for the blood AFGPs is rectal absorption of intestinal AFGPs whose source is the diffuse pancreatic tissue that drains into the anterior intestine and is not digested. It would appear that an alternate source must exist as rectal absorption does not appear to adequately account for the high levels in the blood because of the low rate of intestinal absorption. Possible alternate sites of synthesis that secrete directly into the circulation should be investigated using modern immunohistochemistry approaches. Although voided with the feces APs are not excreted via the urine. The kidneys of notothenioids are aglomerular and while most northern fishes have variable numbers of glomeruli, they are non-functional. The flat fishes are an exception where AFPs are conserved via a charge–charge repulsion mechanism involving the acidic peptide and the negatively charged basement membrane of the glomerulus.

Endogenous ice has been identified in the spleen of some Antarctic fishes but the site of ice entry has not been specifically identified. Unless breached, the integument is an excellent barrier to ice entry, and thus it would appear that the delicate gill and intestinal epithelia are likely entry sites because the blood in these tissues with large surface areas is separated from the environmental ice by a single layer of cells. Although ice only occasionally enters the circulation its growth is quickly inhibited by adsorption of the APs. In the Antarctic notothenioid fishes, AFGP decorated ice crystals are recognized by systemic macrophages that endocytose them and then migrate to the spleen where they reside. Although less than a hundred ice crystals are present in the spleen it melts when fishes enter warm surface water (0 °C) during the summer. Melting is not a viable option for fishes inhabiting the high latitude marine waters that never warm above −1 °C the equilibrium freezing point of the fish blood. For these fishes, the most intriguing question is the fate of the ice that accumulates in the spleen. Are there cellular mechanisms that can melt the ice in the macrophages or does ice continue to accumulate throughout the life of the fish or at some point in its life cycle does it enter water warm enough to melt it? To address these questions novel approaches will be required.

There is a correlation between the amount of blood AP and the severity of the environment. Near ice shelves where the water is at its fp and laden with minute ice crystals, fishes have the highest concentration of AP while in milder climates where water temperatures rise to near 0 °C during the summer, the fish fauna has much lower concentrations of AP. In northern environments, like the southern hemisphere fishes some species are endowed with high levels of AP year round. Others have high winter levels but during the summer synthesis ceases thus conserving energy.

References

Ahlgren JA, Devries AL (1984) Comparison of antifreeze glycoproteins from several Antarctic fishes. Polar Biol 3:93–97

Ahlgren JA, Cheng C-HC, Schrag JD, Devries AL (1988) Freezing avoidance and the distribution of antifreeze glycopeptides in body fluids and tissues of Antarctic fish. J Exp Biol 137:549–563

Antson AA, Smith DJ, Roper DI, Lewis S, Caves LSD, Verma CS, Buckley SL, Lillford PJ, Hubbard RE (2001) Understanding the mechanism of ice binding by type III antifreeze proteins. J Mol Biol 305:875–889

Baardsnes J, Davies PL (2001) Sialic acid synthase: the origin of fish type III antifreeze protein? TIBS 24:468–469

Baardsnes J, Kondejewski LH, Hodges RS, Chao HM, Kay C, Davies PL (1999) New ice-binding face for type I antifreeze protein. FEBS Lett 463:87–91

Baardsnes J, Jelokhani-Niaraki M, Kondejewski LH, Kuiper MJ, Kay CM, Hodges RS, Davies PL (2001) Antifreeze protein from shorthorn sculpin: identification of the ice-binding surface. Protein Sci 10:2566–2576

Bar DM, Bernheim R, Guo AS, Davies PL, Braslavsky I (2016) Putting life on ice: bacteria that bind to frozen water. J R Soc Interface 13:121. https://doi.org/10.1098/rsif.2016.0210

Bielecki A, Rokicka M, Ropelewska E, Dziekonska-Rynko J (2008) Leeches (Hirudinida: Piscicolidae) – parasites of Antarctic fish from *Channichthyidae* family. Wiad Parazytol 54 (4):345–348

Bilyk K, Devries AL (2010) Freezing avoidance of the Antarctic ice fishes (*Channichthyidae*) across thermal gradients in the Southern Ocean. Polar Biol 33(2):203–213

Black VS (1951) Some aspects of the physiology of fish. Univ Tor Stud Biol Ser 71:53–89

Boyd RB, Devries AL (1983) The seasonal distribution of anionic binding sites in the basement membrane of the kidney glomerulus of the winter flounder *Pseudopleuronectes americanus*. Cell Tissue Res 234:271–277

Braslavsky I, Drori R (2013) LabVIEW-operated novel nanoliter osmometer for ice binding protein investigations. J Vis Exp 72:e4189

Brown JD, Sonnichsen FD (2002) The structure of fish antifreeze proteins. In: Ewart KV, Hew CL (eds) Fish antifreeze proteins. World Scientific, Singapore

Burcham TS, Osuga DT, Chino H, Feeney RE (1984) Analysis of antifreeze glyco-proteins in fish serum. Anal Biochem 139:197–204

Bush CA, Feeney RE (1986) Conformation of the glycotripeptide repeating unit of antifreeze glycoprotein of polar fish as determined from the fully assigned proton N.M.R. spectrum. Int J Pept Protein Res 28:386–397

Celik Y, Drori R, Pertaya-Braun N, Altan A, Barton T, Bar-Dolev M, Groisman A, Davies PL, Braslavsky I (2013) Microfluidic experiments reveal that antifreeze proteins bound to ice crystals suffice to prevent their growth. Proc Natl Acad Sci U S A 110:1309–1314

Chakrabartty A, Yang DS, Hew CL (1989) Structure-function relationship in a winter flounder antifreeze polypeptide: II. Alteration of the component growth rates of ice by synthetic antifreeze polypeptides. J Biol Chem 264:11313–11316
Chao HM, Deluca CI, Davies PL (1995) Mixing antifreeze protein types changes ice crystal morphology without affecting antifreeze activity. FEBS Lett 357:183–186
Chao H, Houston ME, Hodges RS, Kay CM, Sykes BD, Loewen MC, Davies PL, Sonnichsen FD (1997) A diminished role for hydrogen bonds in antifreeze protein binding to ice. Biochemistry 36:14625–14660
Chen L, Devries AL, Cheng C-HC (1997a) Evolution of antifreeze glycoprotein gene from a trypsinogen gene in Antarctic notothenioid fish. Proc Natl Acad Sci U S A 94:3811–3816
Chen L, Devries AL, Cheng C-HC (1997b) Convergent evolution of antifreeze glycoproteins in Antarctic notothenioid fish and Arctic cod. Proc Natl Acad Sci U S A 94:3817–3822
Cheng C-HC (1996) Genomic basis for antifreeze glycopeptide heterogeneity and abundance in Antarctic notothenioid fishes. In: Ennion SJ, Goldspink G (eds) Gene expression and manipulation in aquatic organisms. Cambridge University Press, Cambridge
Cheng C-HC (1998) Evolution of the diverse antifreeze proteins. Curr Opin Genet Dev 8:715–720
Cheng C-HC, Detrich HW III (2007) Molecular ecophysiology of Antarctic notothenioid fishes. Phil Trans R Soc Lond B 362:2215–2232
Cheng C-HC, Devries AL (1989) Structures of antifreeze peptides from the Antarctic eel pout, *Austrolycichthys brachycephalus*. Biochim Biophys Acta 997:55–64
Cheng Y, Yang Z, Tan H, Liu R, Chen G, Jia Z (2002) Analysis of ice-binding sites in fish type II antifreeze protein by quantum mechanics. Biophys J 83:2202–2210
Cheng C-HC, Cziko AP, Evans CW (2006) Non-hepatic origin of notothenioid antifreeze reveals pancreatic synthesis as common mechanism in polar fish freezing avoidance. Proc Natl Acad Sci U S A 103:10491–10497
Clarke A (1987) Seasonality in the Antarctic marine environment. Comp Biochem Physiol 90:461–473
Clarke CJ, Buckley SL, Linder N (2002) A new name for antifreeze proteins–ice structuring proteins. CryoLetters 23:89–92
Cohen DM, Inada T, Iwamoto T, Scialabba N (1990) Gadiform fishes of the world. FAO, Rome
Cole AM, Peddrick W, Diamond G (1997) Isolation and characterization of pleurocidin, an antimicrobial peptide in the skin secretions of winter flounder. J Biol Chem 272:12008–12013
Cziko P, Evans CW, Cheng C-HC, Devries AL (2006) Freezing resistance of antifreeze-deficient larval Antarctic fish. J Exp Biol 209:407–420
Cziko PA, Devries AL, Evans CW, Cheng C-HC (2014) Antifreeze protein-induced superheating of ice inside Antarctic notothenioid fishes inhibits melting during summer warming. Proc Natl Acad Sci U S A 111:14583–14588
Davies PL (2014) Ice-binding proteins: a remarkable diversity of structures for stopping and starting ice growth. Trends Biochem Sci 39:548–559
Davies PA, Baardsnes J, Kuiper MJ, Walker VK (2002) Structure and function of antifreeze proteins. Philos Trans R Soc Lond Ser B Biol Sci 357:927–935
Deluca CI, Davies PL, Ye Q, Jia Z (1998) The effects of steric mutations on the structure of type III antifreeze protein and its interaction with ice. J Mol Biol 275:515–552
Demarest JR (1984) Ion and water transport by the flounder urinary bladder: salinity dependence. J Physiol 246:F395
Dempson JB, Kristofferson AH (1987) Spatial and temporal aspects of the ocean migration of anadromous Arctic char. Am Fish Soc Symp 1:340–357
Deng G, Laursen RA (1998) Isolation and characterization of an antifreeze protein from the longhorn sculpin, *Myoxocephalus octodecimspinosis*. Biochim Biophys Acta 1388:305–314
Deng G, Andrews DW, Laursen RA (1997) Amino acid sequence of a new type of antifreeze protein from the long horn sculpin *Myoxocephalus octodecimspinosis*. FEBS Lett 402:17–20

Deng C, Cheng C-H, Ye H, He X, Chen L (2010) Evolution of an antifreeze protein by neofunctionalization under escape from adaptive conflict. Proc Natl Acad Sci U S A 107:21593–21598

Denstad J-P, Aunaas T, Borseth JF, Aaset AV, Zachariassen KE (1987) Thermal hysteresis antifreeze agents in fishes from Spitsbergen waters. Polar Res 5:171–174

Desjardins M, Graham LA, Davies PL, Fletcher GL (2012) Antifreeze protein gene amplification facilitated niche exploitation and speciation in wolffish. FEBS J 279:2215–2230

Devries AL (1970) Freezing resistance in Antarctic fishes. In: Holdgate MW (ed) Antarctic ecology. Academic, New York

Devries AL (1971) Glycoproteins as biological antifreeze agents in Antarctic fishes. Science 172:1152–1155

Devries AL (1982) Biological antifreeze agents in cold water fishes. Comp Biochem Physiol 73A:627–640

Devries AL (1984) Role of glycopeptides and peptides in inhibition of crystallization of water in polar fishes. Philos Trans R Soc Lond B Biol Sci B304:575–588

Devries AL (1986) Glycopeptide and peptide antifreeze - interaction with ice. Methods Enzymol 127:293–303

Devries AL (1988) The role of antifreeze glycopeptides and peptides in the freezing avoidance of Antarctic fishes. Comp Biochem Physiol 90B:611–621

Devries AL, Cheng C-HC (2005) Antifreeze proteins and organismal freezing avoidance in polar fishes. In: Farrell AP, Steffensen JF (eds) The physiology of polar fishes. Elsevier, San Diego

Devries AL, Lin Y (1977) Structure of a peptide antifreeze and mechanism of adsorption to ice. Biochim Biophys Acta 495:388–392

Devries AL, Steffensen JF (2005) The Arctic and Antarctic polar marine environments. In: Farrell AP, Steffensen JF (eds) The physiology of polar fishes. Elsevier, San Diego

Devries AL, Wohlschlag DE (1969) Freezing resistance in some Antarctic fishes. Science 163:1073–1075

Devries AL, Komatsu SK, Feeney RE (1970) Chemical and physical properties of freezing point depressing glycoproteins from Antarctic fishes. J Biol Chem 245:2901–2908

Dieckmann G, Rohardt G, Hellmer H, Kipfstuhl J (1986) The occurrence of ice platelets at 250 m depth near the Filchner Ice Shelf and its significance for sea ice biology. Deep-Sea Res 33:141–148

Dobbs GH, Devries AL (1975) Renal function in Antarctic teleost fishes: serum and urine composition. Mar Biol 29:59–70

Dobbs GH, Lin YL, Devries AL (1974) Aglomerularism in Antarctic fish. Science 185:793–794

Duman JG (2001) Antifreeze and ice nucleator proteins in terrestrial arthropods. Annu Rev Physiol 63:327–357

Duman JG, Devries AL (1974a) The effects of temperature and photoperiod on antifreeze production in cold water fishes. J Exp Zool 190:89–97

Duman JG, Devries AL (1974b) Freezing resistance in winter flounder, *Pseudopleuronectes americanus*. Nature 247:237–238

Duman JG, Devries AL (1975) The role of macromolecular antifreezes in cold water fishes. Comp Biochem Physiol 52A:193–199

Duman JG, Devries AL (1976) Isolation, characterization, and physical properties of protein antifreezes from the winter flounder, *Pseudopleuronectes americanus*. Comp Biochem Physiol 53B:375–380

Eastman JT, Devries AL (1986) Renal glomerular evolution in Antarctic notothenioid fishes. J Fish Biol 29:649–662

Eastman JT, Devries AL (1997) Morphology of the digestive system of Antarctic notothenioid fishes. Polar Biol 17:1–13

Eastman JT, Devries AL, Coalson RE, Nordquist RE, Boyd RB (1979) Renal conservation of antifreeze peptide in Antarctic eelpout, *Rhigophila dearborni*. Nature 282:217–218

Eastman JT, Boyd RB, Devries AL (1987) Renal corpuscle development in boreal fishes with and without antifreezes. Fish Physiol Biochem 4:89–100

Enevoldsen LT, Heiner I, Devries AL, Steffensen JF (2003) Does fish from the Disko Bay area of Greenland possess antifreeze proteins during the summer? Polar Biol 26:365–370

Evans DH (1993) Osmotic and ionic regulation. In: Evans DH (ed) The physiology of fishes. CRC Press, Boca Raton

Evans CW, Devries AL (2017) Coping with ice: freeze avoidance in the Antarctic silverfish (*Pleuragramma antarctica*) from egg to adult. In: Vacchi M, Pisano E, Ghigliotti L (eds) The Antarctic silverfish: a keystone species in a changing ecosystem. Springer, Cham

Evans CW, Gubala V, Nooney R, Williams D, Brimble M, Devries AL (2011) How do Antarctic notothenioid fishes cope with internal ice? A novel function for antifreeze glycoproteins. Antarct Sci 23:57–64

Evans C, Hellman L, Middleditch M, Wojnar JM, Brimble MA, Devries AL (2012) Synthesis and recycling of antifreeze glycoproteins in polar fishes. Antarct Sci 24:259–268

Ewart KV, Fletcher GL (1990) Isolation and characterization of antifreeze proteins from smelt (*Osmerus mordax*) and Atlantic herring (*Culpea harengus harengus*). Can J Zool 68:1652–1658

Ewart KV, Fletcher GL (1993) Herring antifreeze protein: primary structure and evidence for a C-type lectin evolutionary origin. Mol Mar Biol Biotechnol 2:20–27

Ewart KV, Rubinsky B, Fletcher GL (1992) Structural and functional similarity between fish antifreeze proteins and calcium-dependent lectins. Biochem Biophys Res Commun 185:335–340

Fänge R, Nilsson S (1985) The fish spleen: structure and function. Experientia 41:152–158

Fields LG, Devries AL (2015) Variation in blood serum antifreeze activity of Antarctic Trematomus fishes across habitat temperature and depth. Comp Biochem Physiol A 185A:43–50

Fletcher GL, Hew CH, Joshi SB (1981) Isolation and characterization of antifreeze glycoproteins from the frost fish, *Microgadus tomcod*. Can J Zool 60:348–355

Fletcher GL, Addison RF, Hew CL, Slaughter D (1982a) Antifreeze proteins in the Arctic shorthorn sculpin (*Myoxocephalus scorpius*). Arctic 35:302–306

Fletcher GL, Slaughter D, Hew CL (1982b) Seasonal changes in the plasma levels of glycoprotein antifreeze, Na^+, Cl^-, and glucose in Newfoundland Atlantic cod, *Gadus morhua*. Can J Zool 60:1851–1854

Fletcher GL, Haya K, King MJ, Reisman HM (1985a) Annual antifreeze cycles in Newfoundland, New Brunswick and Long Island winter flounder *Pseudopleuronectes americanus*. Mar Ecol Prog Ser 21:205–212

Fletcher GL, Hew CL, Li X, Haya K, Kao MH (1985b) Year-round presence of high levels of plasma antifreeze peptides in a temperate fish, ocean pout (*Macrozoarces americanus*). Can J Zool 63:488–493

Fletcher GL, King MJ, Kao MH, Shears MA (1989) Antifreeze proteins in the urine of marine fish. Fish Physiol Biochem 6:121–127

Fletcher GL, Hew CL, Davies PL (2001) Antifreeze proteins of teleost fishes. Annu Rev Physiol 63:359–390

Foldvik A, Kvinge T (1974) Conditional instability of sea water at the freezing point. Deep-Sea Res 21:169–174

Foster TD (1984) The marine environment. In: Antarctic ecology. Academic, London

Franklen CE, Davison W, Mckenzie JC (1993) The role of the spleen during exercise in Antarctic teleost, *Pagothenia borchgrevinki*. J Exp Biol 174:381–386

Furukawa Y, Nagashima K, Nakatsubo S, Yoshizaki I, Tamaru H, Shimaoka T, Sone T, Yokoyama E, Zepeda S, Terasawa T, Asakawa H, Murata K, Sazaki G (2017) Oscillations and accelerations of ice crystal growth rates in microgravity in presence of antifreeze glycoprotein impurity in supercooled water. Sci Rep 7:43157

Garnham CP, Natarajan A, Middleton AJ, Kuiper MJ, Braslavsky I, Davies PL (2010) Compound ice-binding site of an antifreeze protein revealed by mutagenesis and fluorescent tagging. Biochemistry 49:9063–9071

Garnham CP, Campbell RL, Davies PL (2011) Anchored clathrate waters bind antifreeze proteins to ice. Proc Natl Acad Sci U S A 108:7363–7367

Gauthier SY, Scotter AJ, Lin FH, Baardsnes J, Fletcher GL, Davies PL (2007) A re-evaluation of the role of type IV antifreeze protein. Cryobiology 57:292–296

Gong Z, Ewart KV, Hu Z, Fletcher GL, Hew CL (1996) Skin antifreeze protein genes of the winter flounder, *Pleuronectes americanus*, encode distinct and active polypeptides without the secretory signal and pro-sequences. J Biol Chem 271:4106–4112

Harding MM, Ward LG, Haymet ADJ (1999) Type I antifreeze proteins: structure-activity studies and mechanism of ice growth inhibition. Eur J Biochem 264:653–655

Harding MM, Anderberg PI, Haymet ADJ (2003) Antifreeze glycoproteins from polar fish. Eur J Biochem 270:1381–1392

Haschemeyer AEV, Mathews RW (1980) Antifreeze glycoprotein synthesis in the Antarctic fish *Trematomus hansoni* by constant infusion *in vivo*. Physiol Zool 53:383–393

Haymet ADJ, Ward LG, Harding MM (1999) Winter flounder antifreeze proteins: synthesis and ice growth inhibition of analogues that probe the relative importance of hydrophobic and hydrogen-bonding interactions. J Am Chem Soc 121:941–948

Hew CL, Yip C (1976) The synthesis of freezing-point depression protein of the winter flounder *Pseudopleuronectes americanus* in *Xenopus laevis* oocytes. Biochem Biophys Res Commun 71:845–850

Hew CL, Slaughter D, Joshi S, Fletcher GL, Ananthanarayanan VS (1984) Antifreeze polypeptides from the Newfoundland Ocean pout, *Macrozoarces americanus*: presence of multiple and compositionally diverse components. J Comp Physiol B B155:81–88

Hew CL, Wang N-C, Joshi S, Fletcher GL, Scott GK, Hayes PH, Buettner B, Davies PL (1988) Multiple genes provide the basis for antifreeze protein diversity and dosage in the ocean pout, *Macrozoarces americanus*. J Biol Chem 263:12049–12055

Hinton DE, Snipes RL, Kendall MW (1972) Morphology and enzyme histochemistry in liver of largemouth bass (*Micropterus salmoides*). J Fish Res Board Can 29:531–534

Howe GJ (1991) Biogeography of gadoid fishes. J Biogeogr 18:595–622

Hsiao KC, Cheng C-HC, Fernandes IE, Detrich HW, Devries AL (1990) An antifreeze glycopeptide gene from the Antarctic cod *Notothenia coriiceps* neglecta encodes a polyprotein of high peptide copy number. Proc Natl Acad Sci U S A 87:9265–9269

Hudson AP, Devries AL, Haschemeyer AEV (1979) Antifreeze glycoprotein biosynthesis in Antarctic fishes. Comp Biochem Physiol 62B:179–183

Hunt BM, Hoefling K, Cheng C-HC (2003) Annual warming episodes in seawater temperatures in McMurdo Sound in relationship to endogenous ice in notothenioid fish. Antarct Sci 15:333–338

Hunt Von Herbing I, Miyake T, Hall BK, Boutilier RG (1996) Ontogeny of feeding and respiration in larval Atlantic cod *Gadus morhua* (Teleostei, Gadiformes). I. Morphology. J Morphol 227:15–35

Jin Y (2003) Freezing avoidance of Antarctic fishes: the role of a novel antifreeze potentiating protein and the antifreeze glycoproteins. PhD Thesis, University of Illinois

Jin Y, Devries AL (2006) Antifreeze glycoprotein levels in Antarctic notothenioid fishes inhabiting different thermal environments and the effect of warm acclimation. Comp Biochem Physiol 76B:560–600

Jordan AD, Jurngersen M, Steffensen JF (2001) Oxygen consumption of East Siberian cod: no support for the metabolic cold adaptation theory. J Fish Biol 59:818–823

Jung A, Johnson P, Eastman JT, Devries AL (1995) Protein content and freezing avoidance properties of the subdermal extracellular matrix and serum of the Antarctic snailfish, *Paraliparis devriesi*. Fish Physiol Biochem 14:71–80

Kao MH, Fletcher GL, Wang NC, Hew CH (1985) The relationship between molecular weight and antifreeze polypeptide activity in marine fish. Can J Zool 64:578–582

Kennett JP (1977) Cenozoic evolution of Antarctic glaciation, the circum-Antarctic Ocean, and their impact on global paleoceanography. J Geophys Res 82:3843–3860

Kennett JP (1982) Marine geology. Prentice-Hall, New Jersey

Kitagawa Y, Ogawa M, Fukuchi M (1990) On the kidney of the saffron cod, *Eleginus gracilis*, and its cold adaptation. Proc NIPR Symp Polar Biol 3:71–75
Knight CA, Cheng C-HC, Devries AL (1991) Adsorption of helical antifreeze peptides on specific ice crystal surface planes. Biophys J 59:409–418
Knight CA, Driggers E, Devries AL (1993) Adsorption of fish antifreeze glycopeptides to ice, and effects on ice crystal growth. Biophys J 64:252–259
Kurlansky M (1997) Cod: a biography of the fish that changed the world. Penguin Books, New York
Kurokawa T, Suzuki T (1995) Structure of the exocrine pancreas of flounder (*Paralichthys olivaceus*): immunological localization of zymogen granules in the digestive tract using anti-trypsinogen antibody. J Fish Biol 46:292–301
Leim AH, Scott WB (1966) Fishes of the Atlantic Coast of Canada. Fish Research Board Canada, Ottawa
Lin Y (1979) Environmental regulation of gene expression. J Biol Chem 254:1422–1426
Lin Y, Long DJ (1980) Purification and characterization of winter flounder antifreeze peptide messenger ribonucleic acid. Biochemistry 19:1111–1116
Littlepage JL (1965) Oceanographic investigations in McMurdo Sound, Antarctica. In: Llano GA (ed) Biology of the Antarctic Seas II. American Geophysical Union, Washington, DC
Love MS, Elder N, Mecklenburg CW, Thorsteinson LK, Mecklenburg TA (2016) Alaska Arctic marine fish species accounts. In: Thorsteinson LK, Love MS (eds) Alaska Arctic marine fish ecology catalog: U.S. Geological Survey Scientific Investigations Report 2016-5038 (OCS study)
Low WK, Miao M, Ewart KV, Yang DS, Fletcher GL, Hew CL (1998) Skin-type antifreeze protein from the shorthorn sculpin, *Myoxocephalus scorpius*: expression and characterization of Mr 9700 recombinant protein. J Biol Chem 273:23098–23103
Low WK, Lin Q, Stathakis C, Miao M, Fletcher GL, Hew CL (2001) Isolation and characterization of skin-type, type I antifreeze polypeptides from the longhorn sculpin, *Myoxocephalus octodecemspinosus*. J Biol Chem 276:11582–11589
Marshall CB, Fletcher GL, Davies PL (2004) Hyperactive antifreeze protein in a fish. Nature 429:153
Mazur P (1966) Physical and chemical basis of injury in single-celled micro-organisms subjected to freezing and thawing. In: Meryman HT (ed) Cryobiology. Academic, London
Meister K, Strazdaite S, Devries AL, Lotze S, Olijve LLC, Voets IK, Bakker HJ (2014) Observation of ice-like water layers at an aqueous protein surface. Proc Natl Acad Sci U S A 111:17732–17736
Meister K, Devries AL, Bakker HJ, Drori R (2018) Antifreeze glycoproteins bind irreversibly to ice. J Am Chem Soc 140:9365–9368
Millero FJ (1978) Freezing point of seawater - eighth report of the joint panel of oceanographic tables and standards. In: UNESCO technical papers in marine science, vol 28. UNESCO, Paris
Mochizuki K, Molinero V (2018) Antifreeze glycoproteins bind reversibly to ice via hydrophobic groups. J Am Chem Soc 140(14):4803–4811. https://doi.org/10.1021/jacs.7b1363
Morris HR, Thompson MR, Osuga DT, Ahmed AI, Chan SM, Vandenheede JR, Feeney RE (1978) Antifreeze glycoproteins from the blood of an Antarctic fish: the structure of the proline containing glycopeptides. J Biol Chem 253:5155–5162
Murray HM, Hew CH, Fletcher GL (2003) Spatial expression patterns of skin-type antifreeze protein in winter flounder (*Pseudopleuronectes americanus*) epidermis following metamorphosis. J Morphol 257:78–86
Ng NFL, Hew CL (1992) Structure of an antifreeze polypeptide from the sea raven: disulfide bonds and similarity to lectin binding proteins. J Biol Chem 267:16069–16075
Nishimiya Y, Sato R, Takamichi M, Miura A, Tsuda S (2005) Co-operative effect of the isoforms of type III antifreeze protein expressed in notched-fin eelpout *Zoarces elongatus* (Kner). FEBS J 272:482–492

O'grady SM, Clarke A, Devries AL (1982a) Characterization of glycoprotein antifreeze biosynthesis in isolated hepatocytes from *Pagothenia borchgrevinki*. J Exp Zool 220:179–189
O'grady SM, Ellory JC, Devries AL (1982b) Protein and glycoprotein antifreezes in intestinal fluid of polar fishes. J Exp Biol 98:429–438
O'grady SM, Schrag JD, Raymond JA, Devries AL (1982c) Comparison of antifreeze glycopeptides from Arctic and Antarctic fishes. J Exp Zool 224:177–185
Olla BL, Bejda AJ, Martin D (1975) Activity, movements, and feeding behavior of the cunner, *Tautogolabrus adspersus*, and comparison of food habits with young tautog, *Tautoga onitus*, off long Island, New York. Fish Bull Ill 733:895–900
Pearcy WG (1961) Seasonal changes in osmotic pressure in flounder sera. Science 134:193–194
Petzel DH, Devries AL (1979) Renal handling of peptide antifreeze in northern fishes. Bull Mt Desert Isl Biol Lab 19:17–19
Petzel DH, Devries AL (1980) Renal handling of anionic and cationic antifreeze peptides in the glomerular winter flounder. Bull Mt Desert Isl Biol Lab 20:17
Petzel DH, Reisman HM, Devries AL (1980) Seasonal variation of antifreeze peptide in the winter flounder, *Pseudopleuronectes americanus*. J Exp Zool 211:63–69
Praebel K, Ramløv H (2005) Antifreeze activity in the gastrointestinal fluids of *Arctogadus glacialis* (Peters 1874) is dependent upon food type. J Exp Biol 208:2609–2613
Praebel K, Hunt B, Hunt LH, Devries AL (2009) The presence and quantification of splenic ice in the McMurdo Sound Notothenioid fish, *Pagothenia borchgrevinki* (Boulenger, 1902). Comp Biochem Physiol 154A:564–569
Ramlov H, Devries AL, Wilson PW (2005) Antifreeze glycoproteins from the Antarctic fish *Dissostichus mawsoni* studied by differential scanning calorimetry (DSC) in combination with nanolitre osmometry. CryoLetters 26:73–84
Raymond JA (1992) Glycerol is a colligative antifreeze in some northern fishes. J Exp Zool 262:347–352
Raymond JA, Devries AL (1972) Freezing behavior of fish blood glycoproteins with antifreeze properties. Cryobiology 9:541–547
Raymond JA, Devries AL (1977) Adsorption inhibition as a mechanism of freezing resistance in polar fishes. Proc Natl Acad Sci U S A 74:2589–2593
Raymond JA, Lin Y, Devries AL (1975) Glycoprotein and protein antifreezes in two Alaskan fishes. J Exp Zool 193:125–130
Raymond JA, Wilson PW, Devries AL (1989) Inhibition of growth of non-basal planes in ice by fish antifreeze. Proc Natl Acad Sci U S A 86:881–885
Reisman HM, Kao MH, Fletcher GL (1984) Antifreeze glycoprotein in a southern population of Atlantic tomcod, *Microgadus tomcod*. Comp Biochem Physiol A 78A:445–447
Schauperl M, Podewitz M, Ortner TS, Waibl F, Thoeny A, Loerting T, Liedl KR (2017) Balance between hydration enthalpy and entropy is important for ice binding surfaces in antifreeze proteins. Sci Rep 7:11901
Scholander PF, Van Dam L, Kanwisher JW, Hammel HT, Gordon MS (1957) Supercooling and osmoregulation in Arctic fish. J Cell Comp Physiol 49:5–24
Schrag JD, Devries AL (1982) The effects of freezing rate on the cooperativity of antifreeze glycopeptides. Comp Biochem Physiol 74A:381–385
Schrag JD, O'grady SM, Devries AL (1982) Relationship of amino acid composition and molecular weight of antifreeze glycopeptides to non-colligative freezing point depression. Biochim Biophys Acta 717:322–326
Schrag JD, Cheng C-HC, Panico M, Morris HR, Devries AL (1987) Primary and secondary structure of antifreeze peptides from arctic and Antarctic zoarcid fishes. Biochim Biophys Acta 915:357–370
Shier WT, Devries AL (1975) Carbohydrate of antifreeze glycoproteins from an Antarctic fish. FEBS Lett 54:135–138
Shier WT, Lin Y, Devries AL (1972) Structure and mode of action of glycoproteins from Antarctic fishes. Biochim Biophys Acta 263:406–413

Slaughter D, Fletcher GL, Ananthanarayanan VS, Hew CH (1981) Antifreeze proteins from the sea raven, *Hemitripterus americanus*. J Biol Chem 256:2022–2026

Somero GN, Lockwood BL, Tomanek L (2017) Biochemical adaptation: response to environmental challenges from life's origins to the anthropocene. Oxford University Press, Oxford

Sonawane ND, Thiagarajay JH, Verkman AS (2002) Chloride concentration in endosomes measured using a ratioable fluorescent Cl^- indicator. J Biol Chem 227:5506–5513

Sonnichsen FD, Deluca CI, Davies PL, Sykes BD (1996) Refined solution structure of type III antifreeze protein: hydrophobic groups may be involved in the energetics of the protein-ice interaction. Structure 4:1325–1337

Sørensen TF, Ramløv H (2001) Variations in antifreeze activity and serum inorganic ions in the eelpout *Zoarces viviparus*: antifreeze activity in the embryonic state. Comp Biochem Physiol 130A:123–132

Sørensen TF, Ramløv H (2002) Maternal-fetal relations in antifreeze production in the eelpout *Zoarces viviparus*. CryoLetters 23:183–190

Tachibana Y, Fletcher GL, Fujitani N, Tsuda S, Monde K, Nishimura SI (2004) Antifreeze glycoproteins: elucidation of the structural motifs that are essential for antifreeze activity. Angew Chem Int Ed 43:856–862

Takamichi M, Nishimiya Y, Miura A, Tsuda S (2007) Effect of annealing time of an ice crystal on the activity of type III antifreeze protein. FEBS J 274:6469–6476

Takamichi M, Nishimiya Y, Miura A, Tsuda S (2009) Fully active QAE isoform confers thermal hysteresis activity on a defective SP isoform of type III antifreeze protein. FEBS J 276:1471–1479

Tien R (1995) Freezing avoidance and the presence of ice in shallow water Antarctic fishes. PhD Thesis, University of Illinois, Urbana-Champaign

Turner JD, Schrag JD, Devries AL (1985) Ocular freezing avoidance in Antarctic fishes. J Exp Biol 118:121–132

Valerio PF, Kao MH, Fletcher GL (1990) Thermal hysteresis activity in the skin of the cunner, *Tautogolabrus adspersus*. Can J Zool 68:1065–1067

Valerio PF, Goddard SV, Kao MH, Fletcher GL (1992a) Survival of northern Atlantic cod (*Gadus morhua*) eggs and larvae when exposed to ice and low temperature. Can J Fish Aquat Sci 49:2588–2595

Valerio PF, Kao MH, Fletcher GL (1992b) Fish skin: an effective barrier to ice crystal propagation. J Exp Biol 164:135–151

Van Voorhies WV, Raymond JA, Devries AL (1978) Glycoproteins as biological antifreeze agents in the cod *Gadus ogac* (Richardson). Physiol Zool 51:347–353

Wang X, Devries AL, Cheng C-HC (1995a) Antifreeze peptide heterogeneity in an Antarctic eel pout includes an unusually large major variant comprised of two 7 kDa type III AFPs linked in tandem. Biochim Biophys Acta 1247:163–172

Wang X, Devries AL, Cheng C-HC (1995b) Genomic basis for antifreeze peptide heterogeneity and abundance in an Antarctic eel pout: gene structures and organization. Mol Mar Biol Biotechnol 4:135–147

Wen D, Laursen RA (1992) Structure-function relationships in an antifreeze polypeptide. The role of neutral, polar amino acids. J Biol Chem 267:14102

Wilson PW (1993) Explaining thermal hysteresis by the Kelvin effect. CryoLetters 14:31–36

Wilson PW, Gould M, Devries AL (2002) Hexagonal shaped spicules in frozen fish antifreeze solutions. Cryobiology 44:240–250

Yang DS, Hon W, Bubanko S, Xue Y, Seetharaman J, Hew CL, Sicheri F (1998) Identification of the ice-binding surface on a type III antifreeze protein with a "flatness function" algorithm. Biophys J 74:2142–2151

Zdzitowiecke K (1998) Diversity of digenea, parasites of fishes in various areas of the Antarctic. In: Fishes of Antarctica. Springer, Milan

Zuang X, Yang C, Murphy KR, Cheng C-HC (2019) Molecular mechanism and history of non-sense to sense evolution of antifreeze glycoprotein gene in northern gadids. Proc Natl Acad Sci USA 116:4400–4405

Chapter 6
Insect Antifreeze Proteins

John G. Duman and Samuel S. Newton

6.1 Introduction

Insects have long been known for their abilities to survive and thrive in cold regions, including far northern latitudes (Salt 1936, 1956, 1961, 1966; Scholander et al. 1953). Two basic mechanisms by which insects survive subzero temperatures are generally recognized: freeze tolerance (the ability to survive freezing of body fluids, usually only of the extracellular fluids) and freeze avoidance in freeze-intolerant species (freeze susceptible, die if frozen) (Salt 1936; see Lee 2010 for review). Various levels of freeze tolerance have been recognized in insects, based on considerable differences in lower lethal temperatures and terminology, such as "partial freeze tolerance", have been introduced (Sinclair 1999; Chown and Terblanche 2007; Chown and Sinclair 2010). Likewise, the abilities of freeze-avoiding species also vary considerably. In addition, over time some insects change subzero tolerance mechanisms (freeze tolerance to freeze avoidance and vice versa). Larvae of the beetles *Dendroides canadensis* and *Cucujus clavipes* were freeze tolerant when initially studied in the late 1970s but became freeze avoiding later (Horwath and Duman 1984a; Duman 1984a). Other insects change strategies over the course of a single winter season, for example, switching from freeze tolerance to avoidance (Bale et al. 2000; Brown et al. 2004), and some even vary spatially within the body at a single point in time. For example, the posterior region (abdomen) of overwintering adults of the fungus gnat *Exechia nugatoria* in interior Alaska freeze at −31 °C but the insects survive, while the anterior regions (head and thorax) supercool to −50 °C

J. G. Duman (✉)
Department of Biological Sciences, University of Notre Dame, Notre Dame, IN, USA
e-mail: duman.1@nd.edu

S. S. Newton
Sanford School of Medicine, University of South Dakota, Vermillion, SD, USA
e-mail: Samuel.Sathyanesan@usd.edu

H. Ramløv, D. S. Friis (eds.), *Antifreeze Proteins Volume 1*,
https://doi.org/10.1007/978-3-030-41929-5_6

but death results upon freezing at that temperaure (Sformo et al. 2009). How ice propagation from abdomen to thorax is prevented between −31 and −50 °C is unknown?

Other species are freeze avoiding and supercool to low temperatures when in a dry winter hibernaculum, but if in contact with ice and are inoculated across the cuticle at much higher subzero temperatures they survive (Fields and McNeil 1986; Gehrken et al. 1991; Rozsypal and Košťál 2018). Why have such unusual mechanisms evolved? These are just a few of the many examples illustrating the variations, and extremes, of insect subzero temperature adaptations.

While we typically emphasize the physiological/biochemical mechanisms associated with extensive subzero tolerance, the importance of behavioral adaptations, especially the potential thermal buffering provided by the overwintering microhabitat site should not be forgotten. Many insects winter in the leaf litter, soil, rotting logs, etc. where they are not exposed to the extremes and temporal variations of air temperatures, and they are often further thermally buffered by a blanket of snow (Danks 1991; Chown and Sinclair 2010).

Low temperatures can induce numerous potentially lethal problems including metabolic imbalance, membrane malfunction, oxygen stress (hypoxia, ischemia, as well as oxidative stress requiring antioxidants), at temperatures above those where the insect actually freezes, and which are generally exacerbated as temperatures continue to decrease. Consequently, insects that successfully winter at subzero temperature extremes must have adaptations to overcome these pre-freeze problems, as well as those that pertain to actual freezing, both avoidance and tolerance (for reviews see Ramløv 2000; Lee 2010; Lee and Denlinger 1991; Denlinger and Lee 2010; Storey and Storey 2010; Michaud and Denlinger 2010; Kostal 2010). However, these nonfreezing problems are not the focus of this review.

Given the tremendous diversity and adaptive radiation of insects (approximately 70% of all known animal species) (Wilson 1988), it is not surprising that as a group they have evolved multiple means of surviving subzero temperatures. A proper treatment of the many known adaptations used by insects to survive low temperature extremes would require more space than is available in this chapter. Consequently, the reader is directed to more expansive reviews of insect cold tolerance (Lee and Denlinger 1991; Denlinger and Lee 2010). Here we will concentrate on ice-binding proteins (IBPs). These include antifreeze proteins (AFPs), recrystallization inhibition proteins (RIPs), and ice-nucleating proteins (INPs), and mention these other adaptations as they relate to IBPs and IBP-producing insects.

6.2 Functional Types of Insect Ice-Binding Proteins

This chapter will discuss the three types of insect IBPs, primarily in the context of their roles in subzero temperature tolerance. (1) Antifreeze proteins (AFPs) produce significant body fluid thermal hysteresis (TH) amounting to approximately 2.5 to as much as 13 °C, and are present in freeze-avoiding species where they function to prevent ice crystal growth and/or ice nucleation. (2) Recrystallization inhibition proteins (RIPs) produce significantly less TH (generally 0.2–0.6 °C) in vivo, and function mainly to inhibit recrystallization in freeze-tolerant insects. (3) Ice nucleator proteins (INPs) lack TH activity and inhibit supercooling by initiating freezing at higher subzero temperatures than would otherwise be the case. While INPs are functional in freeze-tolerant species, they are counterproductive in freeze avoidance. Most IBP studies involving insects have concerned AFPs in freeze-avoiding species, and consequently these will be highlighted in this chapter, although RIPs and INPs will also be treated. AFPs and RIPs both produce TH in aqueous solution and were traditionally both labeled as AFPs, although the latter are generally found in freeze-tolerant species that do not appear to require antifreezes and the hemolymph level of activity is insufficient to be of much value as an antifreeze. Their only similarity was TH. INPs have a completely different function, to inhibit supercooling. However, all three of these types of proteins interact with water and ice, and more recent research indicates similarities in structure and ice-binding mechanism(s). Consequently, all three are now commonly labeled as ice active proteins or IBPs, although their physiological functions differ dramatically (Wharton et al. 2005, 2009; Davies 2014). These three IBPs will be discussed primarily in the context of their functions in freeze avoidance and freeze tolerance.

6.3 Freeze Avoidance: Antifreeze Proteins and Other Adaptations

Antifreeze proteins, initially found in marine fish (DeVries and Wohlschlag 1969; DeVries 1971; see Chap. 5), are defined by their ability to lower the freezing point of an aqueous solution below the equilibrium melting point (as determined by solute concentration) by a non-colligative mechanism, thereby producing a thermal hysteresis (TH) that protects the organism from freezing without significantly affecting the osmotic pressure. This AFP-induced freezing point is appropriately termed the hysteretic freezing point (hFP) to differentiate it from the colligative equilibrium melting/freezing point (eqMP). Fish AFPs typically produce a TH of approximately 0.8–2.0 °C and thereby, in conjunction with the normal solute concentration, lower the hFP of the body fluids of these hypoosmotic teleost fish below the freezing point of normal seawater (−1.86 °C), the lowest temperature normally experienced by the fish (DeVries 1986). In addition to lowering the hFP of water below the eqMP, AFPs also raise the melting temperature of an ice crystal above the eqMP to the hMP, although this increase is much smaller (not more, and typically less, than 20%) than the decrease in the freezing temperature of the crystal below the eqMP (Knight and DeVries 1989; DeVries and Cheng 2005; Celik et al. 2010). In contrast to marine fishes, high latitude terrestrial insects routinely experience much lower environmental temperatures, and not surprisingly their antifreezes, both protein and otherwise, must provide greater protection than found in fishes. Insect AFPs typically produce hemolymph TH of 2–6 °C and at times as much as 13 °C, and are therefore sometimes referred to as "hyperactive" AFPs (Davies 2014). However, the TH and hFP resulting from the AFPs may not be the most important parameters of insect AFPs as the functional depression of organismal freezing typically extends well beyond the hFP. To avoid lethal freezing, these freeze-avoiding (freeze susceptible) insects, in contrast to freeze-tolerant species, must (1) inhibit inoculative freezing initiated by contact with the external ice that often surrounds the insects in their winter hibernaculum, and (2) have the ability to promote supercooling and lower their supercooling points (SCPs, the temperature at which spontaneous freezing occurs in the absence of ice, also known as the nucleation temperature) below the colligative and/or hysteretic freezing point of the body fluids. AFPs are critical in overcoming both of these problems (Duman 2001, 2015; Duman et al. 2010).

As suggested by the above brief introduction, there is no single "magic bullet" means of achieving subzero temperature tolerance for an organism, in either freeze-tolerant or freeze-avoiding species, as the multitude of problems to be overcome is too great. In addition, the tremendous diversity of insects has resulted in the evolution of multiple adaptations to achieve overwintering success. Important adaptations include high (often molar) concentrations of various low molecular mass solutes such as polyhydroxy alcohols (polyols, especially glycerol), sugars (trehalose, etc.), amino acids (especially proline), antifreeze glycolipids (AFGLs), removal of ice nucleators in winter, etc., and the main topic of this chapter AFPs and other IBPs (for reviews see Zachariassen 1985; Lee and Denlinger 1991; Denlinger and Lee 2010). These are not mutually exclusive mechanisms, and as we will see, many of these adaptations have additive, or even synergistic, effects on subzero temperature tolerance.

6.3.1 *Inhibition of Inoculative Freezing*

Overwintering insects are routinely in contact with external ice that can inoculate body fluids across the cuticle if the temperature is below the eqMP and/or hFP of those fluids. Antifreezes (AFPs, AFGLs, polyols, etc.) provide protection from inoculative freezing by lowering the eqMP and/or hFP (Husby and Zachariassen 1980; Duman and Horwath 1983; Gehrken 1992; Olsen et al. 1998). Also, for most freeze-avoiding insects, the wax coating on the surface of the cuticle that protects against evaporative water loss also generally provides a reasonably effective physical barrier to propagation of ice across the cuticle—an intuitive example of the importance of the wax-coated cuticle in this regard is a comparison of the wet surface of a frog or an earthworm with that of a typical insect. Seasonal changes in cuticular proteins may also be important in further fortifying the cuticle such that the winter cuticle is more resistant to inoculative freezing than the summer cuticle (Olsen et al. 1998; Michaud and Denlinger 2010; Carrasco and Duman 2011; Carrasco et al. 2011, 2012).

6.3.1.1 Low Molecular Mass Solutes: Polyols, Sugars, and Amino Acids

The seasonal increase of low molecular mass colligative antifreezes such as glycerol and other polyols by overwintering insects often plays a role in inhibiting inoculative freezing by decreasing the eqMP. However, it is important to remember that depression of the eqMP by most solutes relies on the colligative properties of water, meaning that the solutes lower the eqMP strictly by the effect of solute concentration on vapor pressure. The molal freezing point constant for water is only 1.86 °C per Osmol of solute. Therefore to lower the eqMP to a fairly moderate winter temperature of −20 °C requires approximately a 10.75 Osmolar solute concentration in the body fluids. Production and maintenance of such a high solute concentration incur significant energetic and osmotic costs.

Early studies of insect cold tolerance mechanisms in both freeze-tolerant and freeze-avoiding species often concentrated on low molecular mass organic solutes. These were generally polyhydroxy alcohols (especially glycerol, but also sorbitol, mannitol, erythritol, ribotol, threitol, etc.), sugars (trehalose, glucose, etc.), and less often amino acids such as proline and alanine (for reviews see Sømme 1982; Zachariassen 1985; Lee 2010; Storey and Storey 1991; Ramløv 2000). These compounds are often present at very high (molar or multimolar) concentrations. For example, Salt (1958) showed that the freeze-tolerant larvae of the wasp *Bracon cephi* concentrate glycerol in winter to 25% of the body mass. Consequently, these low molecular mass antifreezes and cryoprotectants must be "compatible solutes" capable of very high concentrations without disrupting proteins, membranes, or general cell activities (Yancey et al. 1982; Yancey 2005; Somero et al. 2017). In fact, many of these solutes are not only compatible, but actually are beneficial in protecting macromolecular structure and function at low temperatures (Bolen 2004; Somero et al. 2017). Metabolic pathways leading to accumulation of most of these solutes are rather well known (Storey and Storey 1991), although pathways to others are less obvious (Walters et al. 2009a).

6.3.1.2 Antifreeze Proteins

AFPs lower the eqMP to the hFP and therefore can inhibit inoculative freezing. However, their effects in this regard may extend further than expected based on the measured hemolymph TH. The magnitude of the measured TH of a solution containing AFPs is in some instances dependant on the size of the crystal used in the measurement (Zachariassen and Husby 1982; Ramløv et al. 2005), and this is the case with *Dendroides canadensis* AFPs (DAFPs) (Nicodemus et al. 2006). This point likely has some bearing on the extended ability of AFPs to inhibit inoculative freezing beyond the measured hemolymph TH because the cuticular pores through which ice might propagate are much smaller than even the smallest seed crystal used to measure TH (Zachariassen and Husby 1982). Therefore, AFPs that produce 5 °C of measured TH could be expected to inhibit inoculative freezing by considerably more than 5 °C. This was first demonstrated in *Ips accuminatus* beetles from Norway (Gehrken 1992). Hemolymph AFPs in wintering adult *I. accuminatus* suppressed inoculative freezing well below the hFP, while beetles with depressed hemolymph TH froze much more readily when in contact with ice. In the beetle *Dendroides canadensis* AFPs in the hemolymph, as well as those in/on the layer of epidermal cells underlying the cuticle of overwintering larvae are required to effectively lower the temperature at which ice propagates across a patch of isolated cuticle (Olsen et al. 1998). In these experiments, AFPs producing a TH of just 1.82 °C in the physiological saline solution on the inside of the cuticular patch lowered the inoculative freezing temperature across the isolated cuticle by approximately 8 °C. In fact, the inoculative freezing inhibition effect of the DAFPs may well have been greater than this, since the eventual freezing of the fluid inside the cuticle was equal to the SCP of the fluid, and therefore the inhibition of inoculative freezing may have been greater if the SCP of the fluid had been lower. Also, the TH enhancing effect of glycerol on *D. canadensis* AFPs was not known at this time and therefore was not investigated in this study (i.e., glycerol was not added to the physiological saline). This is likely to be of some importance because it was later determined that glycerol significantly enhances the activity of *D. canadensis* AFPs (Li et al. 1998a), indicating that the presence of glycerol (typically present at 0.5–1.0 M in winter *D. canadensis* hemolymph) is likely to be of value beyond its colligative effect by virtue of enhancing the AFPs.

6.3.1.3 Antifreeze Glycolipids

AFGLs that have specific TH and recrystallization inhibition activities comparable to those of insect AFPs were first identified and characterized in a freeze-tolerant Alaskan Tenebrionid beetle, *Upis ceramboides* (Walters et al. 2009c). While most of the animals and plants subsequently shown to have AFGLs are likewise freeze tolerant, a number of freeze-avoiding insects with AFGLs were also identified (Walters et al. 2011). The physiological function of AFGLs in freeze-avoiding insects is currently unknown. However, while the levels of AFGLs present in the hemolymph of these species are generally not great, the AFGLs are most abundant in cell membranes, and if the AFGLs are present on cells under the cuticle they may be involved in inhibition of inoculative freezing. Also, AFGLs and *D. canadensis* DAFPs synergistically enhance one another's TH (Duman, unpublished data). As AFGLs are best understood in freeze-tolerant species, further discussion of the AFGLs will be covered under the section of this chapter concerned with freeze tolerance.

6.3.2 Promotion of Supercooling

If a freeze-avoiding insect is protected against inoculative freezing, then the next problem to be overcome to further extend freeze avoidance protection is the promotion of supercooling of all body fluid compartments below the lowest temperature experienced by the insect over the course of the winter. A sample of pure water with an eqMP/FP of 0 °C will typically remain in the metastable supercooled state well below 0 °C before freezing (Knight 1967; Lee et al. 1995; Vali 1995). Perhaps surprisingly, if a very small water sample is especially pure it may actually supercool to approximately −40 °C, the homogeneous nucleation temperature of water, although there are volume and time effects that can limit homogeneous supercooling. Small ice-like clusters of water molecules (embryo crystals) form as water is cooled, and as the temperature decreases the clusters grow until they reach a critical size where they seed the supercooled water. In biological systems homogeneous nucleation is generally thought to be rare as ice-nucleating surfaces of varying effectiveness lead to heterogeneous nucleation by providing sites where embryo crystals grow to a critical size at temperatures above the homogeneous nucleation temperature, thereby inhibiting supercooling (Knight 1967; Vali 1995). The efficiency of ice nucleators varies considerably, with the best biological nucleators probably being proteins on the surface membranes of ice-nucleating bacteria and fungi that nucleate at temperatures as high as −2 °C (Lindow 1983; Green and Warren 1985; Wolber and Warren 1989). These ice nucleators can be important to freeze-avoiding insects as ice-nucleating bacteria, etc., are common in the food and water of many insects, potentially providing nucleation sites for gut fluid freezing (Lee 2010). Also, hemolymph proteins with varying ice-nucleating efficiencies are also common. Increased nucleating efficiency of some of these hemolymph ice nucleator proteins (INPs) has apparently been selected for by many freeze-tolerant insects, leading to adaptive INPs, evolved to limit supercooling and thus inhibit lethal intracellular freezing that can accompany freezing following more extensive supercooling (Zachariassen and Hammel 1976; Zachariassen 1982). In contrast, some proteins have some level of ice-nucleating surface as a component of their normal structure that is essential to their natural function (incidental ice nucleators). These are necessarily selected against, or must be masked, in freeze-avoiding insects (Lundheim 1996; Zachariassen et al. 2011). However, Zacchariassen and colleagues have argued that most nucleation below −18 °C in insects is homogeneous and the nucleation temperature is dependant on the water volume of the insect (Zachariassen et al. 2004, 2011). Also important is that the supercooled state is metastable, and consequently there is a time element involved such that the longer an insect is supercooled the more likely it is to freeze, and many insects are supercooled for long periods in winter. In addition, the colder the insect becomes the more likely it is to freeze. Consequently, stabilization of the supercooled state is very important. For a more detailed discussion of these points see Zachariassen et al. (2011).

Freeze-avoiding insects promote supercooling by several, non-mutually exclusive, means: removing ice nucleators and/or production of colligative antifreezes, AFPs, AFGLs, and cryoprotective dehydration.

6.3.2.1 Removal of Ice Nucleators

Removal of ice nucleators from body fluid compartments, either on an evolutionary or seasonal time frame, can extend supercooling. Freeze-avoiding insects typically stop feeding and clear the gut prior to the onset of freezing temperatures (Salt 1953; Sømme 1982; Cannon and Block 1988; Lee et al. 1996), thereby removing ice-nucleating bacteria, etc. that may have otherwise initiated ice nucleation.

Also, some freeze-avoiding species remove or reduce the concentrations of hemolymph incidental ice-nucleating proteins that, because their surface structures inadvertently order water in an ice-like fashion, give these proteins ice-nucleating activity (Zachariassen 1982; Neven et al. 1986; Gehrken 1988). Obviously, this can only happen if the proteins are not essential in winter. Larvae of the beetle *Ceruchus piceus* have a hemolymph lipoprotein with ice-nucleating activity that also functions as a lipophorin to carry lipid from site to site during most of the year. However, in autumn, the larvae enter diapause where the metabolic rate is greatly reduced and therefore the lipophorin is apparently expendable. Consequently, the ice-nucleating lipohorin is removed in winter and therefore the larvae supercool to approximately −25 °C, without having to produce and maintain energetically costly colligative antifreeze or AFPs (Neven et al. 1986). Other freeze-avoiding species may retain incidental INPs (apparently their functions are required in winter), and therefore these INPs must be inhibited, or their effects muted, by antifreezes.

6.3.2.2 Low Molecular Mass Solutes

Polyols and other low molecular mass colligative antifreezes depress the supercooling point, as well as the eqMP. Actually, these solutes appear to lower the SCP by 1–2 times more than they depress the eqMP (Block and Young 1979; Lee et al. 1981; Reid et al. 1985; Duman et al. 1995). However, recall that the colligative depression of the eqMP is only 1.86 °C per osmol, and therefore, the colligative effect of a glycerol concentration that lowered the eqMP by 2 °C (~1 Osmol) would be expected to depress the SCP by just 2–4 °C. In spite of this theoretical restriction, as well as the attendant significant energy requirement and metabolic coordination of producing high concentrations of these antifreezes, there are numerous examples of insects that do so. One of the most extreme examples of this strategy is seen in the willow cone gall, *Rhabdophaga strobiloides* in Alaska (Miller and Werner 1987) where the larvae winter in galls on the tips of willow branches exposed to extreme air temperatures. In winter they produce very high polyol concentrations that lower the eqMP to −19 °C, and this is partially responsible for their ability to supercool to −56 °C. In addition, they feed by sucking phloem from the willows, and therefore they do not normally ingest microbial, or other, ice nucleators. Consequently, in summer they supercool to −26 °C without high polyol concentrations, but even supercooling to −26 °C is insufficient for these non-thermally buffered insects in this extremely cold environment, hence the need for additional high polyol antifreeze protection in winter. At lower latitudes, southern Michigan (~42° N), *R. strobiloides* has hemolymph TH, indicating the presence of AFPs and/or AFGLs (Duman, unpublished data), but Alaskan larvae have not been tested for TH.

6.3.2.3 Antifreeze Proteins

It has been assumed that the seasonal production of AFPs in numerous insects is instrumental in lowering the SCPs of these insects in winter (Duman et al. 1991b, 2010; Duman 2001). However, there is some experimental evidence that counters this idea. AFP-containing hemolymph of *Rhagium inquisitor* beetles did not inhibit the freezing temperature of whole body homogenates containing ice nucleators (Baust and Zachariassen 1983), and hemolymph supercooling points of intact *R. inquisitor* did not correlate with hemolymph TH (Bremdal and Zachariassen 1988). While experimental evidence of the abilities of insect AFPs to depress SCPs is limited, a few experiments have shown that insect AFPs can indeed recognize and bind to ice nucleators, or to embryo crystals on their surfaces, thereby inhibiting their growth to the critical size required for nucleation. For example, although *D. canadensis* larvae decrease the concentrations of hemolymph proteins and lipoproteins with ice-nucleating activity in winter, these ice nucleators are not completely removed. Addition of *D. canadensis* AFPs (DAFPs) to these purified ice nuleators inhibited ice nucleator activity and promoted supercooling, although the decrease in nucleation temperature, while statistically significant, was not great (Olsen and Duman 1997a). This was also the case when DAFPs were added to solutions containing the ice-nucleating bacteria *Pseudomonas syringae*, a common bacteria present in the larval gut during much of the year (Olsen and Duman 1997b). However, later studies showed that addition of DAFPs to solutions containing either these same hemolymph INPs or ice-nucleating *P. syringae* bacteria eliminated ice nucleator activity if physiological concentrations (0.5 M) of the DAFP enhancer glycerol were also present (Duman 2002). In addition, various crystals that often form in the urine of overwintering insects (calcium phosphate, potassium phosphate, sodium urate, uric acid) can induce ice nucleation (Mugnano et al. 1996). This IN activity is also inhibited by DAFPs in the primary urine of *D. canadensis* larvae (Nickell et al. 2013).

In spite of this evidence of the ability of DAFPs to inhibit ice nucleators, numerous plots of winter *D. canadensis* hemolymph TH versus larval SCPs over multiple years failed to show a correlation between the magnitude of TH and SCPs (Duman, unpublished data). In fact, larvae with rather low hemolymph TH often had the lowest SCPs in a group of larvae, and vice versa. Why? A potential answer may be that the ratio of DAFPs to INs is likely to be critical. Along with providing protection from inoculative freezing, inhibition of ice nucleators to promote

Table 6.1 Effects of the addition of ice-nucleating bacteria (*Pseudomonas syringae*) at two concentrations on the thermal hysteresis of an aqueous solution of DAFP-1 plus glycerol (0.5 M)

Sample	TH (°C)
DAFPs + Glycerol	2.20 ± 0.28
DAFPs + Glycerol + Bacteria (Lo: 2.8 × 10^4/ml)	1.85 ± 0.10
DAFPs + Glycerol + Bacteria (Hi: 2.8 × 10^6/ml)	0.42 ± 0.00

Based on data from Duman (2002)

supercooling is probably the most important function of DAFPs in the freeze-avoiding larval *D. canadensis*, whether in the hemolymph, gut or urine. When DAFPs bind to and inhibit INs it is likely that they are then unavailable to bind to the ice crystal used in the measurement of TH of that particular sample. Consequently, those larval hemolymphs with greater levels of INs should exhibit less TH than those with the same DAFP concentrations but lower levels of INs. Obviously, to keep SCPs low there must be sufficient DAFP to bind to and inhibit the INs, but as IN concentrations increase, the unbound DAFP available to bind to the seed crystal used for TH measurement decreases. This is evident in Table 6.1, based on previously described data showing that purified DAFPs inhibited the ice-nucleating activity of *Pseudomonas syringae* bacteria (Duman 2002). Note the inverse relationship between TH and bacterial concentration. As the numbers of ice-nucleating bacteria are increased the TH decreases. Another potentially important factor is that *D. canadensis* produce some 30 DAFP isomers and there is considerable tissue specificity in the location of these isomers, suggesting that various isomers, or groups of isomers, are better than others at preventing freezing in certain locations (Nickell et al. 2013). Also, as will be discussed later, some of the DAFP isomers enhance one another's activity (Wang and Duman 2005, 2006) as do certain low molecular mass solutes, such as glycerol, accumulated in winter (Li et al. 1998a). Laboratory experiments concerning functions of AFPs (retardation of inoculative freezing, inhibition of ice nucleators) generally use only one, or at best two or three, of the AFP isomers present in the given insect and often do not include potentially important low molecular mass solutes.

6.3.2.4 Antifreeze Glycolipids

AFGLs have TH and RI activities comparable to those of insect AFPs on a per weight basis (Walters et al. 2009c). Some of the AFGL producing insects are freeze avoiding (Walters et al. 2011), although the levels of AFGL present in these species are probably not sufficient to significantly promote supercooling directly. However, AFGLs and *D. canadensis* DAFPs synergistically enhance one another's TH (Duman, unpublished data), suggesting that AFGLs may enhance the supercooling abilities of AFPs. The ability of AFGLs to inhibit ice nucleators, alone or in concert with AFPs, has not been tested.

6.3.2.5 Cryoprotective Dehydration

Unless the solute concentration is very high, the vapor pressure of air in contact with liquid water (in the insect) is higher than that of air in contact with external ice at the same temperature, and this vapor pressure difference (deficit) of the insect becomes greater as the temperature decreases. Consequently, there is a net flux of water molecules from the liquid (hemolymph) to the air and onto the external ice surface. Therefore, freeze-avoiding insects tend to lose water in frozen environments, by both evaporation across the integument and respiration (Lundheim and Zachariassen 1993; Worland and Block 2003). This process, while potentially leading to lethal dehydration, can, in some cases, actually be functional in freeze-avoiding species, resulting in a concentration of antifreezes that promotes supercooling, and also leaving less water available for freezing, thereby making the metastable supercooled water less likely to freeze. For example, larvae of the beetle *Pytho deplanatus* from Alaska and Canada supercool to −54 °C after extensive dehydration resulting in loss of nearly 30% of their body water, and even after freezing at such a low temperature they are freeze tolerant (Ring 1982).

If dehydration proceeds to vapor pressure equilibrium between the organism and the surrounding air in contact with ice the organism is essentially rendered unfreezable. The best examples of this are some small noninsect invertebrates with high surface/volume ratios that lack the cuticular wax coating of insects. Consequently, these organisms lose water quickly when in water vapor pressure deficit, allowing them to respond quickly to decreased temperatures and readily employ this cryoprotective dehydration to avoid freezing. The best-known examples are certain earthworm cocoons (Holmstrup and Westh 1994; Holmstrup and Zachariassen 1996), soil nematodes (Wharton et al. 2003), an Arctic enchytraeid worm (Pedersen and Holmstrup 2003) and polar collembola (Holmstrup and Sømme 1998; Worland and Block 2003). Perhaps because of the slower rate of water loss from insects due to the wax coated cuticle, cryoprotective dehydration is not common in insects, however, larvae of the Antarctic midge *Belgica antarctica* use this freeze avoidance mechanism if the soil they inhabit is dry (Elnitsky et al. 2008). In contrast, if the moisture content of the soil is high the midge larvae freeze at high temperatures, but they are freeze tolerant when inoculated by external ice at higher subzero temperatures. A similar situation occurs in the freeze-avoiding larvae of the Alaskan subspecies (*C. c. puniceus*) of the beetle *Cucujus clavipes* (Bennett et al. 2005) where the dehydration, along with additional adaptations, can lead to vitrification (glass formation) of the body fluids (Sformo et al. 2010, 2011).

6.3.2.6 Beetle Examples of Freeze-Avoidance Mechanisms

To highlight the variety and range of mechanisms that have evolved in freeze-avoiding insects, we will discuss a few examples provided by beetles from the area around South Bend, Indiana, USA (42°N latitude, 86.2°W longitude), northern

Table 6.2 Comparisons of the supercooling abilities of five beetle species that overwinter in the same microhabitats in decomposing logs, in both summer and winter

Species	Season	eqMP	hFP	TH	SCP	Undercooled
Ceruchus piceus	Summer	−0.5	−0.5	0	−7	6.5
	Winter	−1.1	−1.1	0	−26	24.9
Meracantha contracta	Summer	−0.8	−0.8	0	−4	3.2
	Winter	−1.3	−5.0	3.7	−11	6.0
Dendroides canadensis	Summer	−0.6	−1.7	1.1	−9	7.3
	Winter	−2.5	−5.7	3.2	−27	21.3
Cucujus piceus	Summer	−0.5	−0.7	0.2	−7	6.3
	Winter	−2.6	−5.7	3.1	−30	24.3
Uloma impressa	Summer	−0.9	−2.0	1.1	−6	4.0
	Winter	−9.9	−14.7	4.8	−21	6.3

The equilibrium melting point (eqMP) is a colligative property dependent on the hemolymph solute concentration, with seasonal variations dependent largely on winter polyol levels. The hysteretic freezing point (hFP) is the temperature at which a small ice crystal in the hemolymph grows. It is a product of the eqMP depression and the thermal hysteresis (if any). Thermal hysteresis (TH) is the difference between the eqMP and hFP, if any, and is dependent on AFPs. The supercooling point (SCP) is the temperature of spontaneous freezing of the whole, intact insect. Undercooling reflects the supercooling of the insect below the hFP. All values are °C. Data are derived from several sources (Duman 1977a, 1979, 1984a, b; Horwath and Duman 1984a, b; Bennett et al. 2005)

Indiana and southwest Michigan, during winters in the early 1980s (Table 6.2). These beetles all inhabit the same hardwood forests and they are typically present in the same fallen, partially decomposed trees. In addition, they all cease feeding and clear the gut prior to the onset of winter.

The stag beetle *Ceruchus piceus* winters mainly as last instar larvae, although some adults winter as well. As mentioned earlier, this is an interesting species because it does not produce antifreeze (Neven et al. 1986). It lacks both high hemolymph concentrations of polyols or other low molecular mass solutes in winter (note the high winter melting point) and hemolymph TH indicating the absence of AFPs. Yet the larvae supercooled to −26 °C, undercooling 25 °C below the equilibrium MP/FP. This is accomplished by the removal in autumn of a hemolymph lipoprotein with ice nucleator activity. In summer this lipophorin shuttles lipid from the gut to fat body to other cells as needed. However, in winter the larvae are in metabolic diapause and apparently do not need the lipophorin, so they can remove it without consequence and therefore they supercool nicely without expending energy on antifreezes. Adipokinetc hormone stimulates production of the lipophorin ice nucleator in both larvae and in cultured fat bodies, resulting in decreased supercooling when applied to winter larvae (Xu et al. 1990). Although *C. piceus* lacks hemolymph TH it is now known to have the antifreeze glycolipid, perhaps on the inside surface of the cuticle where it may inhibit inoculative freezing (Walters et al. 2011).

The Tenebrionid beetle *Meracantha contracta* was the first insect that we studied in the field that showed hemolymph TH (Duman 1977a, b), but note that the winter eqMP was high as they do not accumulate polyols, etc. (Table 6.2). The wintering larvae only supercool to approximately −11 °C, just 6 °C below the hFP. They successfully overwinter without more extensive supercooling because in late autumn they move to the underside of the fairly large logs they inhabit and there are insulated by the overlying log, soil, detritus and snow.

Another beetle example from these same logs is the Pyrochroid beetle *Dendroides canadensis*, an insect that we have studied for over 40 years. *D. canadensis* winter in multiple larval instars and take 2–3 years to complete their lifecycle. They produce polyols, mainly glycerol, at hemolymph concentrations of 0.5–1.0 M, along with AFPs in the hemolymph, gut, urine, and on epithelial cells underlying the cuticle (Duman 1979; Olsen and Duman 1997a, b; Olsen et al. 1998; Nickell et al. 2013). In addition, *D. canadensis* larvae produce antifreeze glycolipids in winter (Walters et al. 2011). These multiple antifreeze types resulted in the larvae supercooling to −27 °C, 21 °C below the hFP of the hemolymph. High antifreeze levels are perhaps necessary because the winter larvae do not completely remove protein ice nucleators from the hemolymph, although these nucleators are reduced in winter (Olsen and Duman 1997a). During most of the more recent warmer winters in this area, the larvae generally did not supercool as much as they did during the much colder 1970s, exhibiting mean SCPs of −17 to −22 °C (Nickell et al. 2013). The *D. canadensis* AFPs are composed of a protein family of some 30 isoforms with tissue-specific expression that we will discuss later. Note that summer *D. canadensis* hemolymph has TH, although much less than is present in winter. This is a fairly common situation in TH-protein producing insects.

Another beetle worthy of mention here is the Tenebrionid *Uloma impressa*. Overwintering adults have a very low hemolymph eqMP (−9.9 °C), due to very high polyol concentrations (Table 6.2). This, along with approximately 5 °C of TH, lowers the hemolymph hFP to nearly −15 °C. However, in spite of such high antifreeze levels these beetles supercooled only to −21 °C, just 6 °C below the hysteretic freezing point (Duman 1979). This suggests that they have potent ice nucleators that the antifreezes cannot efficiently inactivate, or that they are subject to inoculative freezing.

Cucujus clavipes beetles (family Cucujidae) winter mainly as larvae (multiple instars requiring 2–3 years for completion of their lifecycle) in the same microhabitat as the others mentioned. This is a widespread species in North America, ranging from North Carolina (~35°N) through Canada and into Alaska north to the limit of the boreal forest (~67°30′N). There are thought to be two subspecies—*C. c. clavipes* in the east and *C. c. puniceus* in the west (Thomas 2002). In the region we are discussing here the *C.c. clavipes* subspecies is present, and employs subzero adaptations similar to those of *D. canadensis*, namely high polyol concentrations, AFPs and AFGLs that promote supercooling to as low as −30 °C (Table 6.2). However, as might be expected the western subspecies, *C.c. puniceus*, in Alaska supercool to much lower temperatures (Bennett et al. 2005).

To summarize this section, note again that these data from the five beetles were gathered from species that are found in the same forests and in the same decomposing logs, yet they demonstrate considerable variation in the adaptations that have evolved. All clear their guts in winter to remove gut ice nucleators. One completely removes hemolymph protein ice nucleators in winter (*C. piceus*), while others reduce and mask them with AFPs (e.g., *D. canadensis*). All, except *C. piceus*, produce significant levels of low molecular mass antifreezes (especially *U. impressa*), and antifreeze proteins (except *C. piceus*). Also, the behavioral choice of the logs provides thermal buffering from the extremes and variations in air temperatures, especially important for *M. contracta* with limited supercooling abilities necessitating their movement to the underside of the logs in winter.

Another unusual point of interest concerning two of these beetles is that both *C.c. clavipes* and *D. canadensis* larvae in northern Indiana were freeze tolerant when we first studied them during the very cold winters of the 1970s, exhibiting mean lower lethal temperatures of approximately −25 °C and high SCPs (generally above −10 °C), the latter indicating the presence of ice nucleators that the AFPs could not inhibit (Duman 1979, 1980). However, both species were freeze avoiding when winters became warmer in the early 1980s (Horwath and Duman 1984a; Duman 1984a), and have, for the most part, remained freeze avoiding since that time. Freeze-tolerant *D. canadensis* larvae in the winter of 1978–1979 had SCPs of −7 to −10 °C, and lower lethal temperatures of approximately −28 °C. However, in the warmer winter of 1981–1982, they died if frozen, but their mean SCPs were −27 °C, so the lower lethal temperatures did not change. Both sets of larvae (freeze tolerant and freeze avoiding) had comparable polyol concentrations and hemolymph TH, but the freeze-avoiding larvae lacked the hemolymph ice nucleator activity exhibited by the freeze-tolerant larvae just a few years earlier.

To further demonstrate the powerful effects of a combination of the above mentioned adaptations (AFP, AFGLs, polyols, cryoprotective dehydration, etc.), we will more closely examine *Cucujus clavipes punicius* in interior Alaska where the larvae can be exposed to winter temperatures of −60 °C or lower. Recall that in northern Indiana the freeze-avoiding larvae can have SCPs of −30 °C in winter (Table 6.2) (Duman 1984a, b), although in the warmer recent winters the SCPs have generally ranged from −18 to −25 °C (Bennett et al. 2005). In Indiana, the winter larvae produce AFPs, AFGLs, and polyols (mostly glycerol at 0.5–1.0 M glycerol). In contrast, in early winter the Alaska larvae supercool to −35 to −42 °C while producing approximately the same level of hemolymph TH (~ 4–5 °C) found in the Indiana larvae in mid-winter, although the glycerol concentrations are higher in the Alaska larvae (2.0–2.5 M) (Bennett et al. 2005). In addition, the Alaska larvae undergo a metabolic diapause, while the Indiana larvae do not (Bennett et al. 2005). However, the biggest difference between the two populations is that when the temperatures drop consistently to approximately −20 °C the Alaska larvae dehydrate, and mean body water then decreases from 65 to 24–40% (0.40–0.68 g H_2O g^{-1} dry weight). This cryoprotective dehydration greatly increases the concentrations of antifreezes, raising glycerol concentrations as high as 6–10 M and hemolymph TH to near 13 °C. As a consequence, SCPs are lowered further and many larvae cannot be frozen even if cooled to −150 °C. At temperatures between −58 and −76 °C the remaining unfrozen body water of those larvae with low body water vitrifies (Sformo et al. 2010, 2011). The larvae are so dehydrated at this time that their hemolymph cannot be collected. However, if hemolymph is sampled from cold acclimated, but non-dehydrated individuals (mid-November), and then concentrated to levels similar to those of the mid-winter desiccated individuals, extremely high TH (13 °C) results. When this concentrated, high TH hemolymph is cooled to sufficiently low temperatures (glass transition temperature) it vitrifies (turns glassy), rather than freezes. The increased viscosity caused mainly by the high glycerol concentrations, along with the inhibition of ice nucleators by the AFPs promotes supercooling to the glass transition temperature. Proteomics study comparisons of Indiana and Alaska *C. clavipes* in winter and summer identified numerous additional factors that are undoubtedly involved in the cold tolerance differences between these two populations (Carrasco and Duman 2011; Carrasco et al. 2012), but the cryoprotective dehydration of the Alaska larvae is a major factor.

6.3.3 Antifreeze Protein Ice-Binding Mechanism(s)

While the structures of TH-producing AFPs and RIPs vary greatly, one common characteristic of all of these proteins, no matter the organism, is that they are able to recognize and bind to ice. DeVries (1971) initially proposed this adsorption–inhibition model of AFP depression of the hFP and attendant TH to explain the TH produced by the Antarctic fish antifreeze glycoproteins. Since then numerous studies have demonstrated that these AFPs do indeed bind to ice (Duman and DeVries 1972; Raymond and DeVries 1977), and that the crystal plane(s) to which the AFP adsorbs depends on the particular AFP (Raymond et al. 1989; Knight et al. 1991; Pertaya et al. 2008; Davies 2014). The preferred binding plane(s) (basal, pyramidal, prism) is important in determining the varying efficiencies of the IBPs as regards TH, RI, etc. (Olive et al. 2016).

As a consequence of the AFP adsorption onto the ice surface, water molecules are prevented from joining the ice lattice at the IBP-binding sites, thereby forcing crystal growth between the IBPs in high radius of curvature (high surface free energy) fronts rather than in the preferred low radius of curvature (low surface free energy) fronts. Consequently, according to the Kelvin effect, the temperature must be lowered to the hFP before significant growth occurs (Raymond and DeVries 1977; Raymond et al. 1989). Also, as mentioned previously, the presence of IBPs on ice also raises the hysteretic MP (hMP) above the normal colligative eqMP, although the magnitude of the mp increase is much less than the depression of crystal growth to the hFP (Knight and DeVries 1989; Celik et al. 2010).

The mechanism(s) of this binding is a matter of some debate, and appears to vary somewhat between different IBPs. The binding mechanism will be discussed in detail in Chap. 4 of Vol. II, but a brief introduction is provided here. The first fish AFPs characterized were amphiphilic with regularly arrayed hydroxyl groups, either from saccharide hydroxyls of AFGPs (DeVries 1971; DeVries et al. 1970, 1971; Shier et al. 1975) or amino acid side chains from AFPs lacking carbohydrate components (DeVries and Lin 1977; Sicheri and Yang 1995; DeVries and Cheng 1992) projecting from one side of the protein, and these were thought to hydrogen bond to oxygens in the ice lattice. Likewise, insect AFPs typically have regular repeat sequences that constitute the ice-binding site, usually with regularly spaced threonine residues with projecting hydroxyls that may hydrogen bond to the ice crystal (Graham et al. 1997; Duman et al. 1998; Gauthier et al. 1998; Davies 2014). Consequently, the adsorption–inhibition model based on hydrogen bonding of the threonine hydroxyls to ice appeared to fit these AFPs as well.

A second ice-binding mechanism was later proposed by Davies and colleagues to apply to fish, insect, and other IBPs. See the review by Davies (2014) and the chapter on ice-binding mechanisms in this book. This view is bolstered by considerable experimental and theoretical evidence (Jia and Davies 2002; Graether and Sykes 2004; Garnham et al. 2011a; Hakim et al. 2013; Sun et al. 2014; Davies 2014) that involves a highly structured ice-like hydration sphere of anchored clathrate waters around the ice-binding site. These waters are structured by hydrophobic interactions at the binding site, largely as a result of the methyl groups of the threonine residues. This ice-like water integrates into the semiliquid/crystalline region at the surface of the solid ice, thereby providing an ice-binding mechanism of the AFPs. It now appears that both ice-binding mechanisms (hydroxyls hydrogen bonding to ice, and ice-like water that forms at the ice-binding site that fits into the ice surface) are involved, with the predominant mechanism dependent on the specific AFP (Ebbinghaus et al. 2010, 2012; Meister et al. 2013, 2015). Cosolvents can extend the long-range hydration sphere around the AFP and thereby enhance the level of TH activity (Meister et al. 2013, 2014).

While most studies dealing with the AFP ice-binding mechanism(s) have concerned the ice-binding site, the non-ice-binding face(s) of the protein is also important. In an interesting experiment (Liu et al. 2016), three AFPs (fish, insect, bacterial) were tethered to a solid surface such that either the ice-binding face (IBF) or the non-ice-binding face (NIBF) of the AFP was tethered, leaving the other face exposed to water. The investigators then determined the effect of the exposed surface on supercooling, demonstrating that exposure of the IBF to water facilitated ice nucleation while exposure of the NIBF inhibited it. Associated molecular dynamics simulations indicated that the regular arrangement of hydroxyl and methyl groups on the IBF of the AFP was responsible for the ice-like arrangement of water that facilitated ice nucleation, while the absence of such an arrangement on the NIBF inhibited nucleation. It should be noted that under natural conditions, the volume of ice-like interfacial water associated with the IBF is too small to act as a functional embryo crystal: otherwise the AFP would nucleate rather than act as an antifreeze. Therefore, in the organism (or in a TH measurement) when the AFP binds to ice or to an ice nucleator, the exposed NIBF serves to inhibit ice formation at its surface, thereby inhibiting overgrowth of the AFP by ice.

6.3.4 Insect Antifreeze Protein Structures

The first non-fish AFPs sequenced were insect AFPs, initially those of two beetles, *Tenebrio molitor* (Graham et al. 1997) and *Dendroides canadensis* (Duman et al. 1998), followed closely by AFPs of a lepidopteran, the spruce budworm, *Choristoneura fumifierana* (Gauthier et al. 1998). Additional AFPs have been described from other beetles (Kristiansen et al. 1999, 2011; Qiu et al. 2013; Mao et al. 2011; Ma et al. 2012) another lepidopteran (Lin et al. 2011), a dipteran (midge) (Basu et al. 2015) and a hemipteran (Guz et al. 2014). Perhaps unexpectedly, the insect AFPs that have been described have all been different from the AFPs of fish or other organisms, and there is considerable sequence variation between the AFPs from the different insect families and even between most the beetle AFPs and those of *Rhagium* beetle AFPs. However, in spite of these sequence differences there is considerable similarity between the flat ice-binding surfaces of the insect AFPs. Short descriptions are provided here as more expanded information is described in Chap. 2 of Vol. II.

6.3.4.1 Beetle AFPs

Tenebrio molitor AFPs were the first insect AFPs to be purified and characterized as to composition (Patterson and Duman 1979, 1982; Schneppenheim and Theede 1980; Tomchaney et al. 1982), although some of these reports are now questionable. While the eventual sequencing of *T. molitor* AFPs showed that they had high cysteine contents of 18–19% (Graham et al. 1997), the earlier compositional analyses sometimes found either little or no cysteine (Patterson and Duman 1979; Tomchaney et al. 1982) or different cysteine contents, ranging from 12% to as high as 28% (Schneppenheim and Theede 1980; Patterson and Duman 1982), suggesting either that the initial AFP samples were not pure or that additional AFP types may be present in *T. molitor*, perhaps due to population variations. The originally sequenced *T. molitor* AFPs (Graham et al. 1997) consisted primarily of a variable number of 12-mer repeats in which approximately every sixth residue is a cysteine, flanked by threonines that form the now familiar -T-C-T- or T-X-T units that combine to form the ice-binding surface common in beetle and many other insect AFPs, plus a 14 residue N-terminal cap (Fig. 6.1a–c). *D. canadensis* AFPs have very similar sequences and structures (Fig. 6.1d–f) consisting of variable numbers of 12- and 13-mer repeats plus a 14-residue N-terminal cap that combined yielded a molecular mass of 8.7 kDa for the most abundant hemolymph DAFP (DAFP-1), although isomers varied in molecular mass from approximately 5.4 to 14 kDa (Duman et al. 1998; Nickell et al. 2013). Signal peptides are present in most of these AFPs indicating they are secreted, and the mature proteins often have blocked N-termini of pyroglutamine, and one or more prolines at or near the C-termini that protect the AFPs from C-terminal proteases (Duman et al. 1998; Andorfer and Duman 2000; Nickell et al. 2013) and/or perhaps edge-to-edge inter-

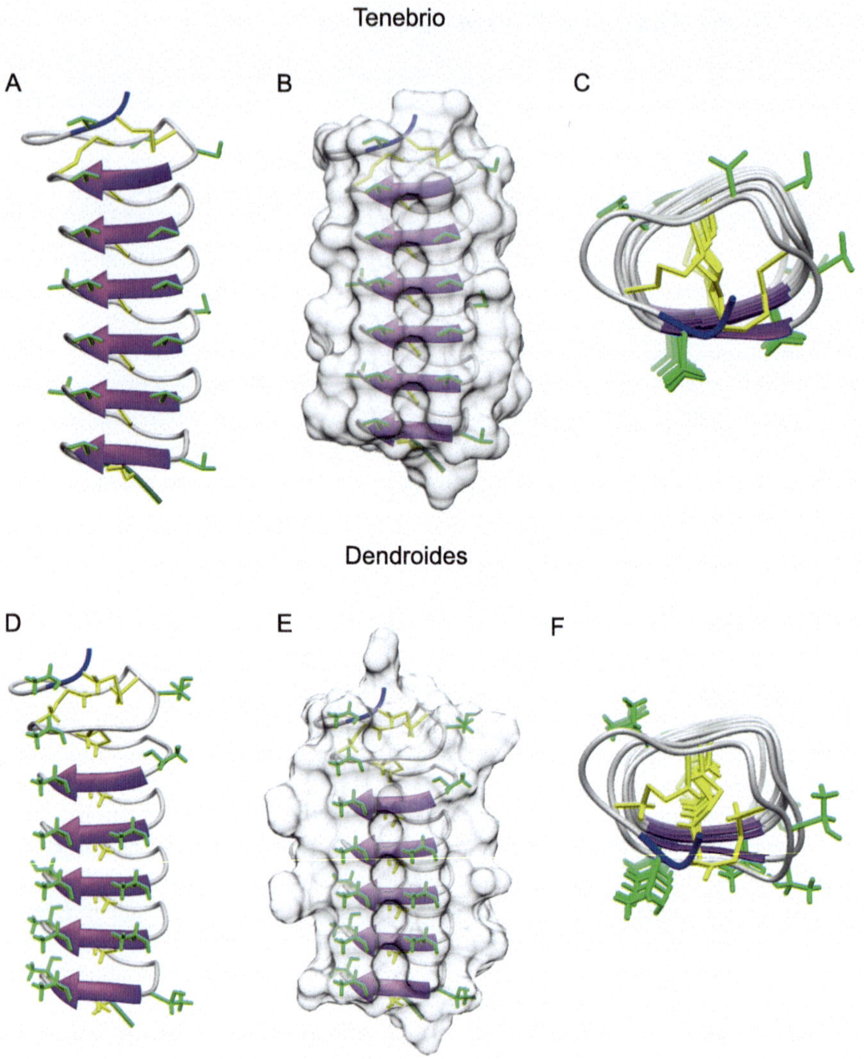

Fig. 6.1 Structures of two beetle AFPs. (**a–c**) Crystal structures of the mealworm, *Tenebrio molitor*, *Tm*AFP (PDB 1EZG). (**a**) Secondary structure is shown with β-strands in purple and loops in gray. The N- and C-termini are colored blue and dark green, respectively. Threonine and cysteine residues are colored green and yellow. (**b**) Semitransparent molecular surface representation. (**c**) View in A is rotated 90° toward the reader. (**d–f**). Homology models of the beetle, *Dendroides canadensis* DAFP-2. Model was generated using the SWISS-MODEL server (Waterhouse et al. 2018). (**d**) Secondary structure is shown with β-strands in purple and loops in gray. Threonine and cysteine residues are in green and yellow, respectively. The N-terminus is in blue and the C-terminus in dark green. (**e**) Semitransparent molecular surface representation is shown. (**f**) The view in (**d**) is rotated 90° toward the reader

AFP associations that could lead to aggregation and amyloid formation (Richardson and Richardson 2002). The cysteine residues are disulfide bridged, usually between cysteines within each repeat (Li et al. 1998b), providing structural stability and aligning the threonine residues on one side of the protein to form the ice-binding site. In both *T. molitor* (Liou et al. 1999) and *D. canadensis* (Andorfer and Duman 2000; Nickell et al. 2013) beetles, the AFPs consist of multiple isoforms of similar proteins that, although maintaining certain highly conserved residues, vary at other positions (Liou et al. 1999; Andorfer and Duman 2000). Some *T. molitor* AFPs are N-glycosylated, although these moieties are not required for TH activity (Liou et al. 1999). Southern blots indicated some 30–50 tandem linked copies of *T. molitor* AFP genes (Liou et al. 1999). The resulting, generally modest, sequence variations beg the question of whether the variable isomers evolved to serve special functions. The original 13 *D. canadensis* AFPs, ranging in mass from 5.4 to 14 kDa due to variations in the number of repeats, were initially separated into three groups based on these sequence variations (Andorfer and Duman 2000), and later it was determined that these three DAFP groups, along with several additional more recently discovered DAFPs (Nickell et al. 2013) exhibited tissue-specific expression (Duman et al. 2002). While just four DAFPs (Group I) are typically present in the hemolymph, the nine Group II and III DAFPs are found in the midgut and epidermis and others of the 30 DAFP isoforms are produced by the Malpighian tubule epithelium and hindgut (Nickell et al. 2013). In contrast to the speculation that the various DAFP isomers evolved to fit slightly different functions in different tissues (Duman et al. 2002; Nickell et al. 2013), it has been argued that these isomer variations in *D. canadensis* and *T. molitor* AFPs might be explained by selection for greater AT content in the third codon position (Graham et al. 2007), however, this does not explain the tissue specific expression in *D. canadensis*.

Crystallization and X-ray analysis along with the NMR solution structure of *Tm*AFP (Liou et al. 1999; Daley et al. 2002) demonstrated the β-barrel structure of these beetle AFPs. The *T. molitor* and *D. canadensis* AFPs both form a somewhat flattened cylinder consisting of a β-sheet face with six or more right handed β-strands (Fig. 6.1). The disulfide bonds are found across the inside of the cylinder that also contains conserved serines and alanines. The highly regular threonines of the conserved -T-C-T- repeats form one flattened, fairly hydrophobic side of the structure that is the ice-binding site with the spacing of the threonine hydroxyls matching oxygens in the ice lattice and methyl groups positioned to structure water in an ice-like fashion. The "pits" between the β-helical strands evident in the semitransparent molecular surface representation (Fig. 6.1b, e) indicate where much of this surface structured water is located. The non-ice-binding faces of these AFPs, as mentioned earlier, may be quite important in generating the high specific TH activity of the AFPs by making it more difficult for ice to overgrow the AFPs adsorbed to the ice surface (Liu et al. 2016).

Except for the *Rhagium* spp. AFPs, the sequences of the other described beetle AFPs are similar to those of *T. molitor* and *D. canadensis,* and are therefore expected to have similar higher order structures. These are AFPs of two additional Tenebrionids from high desert regions of China *Microdera punctipinnis* (Qiu et al. 2013) and *Anatolica polita* (Mao et al. 2011; Ma et al. 2012). Also, sequences of probable AFPs from two additional beetles based on transcriptomics have been submitted to NCBI: a Lucanidae, *Dynodorcus curvidens binodulosus*, (Accession #A1IIC7_9SCAR) (Nishimiya et al. 2006); and a Curculionidae, the mountain pine beetle *Dendroctonus* ponderosae (multiple AFPs such as Gen Bank Accession #ABB03885.1) (Keeling et al. 2012). Both contain repeating units containing -T-C-T- sequences as known from *T. molitor* and *D. canadensis* AFPs. Also, larvae of *C. clavipes* beetles, subspecies from both Alaska and Indiana/Michigan produce AFPs quite similar to those of *T. molitor* and *D. canadensis* (Sformo, Barnes, Duman, unpublished data).

The AFPs of two species of *Rhagium* beetles (*R. inquisitor* and *R. mordax*) have been sequenced and their conformations described and found to differ from the other known beetle AFPs (Fig. 6.2). This difference is perhaps not surprising as the other beetle AFPs are from species belonging to the superfamily Tenebrionoidea while the *Rhagium* spp. are in a distantly related superfamily, the Chrysomeloidea. *R. inquisitor* is found throughout Europe wintering under the bark of dead spruce and pine trees where the freeze-avoiding insects supercool to −25 °C or lower, in part due to the *Ri*AFPs (Zachariassen et al. 2008). Not only are the *Ri*AFPs found in the hemolymph (Zachariassen and Husby 1982), some are also present intracellularly (Kristiansen et al. 1999). The *Ri*AFPs have even greater TH activities than other insect AFPs, and therefore they were expected to differ somewhat in sequence and structure (Kristiansen et al. 2011). A 12.8-kDa hemolymph *Ri*AFP has only one disulfide bridge, and contains six irregularly spaced 13-mer repeat segments that each contain a central highly conserved sequence of -T-x-T-x-T-x-T- with interspersed non-repeating sequences of varying length (Kristiansen et al. 2011). Note the greater number of threonine residues per repeat in *Ri*AFP relative to the repeats of the other beetle AFPs (four versus two). A subsequent study of the sister species, *R. mordax*, identified AFP isoforms similar to those of *R. inquisitor* (Kristiansen et al. 2012). Using molecular dynamics modeling the authors showed that the *Rhagium* AFPs have a much flattened β-helix that is considerably different than those of the other beetle AFPs (Fig. 6.2). The predicted β-sheet conformation of the *Ri*AFP is such that on one side of the protein the repeated threonines form a flat ice-binding surface similar to the -T-C-T- binding surface of the other beetle AFPs. However, the increased number of threonines appears to provide a stronger ice-binding surface that translates to increased TH activity, or alternatively this may increase the solubility of the protein, also leading to higher TH activity (Kristiansen et al. 2011, 2012). *R. mordax* is, like *R. inquisitor*, present throughout northern Europe. Unlike *R. inquisitor*, *R. mordax* does not accumulate high concentrations of polyols or other small colligative antifreezes in winter, but its hemolymph TH activity routinely reaches 8 °C or more in winter (Zachariassen et al. 2008; Wilkens and Ramløv 2009) and permits the beetles to supercool well below the *h*FP of the hemolymph. One of the eight isoforms of *Rm*AFPs lacked a signal peptide indicating an intracellular location (Kristiansen et al. 2012).

A crystal structure study of *Ri*AFP, along with molecular dynamics simulations and ice etching, confirmed the flattened β-helical *Ri*AFP structure, and added

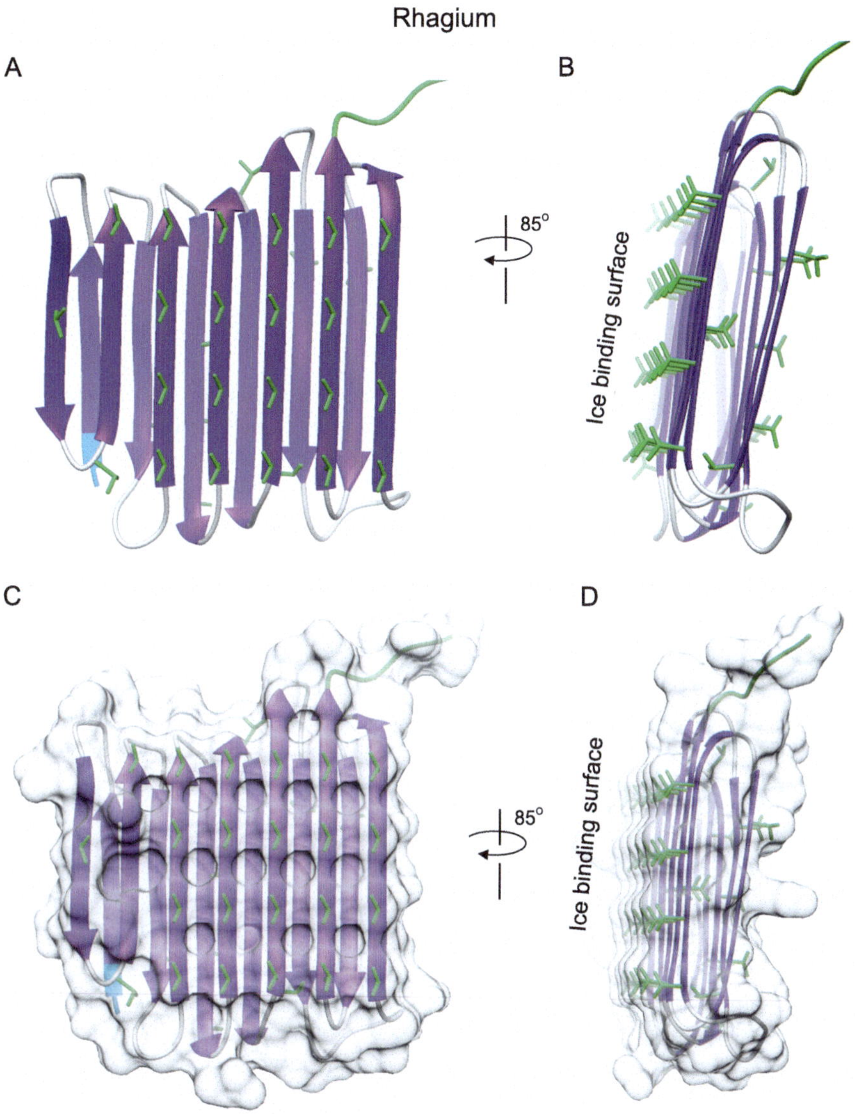

Fig. 6.2 Crystal structure of the *Rhagium inquisitor Ri*AFP (PDB 4DT5). (**a**) Secondary structure is shown with β-strands in purple and loop in gray. Threonine side chains are in green. N-terminus is colored blue and C-terminus is in green. (**b**) The structure in (**a**) is rotated toward the left to show the location of threonine residues. (**c**) Semitransparent molecular surface representation of the view in (**a**). (**d**) Rotation of view in (**c**) to indicate the flatness of the ice-binding surface

additional information on these, especially active AFPs (Hakim et al. 2013). The highly ordered crystallographic water molecules associated with the threonine residues match the prism plane of ice, and these, based on molecular dynamics simulations, appear to recruit additional waters, and some of these matches the basal plane of ice. Ice-etching experiments also indicated that the *Ri*AFP binds to the prism, basal, and probably additional ice planes. Interdigitation of short side chains of amino acid residues located in the top and bottom β-sheets stabilize the flattened structure of the *Rhagium* AFPs in lieu of the numerous disulfide bridges seen in the other beetle AFPs. Molecular dynamics simulations verified the positions of structured waters shown by the crystal structure (indicated by the circular pits in Fig. 6.2c) that, along with associated less well-structured water molecules, assist in binding the AFP to the ice surface.

6.3.4.2 Lepidopteran AFPs

The AFPs of two species of Lepidoptera, larvae of the eastern spruce budworm (*Choristoneura fumiferana,* family Tortricidae) and an inchworm (*Campaea perlata*, family Geometridae) have been characterized. A 9-kDa spruce budworm AFP (*Cf*AFP) consists of a left-handed β-helix containing 15-mer repeats with -T-X-T- motifs and a solenoid with a three-sided, triangular, prism shaped cross section stabilized by a few disulfide bridges (Fig. 6.3a–c) (Tyshenko et al. 1997; Gauthier et al. 1998; Graether et al. 2000). An improved NMR solution structure (Graether et al. 2003) confirmed the earlier findings and showed that the *Cf*AFP is more structured at low (5 °C) than at high temperature (30 °C). Also, the organized waters associated with the ice-binding surface are more ordered at low temperature, resulting in improved binding of the AFP to ice, while at the same low temperature the water structure on the other two non-ice-binding surfaces becomes more disassociated, thereby impeding ice from overgrowing the AFP (Nutt and Smith 2008). As is the case in other AFP-producing insects, multiple copies of the *Cf*AFPs of various masses ranging from approximately 9–12 kDa are present in spruce budworm larvae (Doucet et al. 2002). Expression of the 17 genes encoding the *Sf*AFPs is developmentally controlled, but the *Cf*AFPs are most abundant in the second instar overwintering stage (Qin et al. 2007).

The pale geometer (inchworm) moth, *Campaea perlata*, is common in coniferous, deciduous, and mixed forests of North America ranging from Alaska, across Canada, and south to the southern United States, overwintering as third and fourth instar larvae in exposed positions on trees. Wintering larvae produce an AFP that is quite different than that of the spruce budworm (Lin et al. 2011). Two variants of the *Cf*AFP were characterized, a 3-kDa peptide and a larger 8 kDa variant. The two genes both encode for larger, approximately 40 kDa, proteins from which most of the encoded protein is cleaved to produce the smaller functional *Cf*AFPs. Although the inchworm *Cf*AFPs lack homology to any known proteins, they have some structural similarities to the *Rhagium* beetle AFPs in that both have stretches of alanine and threonine dipeptide repeats (although the repeats of *Cf*AFPs are shorter than those of *Rhagium*), and the repeats are separated by discontinuities. Circular dichroism of the larger *Cf*AFP indicated a mostly β-strand secondary structure, and modeling suggested seven parallel β-helices that form two β-sheets with a thin core made up of alternating serines and alanines, similar to that of silk and the thin core of

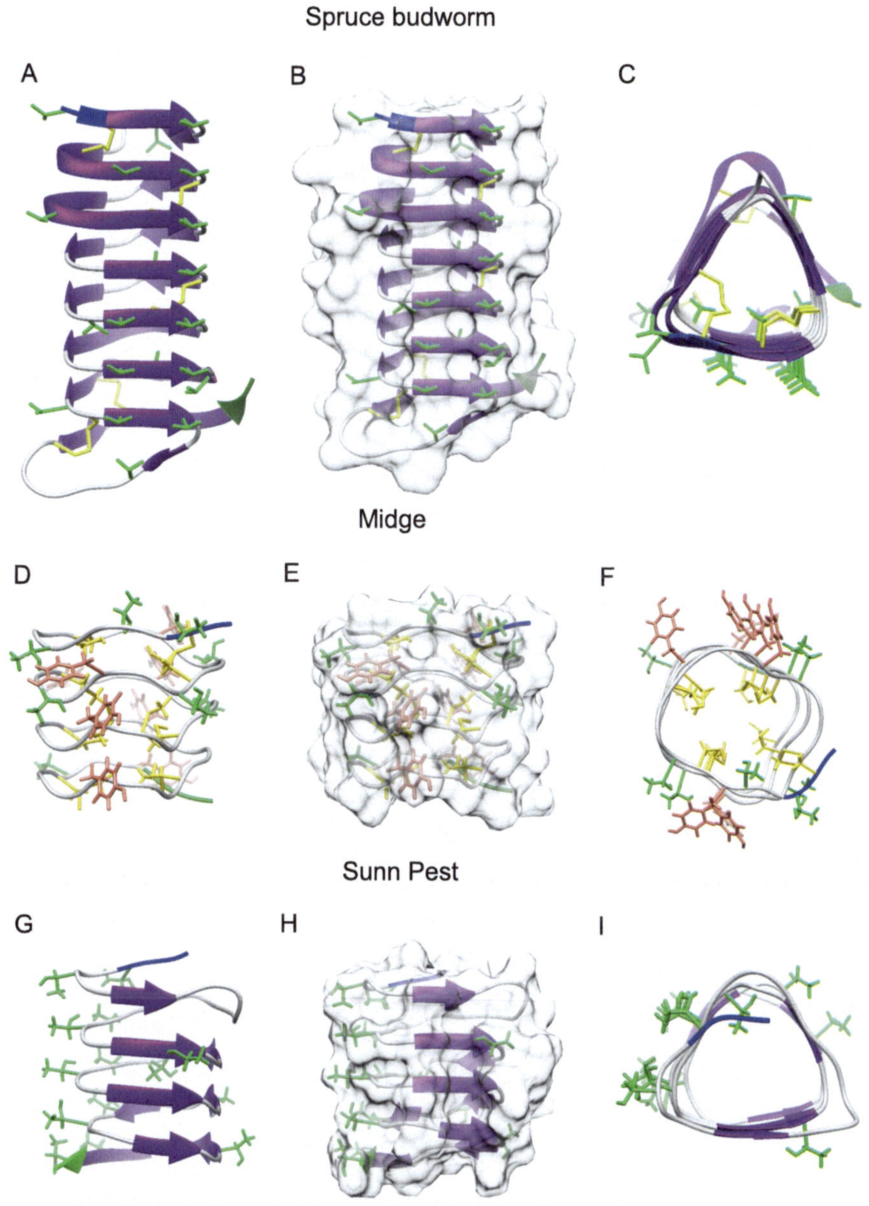

Fig. 6.3 Structures of AFPs of spruce budworm (**a–c**), midge (DEF) and Sunn pest (**g–i**). Crystal structure of the spruce budworm, *Choristoneura fumiferana*, *Cf*AFP (PDB 1M8N). (**a**) β-strands are shown in purple and loops in gray. The N- and C-termini are colored blue and dark green, respectively. Threonine residues are in green and cysteine residues in yellow. (**b**) Semitransparent molecular surface representation is shown. (**c**) The view in A is rotated 90° toward the reader. (**d**) Predicted model of midge AFP. The model was generated using the I-TASSER server (Yang and Zhang 2015). Threonine, cysteine, and tyrosine residues are in green, yellow, and salmon, respectively. (**e**) Semitransparent molecular surface representation. (**f**) The view in D is rotated 90° toward the reader. (**g**) Homology model of the sunn pest, *Eurygaster maura, Em*AFP is shown with threonine side chains in green. The model was generated using the I-TASSER server (Yang and

the *Rhagium* AFPs. The resulting flat *Cf*AFP ice-binding surface is similar to those of the beetle and spruce budworm AFPs.

6.3.4.3 Midge (Dipteran) AFPs

Although dipterans (flies, midges, and mosquitoes) are well represented in high latitudes, thermal hysteresis is rare in this insect family (Duman et al. 1991a, b, 2004) and the AFPs of just two members of this large insect family have been characterized. The gall midge *Thecodiplosis japonica* overwinters as third instar larvae near the soil surface and concentrates glycerol and trehalose as well as AFPs (Li et al. 2000). The AFPs were purified, but only characterized as to molecular mass (34.9 and 37.8 kDa), which is quite large compared to other known insect AFPs. An extremely high thermal hysteresis of ~11 °C was reported for this AFP, although this was at a very high, probably nonphysiological, concentration of 50 mg/ml. The other characterized dipteran AFP came from the emerged adults of an unidentified chironomid midge (Basu et al. 2015). Freshwater organisms do not generally have AFPs, however, these AFP-producing chironomids had emerged from a lake in southern Ontario, Canada where they winter in the benthos as larvae, and then emerge in March at a time when subzero air temperatures still occur, hence the need for AFPs. The midge AFPs form a family of isomers that are in some ways quite different from other known AFPs, yet similar in other respects. The primary isomer is a 9.1-kDa protein that is glycosylated, with a 10-mer repeating unit consisting of xxCxGxCYx forming approximately half of the protein. Regularly spaced tyrosine residues provide the main feature of the binding site (Fig. 6.3d–f), rather than the usual threonines of many other AFPs. The authors state "... the tyrosine ladder plays a role in ice binding through formation of chlathrate waters around the phenyl groups with 'anchoring' to the Tyr hydroxyls and backbone peptide groups." Disulfide bonds in the core strengthen the structure, as in the beetle AFPs, but the midge AFP lacks the β-strand structure of the *T. molitor* and *D. canadensis* AFPs.

6.3.4.4 True Bug (Hemipteran) AFPs

The first AFP that was purified and described from this important insect family is that of the milkweed bug, *Oncopeltus fasciatus* (Patterson et al. 1981). This AFP, purified from a cold acclimated laboratory colony originating from Iowa (USA), consisted of 30.5 mol% serine. Unfortunately, the sequence of the AFP was not

---

Fig. 6.3 (continued) Zhang 2015). Energy minimization was performed in Yasara Structure (Krieger et al. 2004) followed by molecular dynamics simulations in water at 298 K. (**h**) Semi-transparent molecular surface representation. (**i**) The view in (**d**) is rotated 90° toward the reader. All figures were generated using UCSF Chimera (Pettersen et al. 2004)

determined. Interestingly, winter field collected *O. fasciatus* from near South Bend, Indiana (42°N) did not have hemolymph TH, indicating population variations in the ability to produce AFPs.

More recently putative AFPs were described from another hemipteran, the sunn pest *Eurygaster maura*, from Turkey (Guz et al. 2014). *E. maura* is a widespread and important cereal crop pest in southern and eastern Europe, north Africa, and central and western Asia, as well as the central and near-East. The bugs used in this study winter as diapausing adults in the mountains of Turkey where they encounter temperatures as low as −30 °C. In spring, the adults fly to low lying fields where they reproduce and the juveniles complete their development before returning to the mountains. The sequence of the *Em*AFP, derived from a fat body cDNA library, shows that the 96 residue *Em*AFP consists of 12–13-mer repeats with a predominance of -T-X-T- residues where X may be any amino acid. The sequence had the most homology (52% identity) to the spruce budworm AFP. Modeling by the authors indicated a left-handed β-helical coil with the four cysteine residues forming disulfide bridges across the center of the solenoid, and the Thr residues aligned in a regular fashion on the flat presumptive ice-binding site (Fig. 6.3g–i).

6.3.5 AFP Tissue Specificity

TH activity was initially identified in the hemolymph, hindgut fluid, and perirectal space, but not in the primary urine, of *T. molitor* larvae (Ramsay 1964; Grimstone et al. 1968). In *D. canadensis* beetle larvae TH was first measured in hemolymph (Duman 1979), then in midgut fluid (Duman 1984a, b), and later in hindgut and primary urine (Nickell et al. 2013) of winter larvae. The absence of TH in primary urine of *T. molitor* studied by Ramsay may be because these larvae were not cold adapted, but this is purely conjecture as the urine of cold adapted *T. molitor* has not been tested for TH. In the summer, *D. canadensis* have low levels of hemolymph TH (Table 6.2), but lack TH in the gut and urine at this time.

D. canadensis larvae produce 30 known DAFP isomers. The tissue-specific expression of the isomers (Duman et al. 1998, 2002; Andorfer and Duman 2000; Nickell et al. 2013) suggests that their functions may vary somewhat. Twelve isomers were initially described and these were separated into three groups based on sequence variations (Andorfer and Duman 2000). Group-I (DAFPs-1, -2, -4, and -6) are produced in the fat body and secreted into the hemolymph while groups-II and -III are present in the midgut and produced both in the fat body and midgut epithelia (Duman et al. 2002). As mentioned previously, the group-I hemolymph DAFPs function to inhibit inoculative freezing across the cuticle (Olsen et al. 1998) and promote supercooling by inhibiting incidental hemolymph INPs (Olsen and Duman 1997a). Individual larvae produce varying combinations of certain groups-I, -II, and III DAFPs in the single cell layer of epidermal cells underlying the cuticle, some of which lack signal peptides and are therefore present in and/or on the cells where they can be visualized by immunofluorescence. These DAFPs, along with the hemolymph DAFPs, are essential in inhibiting inoculative freezing (Olsen et al. 1998). Gut DAFPs inhibit microbial, etc., ice nucleators ingested by the larvae. While the larvae cease feeding and clear their guts of ice-nucleating bacteria, etc. by the onset of prolonged cold weather (usually sometime in late November) the gut DAFPs are produced in early October and persist well into the spring, thereby allowing the larvae to feed later in autumn and earlier in spring, and to drink during winter thaws (Nickell et al. 2013). In addition to the originally identified 13 DAFP isomers, numerous additional isoforms (24 total) are produced by the Malpighian tubule epithelium and secreted into the primary urine, probably to inhibit ice-nucleating activity of various crystals that form in insect urine over the winter (Nickell et al. 2013). The primary urine is released into the hindgut, thereby positioning the contained DAFPs to assist gut produced DAFPs in providing freeze protection of hindgut and rectal fluid.

The freeze-tolerant alpine cockroach *Celatoblatta quinquemaculata* is particularly interesting as regards AFP tissue specificity. Although this species lacks hemolymph TH, activity of approximately 2 °C is present in the gut tissue and 0.6 °C in the gut contents (Wharton et al. 2009). The gut AFP is thought to protect gut tissue from freezing damage by inhibiting recrystallization.

A recent study using fluorescent *Rm*AFP antibodies showed that *R. mordax* larvae had significant levels of *Rm*AFP in the cuticle and gut in summer, actually more than in winter (Buch and Ramløv 2017). The authors suggested that the larvae store the AFP in and/or around the fat body in summer for use when cold weather returns. Also, as previously mentioned, the related *R. inquisitor* beetles not only have hemolymph AFPs (Zachariassen and Husby 1982), but also intracellular AFPs are present (Kristiansen et al. 1999).

6.3.6 Insect AFP Activity and Enhancers

Insect AFPs from freeze-avoiding species typically have higher specific activities (TH/AFP concentration) than those of most other organisms: fish, plants, microorganisms, or the RIPs of freeze-tolerant insects (Duman 2015). Mid-winter hemolymph concentrations of AFPs in the freeze-avoiding larvae of the beetle *D. canadensis* are approximately 2–4 mg/ml, producing TH of approximately 2.5–6 °C, occasionally as much as 9 °C. TH of fish blood serum is rarely as high as 2.5 °C, even though the AFP concentrations are much higher than those of *D. canadensis*. In contrast to these freeze-avoiding species, the specific TH activities of RIPs from freeze-tolerant insects have yet to be determined (none have been purified), the hemolymph TH of these species is low, generally 0.2–0.6 °C, and consequently it is likely that their specific TH activities are low.

In comparisons of AFPs representing different organisms (fish versus insects versus plants, etc.), different species within these groups, or of different AFP isomers it is important to realize that a number of factors can influence the magnitude of TH measurements. These include the annealing time (the time the sample is permitted to interact with the seed crystal), the cooling rate after determination of the eqMP and/or hMP, the size of the seed crystal, and the technique used to measure TH: capillary, nannoliter osmometer, differential scanning calorimeter, sonocrystallization, etc. (DeVries 1986; Ramløv et al. 2005; Nicodemus et al. 2006; Olive et al. 2016). This is especially pertinent when comparing TH values produced by different research groups. Also, the ice crystal adsorption plane of a given AFP influences the magnitude of the measured TH and RI (Olive et al. 2016).

While the hemolymph TH of mid-winter collected *D. canadensis* larvae is typically 2.5–6 °C, and as high as 9 °C in some individuals (Duman 1979, 1980), TH measurements of purified DAFPs were much less, even at very high nonphysiological concentrations (Wu et al. 1991; Duman et al. 1998). The first clue to this paradox was provided by chance when graduate student Ding Wen Wu attempted to titrate out the TH activity of the DAFP she had purified (Wu and Duman 1991) by adding a polyclonal anti-DAFP immunoglobulin she had prepared. Addition of the antibody increased the TH from 1.20 °C with DAFP alone to 1.72 °C with addition of the antibody, rather than decreasing TH as expected (Wu and Duman 1991). The explanation for the enhanced TH was that the DAFP–antibody complex was much larger than the DAFP alone and therefore it covered a larger surface area of the seed crystal used in the TH measurement, plus it rose higher above the crystal surface making it harder for the ice to overgrow it. Further proof was provided by the addition of a goat–anti-rabbit IgG to the anti-DAFP–rabbit antibody–DAFP complex that further increased the TH (from 1.72 to 2.86 °C). Of course, insects do not produce antibodies, but this serendipitous result suggested that other naturally occurring proteins might also bind to, and thereby enhance, the DAFPs. Therefore, *D. canadensis* hemolymph protein fractions that lacked TH were added to purified DAFPs and the resulting TH measured. While most protein fractions did not enhance the TH, others did, as did other non-endogenous proteins (Duman et al. 1993). A modified immunoblot technique indicated direct binding of the enhancing proteins to the DAFPs, as occurred with the DAFP–antibody enhancement. This binding was confirmed when yeast-two hybrid studies, in combination with immunoprecipitation, were later used to identify proteins that bind to, and thereby might enhance, the DAFPs (Wang and Duman 2005). There are four hemolymph DAFPs: DAFP-1, -2, -4, and -6. Each of these was used individually as "bait" to identify other proteins that bind to the bait. Surprisingly, certain of the four hemolymph DAFPs bound to one another, but not to others or to themselves. When the interacting DAFPs were mixed (while holding total DAFP concentration constant and physiological at 3 mg/ml) the TH increased synergistically, while mixtures of the noninteracting DAFPs did not exhibit increased activity. Interestingly, two enhancing DAFPs, DAFP-1 and -2, differ at just two positions: an Ile at position 35 and an Ala at position 75 in DAFP-1, while DAFP-2 has an asparagine at position 35 and a Thr at position 75. Point mutations at either of these positions in DAFP-2 eliminated both the binding to one another and the enhanced TH (Wang and Duman 2005). The only non-DAFP protein identified by the yeast-two-hybrid was a thaumatin-like protein that interacted with and increased TH when mixed with either DAFP-1 or -2, but not with DAFPs -4 or -6 (Wang and Duman 2006).

Initial investigations of potential DAFP enhancers also demonstrated that glycerol (normal winter hemolymph concentrations of ~0.5–1.0 M), sorbitol and to a lesser extent NaCl also had positive effects on TH activity of DAFPs (Duman et al. 1993). An expanded study of various low molecular mass solutes (Li et al. 1998a) showed that various sodium salts, polyhydroxy alcohols (glycerol, sorbitol, etc.), amino acids and amine compounds all had an enhancing effect on DAFP-4. Of the solutes tested, citrate was the best, raising TH from 1.2 °C in its absence to 6.8 °C with a trisodium citrate concentration of 1 M (an obviously nonphysiological concentration). At the time the mechanism of enhancement by small solutes was not understood. However, terahertz absorption spectroscopy and molecular dynamics simulations later indicated the gradient of hydrogen bond dynamics was more pronounced (the hydrogen bond dynamics were greatly slowed relative to the rest of the protein) toward the ice-binding region following addition of sulfate enhancer (Meister et al. 2014), suggesting that the enhancer increased the ice-like ordering of water near the ice-binding face.

Dr. Xin Wen and colleagues have systematically investigated a large number of low molecular mass enhancers of DAFP-1. Using a series of polycarboxylates they found that the number of carboxylates and/or hydroxyl groups present in an enhancer positively correlated with TH (Amornwittawat et al. 2008). Likewise, TH increased with the number of hydroxyls on polyhydroxyl compounds (polyols and carbohydrates) up to five, with trehalose being the best enhancer (Amornwittawat et al. 2009). Trehalose is the naturally occurring blood sugar of insects and it increases from 0.3 mM in summer *D. canadensis* hemolymph to 90 mM in winter (Wen et al. 2016). Modification of arginine-9 of DAFP-1 with 1, 2-cyclohexanedione resulted in a loss of the TH enhancement provided by these small organic enhancers, while reversal of the modification with hydroxylamine treatment restored enhancement, indicating that the guanidinium group of this arginine is critical for enhancement by these compounds (Wang et al. 2009a). Also, a study of monovalent inorganic salts demonstrated that while there was a salting in effect at low concentrations, at higher concentrations the salts had a salting out effect on the DAFP-1 leading to increased TH, presumably as more DAFP then adsorbed onto the ice crystal surface. The magnitude of the salting out, and therefore the TH, followed a Hofmeister series effect (Wang et al. 2009b), described in detail in Chap. 6 of Vol. II.

Manipulations of winter *D. canadensis* hemolymph collected from three different years demonstrated that endogenous proteins (containing both protein enhancers such as thaumatin as well as ice nucleators that would lower TH) and low molecular mass enhancers increased TH by 60–97% and 35–55%, respectively (Duman and Serianni 2002). Also, mixtures of the four naturally occurring hemolymph DAFPs (1, 2, 4, and 6) at equal concentrations that together yielded a physiological total DAFP concentration of 3 mg/ml, plus the thaumatin-like protein and 0.5 M glycerol resulted in TH of 8–10 °C, equaling the highest TH measured in winter *D. canadensis* hemolymph (Wang and Duman 2006). Note that these latter studies were done with expressed and purified proteins, and therefore, unlike the experiments done with natural hemolymph, there were no incidental ice nucleators included. The relative concentrations of DAFPs in hemolymph are not equal (Duman et al. 2002), as was the case in this experiment. While these studies demonstrate that the DAFPs and other proteins (thaumatin-like protein) bind to one another, the nature of this binding and the composition of the hemolymph protein aggregates are unknown. Also, recall that while there are only four DAFPs in *D. canadensis* hemolymph, the midgut fluid has nine different DAFPs (Duman et al. 2002) along with a gut-specific thaumatin-like protein (Wang and Duman 2006), and the Malpighian tubule epithelia produce 24 DAFP isoforms that are presumably released into the urine and subsequently enter the hindgut (Nickell et al. 2013). Consequently, identification of the potential interactions of gut and urine DAFPs is a daunting task.

In addition to their effects on TH, DAFP enhancers are also essential to optimize other critical functions of DAFPs, such as inhibition of ice nucleators to improve supercooling (Olsen and Duman 1997a, b; Duman 2002; Nickell et al. 2013) and inhibition of inoculative freezing across the cuticle by external ice (Olsen et al. 1998).

6.3.7 Seasonal Variations and Controls of Antifreeze Protein Activity

AFP activity in various fluid compartments is seasonal. For example, in *D. canadensis* larvae hemolymph TH begins to increase in early autumn, peaks in winter and decreases gradually to very low summer values over several weeks through the spring (Duman 1977b, 1980; Nickell et al. 2013). DAFP transcript production is largely in tune with hemolymph TH, with transcript for the primary hemolymph DAFP (DAFP-1) peaking in late November and December just prior to the winter maximum in TH (Andorfer and Duman 2000). The autumn hemolymph TH increase precedes that of the gut by about 2 weeks and that of the urine by 3–4 weeks (Nickell et al. 2013), while the hemolymph TH decrease in the spring lags behind that of the gut and urine by an equal time period (Nickell et al. 2013). This timing of early autumn induction and late spring loss of AFPs provides the insects with AFPs that contribute to freeze-avoidance protection that permits larval activity, including feeding and growth, to continue longer into late autumn/early winter and to resume earlier in late winter/early spring, especially when the AFPs are expressed in the gut (Duman 1984b) and urine (Nickell et al. 2013). *D. canadensis* larvae near South Bend, IN (USA) typically begin feeding in late winter (generally early to mid-March or even late February in warm winters). In contrast, polyols are rapidly removed with the advent of warmer weather in late winter/early spring (Duman 1980; Nickell et al. 2013). In some species, hemolymph TH is absent in summer, while in others a low level of activity, generally 0.2–0.4 °C is maintained in summer (Table 6.2), perhaps suggesting that the AFPs may degrade very slowly or that they have a role in addition to cold tolerance (to be discussed later in this chapter). In

D. canadensis larvae certain AFP transcripts are produced throughout the summer, apparently resulting in the low summer TH (Andorfer and Duman 2000). *Microdera punctipennis*, a beetle from high desert regions of China with AFPs similar to those of *D. canadensis* also synthesizes certain of its AFPs in summer (Qiu et al. 2013). Also, recall that while *R. mordax* larvae have hemolymph TH as high as 8 °C in winter (Wilkens and Ramløv 2009) significant levels of *Rm*AFP are present in the cuticle and gut in summer, actually more than in winter (Buch and Ramløv 2017). The authors suggested that the larvae store the AFP in and/or around the fat body in summer for use when cold weather returns.

Increased hemolymph TH was induced in the laboratory by low temperatures or short photoperiods in the Tenebrionid beetles *Meracantha contracta* (Duman 1977a, c) and *Tenebrio molitor* (Patterson and Duman 1978; Horwath and Duman 1982). Low relative humidity also induced increased hemolymph TH in *T. molitor*. The larvae did not decrease body water during the low humidity acclimation period, so the increased TH was not due to decreased volume of the hemolymph (Patterson and Duman 1978). However, another study of *T. molitor* larvae concluded that while cold (4 °C), desiccation or starvation increased TH and *Tm*AFP transcript, short photoperiod (at 22 °C) did not (Graham et al. 2000). This latter work concluded that any environmental condition that inhibited growth (cold, desiccation, starvation) resulted in the larvae becoming quiescent with attendant induction of AFP.

A series of experiments with *D. canadensis* larvae demonstrated that temperature, photoperiod and thermoperiod were all important cues controlling TH. Low temperature in the autumn is inductive (critical temperature between 10 and 15 °C), as is short photoperiod (critical photoperiod between 10 and 11 h of light per day), but high temperature (above 25 °C) inhibited short photoperiod induction (Horwath and Duman 1983b). Photoperiodism provides an organism with *the* most reliable mechanism for determination of the season, and consequently it is an important cue in the timing of many physiological responses. "Resonance" and "T" experiments showed that the photoperiodic induction of TH in *D. canadensis* larvae involves the circadian system (Horwath and Duman 1983b, 1984b). In addition, *D. canadensis* larvae can perceive and use short thermoperiods to induce DAFP production in the autumn, even when neither the thermophase nor scotophase of the 24-h thermocycle are below the critical temperature (Horwath and Duman 1986). Use of thermoperiod to cue DAFP production is likely useful in the larval microhabitat under the bark of fallen partially decomposed trees where the magnitude of variations in air temperature is dampened and photoperiod may not always be perceptible, depending on the site of the larva in the log. In spring, a combination of both long photoperiod and high temperature appears to be necessary for reduced hemolymph TH. This "failsafe" combination of photoperiod and temperature cues provides a safeguard for the larvae, protecting them from yearly variations in weather conditions (early autumn or late spring frosts, mid-winter thaws) that might otherwise render the larvae without AFP protection (Horwath and Duman 1983b). Unfortunately, these studies only looked at hemolymph TH, because at the time the presence of DAFPs in gut, Malpighian tubules, and epidermis was not known. Also, as discussed earlier both polyols and thaumatin-like protein enhance DAFP activity, but when these studies were done this enhancement was not known. Consequently, there is the possibility that induction or loss of these enhancers may have contributed to the hemolymph TH changes.

Juvenile hormone (JH) plays a role in AFP induction, both in *D. canadensis* (Horwath and Duman 1983c; Xu and Duman 1991) and *T. molitor* (Xu et al. 1992). *D. canadensis* larvae collected from the field in early autumn prior to production of AFPs and held in non-inducing conditions (long photoperiod and 22 °C) increased hemolymph TH when topically treated with JH, while larvae held under TH inducing conditions (short photoperiod) failed to increase hemolymph TH if treated with the anti-JH drug precocene (Horwath and Duman 1983c). Isolated fat bodies of larvae produced AFPs when JH was added to the incubation medium, but only if the fat bodies were taken from larvae that had been pre-treated with JH prior to fat body removal (Xu and Duman 1991). Similar results occur with JH induction of vitellogenin production in cultured fat bodies in other insects: pre-treatment of the larvae is necessary prior to the removal of the fat body (Koeppe et al. 1985). Likewise, in *T. molitor* topical application of JH to larvae induced AFP production, as did the addition of JH to the culture medium bathing previously JH induced fat bodies (Xu et al. 1992). In addition, JH titers were increased in *T. molitor* larvae held under DAFP inducing conditions (low temperature or short photoperiod), further implicating JH in DAFP induction.

A comparison of the regulation of spruce budworm, *Choristoneura fumiferana*, AFP production once again demonstrates the variability of insect wintering mechanisms and their controls. Unlike *D. canadensis* larvae that overwinter in quiescence in multiple larval instars and require 2–3 years for development to adults, *C. fumiferana* winter in cocoons as diapausing second instar larvae, and then complete development after breaking diapause in the spring. They produce 17 *Cf*AFP isoforms and production of these is not directly influenced by environmental conditions, but rather is controlled developmentally (Qin et al. 2007). While low levels of certain *Cf*AFP transcripts are present in eggs and first instars, high transcript levels and *Cf*AFPs are not produced until the larvae enter the second instar, and diapause induction by low temperature and short photoperiod are not required. Perhaps surprisingly, since the *Cf*AFPs are developmentally regulated, hormones are not involved. In contrast, glycerol accumulation in the wintering larvae requires low temperature for induction (Han and Bauce 2000). *Cf*AFP transcripts disappear shortly after diapause is broken and the *Cf*AFPs decrease thereafter in approximately 1 week, although low levels of certain transcripts continue into the summer in later instars (Qin et al. 2007).

While TH is dependant on AFP concentration and specific activity, when considering the annual variations in insect TH it is important to remember that enhancers of AFP activity may also be important and that these may change seasonally as well.

6.4 Adaptations to Freeze Tolerance

As with freeze avoidance, there are multiple potential adaptations to achieve freeze tolerance, and various freeze-tolerant insects exhibit different combinations of these adaptations. Undoubtedly, there are numerous adaptations yet to be discovered. Two types of IBPs (RIPs and INPs), along with AFGLs, are appropriate to this review, and these, along with a few other adaptations will be discussed in this section.

6.4.1 Ice Nucleator Proteins

Adaptive INPs are present in many freeze-tolerant insects (Zachariassen and Hammel 1976; Zachariassen 1982, 1985; Ramløv 2000; Duman et al. 2010). As initially reported in beetles from mountains in southern California (Zachariassen and Hammel 1976), the INPs are typically located in the hemolymph where they function to limit supercooling of the extracellular fluid, and thereby, paradoxically, prevent lethal freezing of the cytoplasm. When freezing is initiated following more extensive supercooling, the subsequent rapid ice growth can lead to lethal propagation of ice across the cell membrane (Mazur 1984). In contrast, with more controlled freezing at higher subzero temperatures (generally −5 to −10 °C in species with extracellular INPs) the cell membrane is better able to prevent inoculation of the cytoplasm by extracellular ice. Subsequently, as the extracellular ice excludes solutes from the crystal lattice the resulting increased osmotic concentration in the unfrozen fraction of the hemolymph results in the osmotic outflux of water from the cells, thereby lowering the freezing temperature of the cytoplasm (Mazur 1984). Another advantage of initially freezing at higher temperatures is that the resulting osmotic imbalance is decreased and consequently is easier to control (Zachariassen 1992). Also, there is likely to be an energy savings aspect to freezing at higher temperatures because frozen insects exhibit a greatly reduced metabolic rate relative to unfrozen insects at the same temperature (Scholander et al. 1953). Early freezing also decreases water loss for the insect, mitigating a potentially serious water balance problem over a long winter. Zacchariassen and colleagues proposed that many freeze-tolerant organisms, including insects, have high rates of transcuticular water loss relative to freeze avoiding insects, and that this provided a stimulus for the evolution of freeze-tolerant adaptations in these organisms, including hemolymph INPs (Zachariassen 1992; Zachariassen et al. 2008). Freezing at higher temperatures produces vapor pressure equilibrium between the extracellular fluids and the surrounding atmosphere thereby reducing water loss. Some adaptive INPs are only produced in winter and consequently ice nucleation may be their primary, or only, function (Baust and Zachariassen 1983).

INPs have been extensively studied in ice-nucleating bacteria and fungi, mainly because of the extensive damage they cause to agricultural crops and their influence on precipitation patterns due to ice nucleation by bacteria in the atmosphere (see Lee et al. 1995 for reviews). Ice-nucleating bacteria, such as *Pseudomonas syringae*, are common in soil, detritus, and as epiphytes on the surface of plants where they initiate ice formation in condensed water (dew) that then can inoculate and kill sensitive plants at temperatures just a few degrees below 0 °C (Lindow 1983, 1995). These bacteria are responsible for much of the agricultural damage done to frost-sensitive crops as a result of fairly minor early autumn and late spring frosts that otherwise would not damage the plants were it not for the presence of the epiphytic bacteria. These are the same microorganisms mentioned earlier in this chapter that, if ingested by freeze avoiding insects along with their normal food, can cause lethal ice formation in the gut fluid, necessitating gut evacuation, and cessation of feeding prior to the onset of subzero temperatures in the autumn (Salt 1953; Sømme 1982; Zachariassen 1985; Cannon and Block 1988; Lee et al. 1996). In contrast, this cessation of feeding and gut evacuation is not necessary for freeze-tolerant insects such as larvae of the cranefly *Tipula trivitta* that maintain full guts throughout the winter (Duman et al. 1985).

In *P. syringae*, the 150-kDa proteins responsible for ice nucleation are present on the outer membrane of the bacteria and have a sequence consisting primarily (81%) of a lengthy central region made up primarily of 16-mer tandem repeats (Green and Warren 1985; Wolber and Warren 1989). Similar configurations are present in the other bacterial INPs (Warren and Corotto 1989). To exhibit their activity, bacterial INPs must be attached to the outer membrane (reviewed in Fall and Wolber 1995), and as a consequence structural information on these proteins is limited. However, there are indications that the central repeating region forms a β-solenoid somewhat similar to those of certain insect AFPs, but on a much larger scale, such that the associated ice-like organized water is sufficiently large to nucleate water (Graether and Jia 2001; Garnham et al. 2011b). The apparent similarities between some insect AFPs and the bacterial INP were demonstrated when a 96-residue recombinant protein based on a portion of the *P. syringae* INP actually had low-level TH activity (Kobashigawa et al. 2005). Given this interesting finding, it is not clear why the native *P. syringae* INP does not produce TH.

In contrast to the great amount of information on bacterial INPs, much less structural information is known about endogenous insect hemolymph INPs, and none of these has been sequenced. A 74-kDa INP with an unusually high glutamine/glutamate content (20 mol%) was purified from the hemolymph of freeze-tolerant queens of the white-faced hornet *Vespula maculata* (Duman et al. 1984), but this INP was not sequenced and further characterized. A large globular 800 kDa ice nucleator lipoprotein (INLP) with a diameter of approximately 135 Å was purified and characterized from the hemolymph of the freeze-tolerant larvae of the cranefly *Tipula trivittata* (Duman et al. 1985; Neven et al. 1989). The *T. trivittata* INLP consisted of (by weight) 45% protein (two apolipoproteins: Apo-I, 265 kDa and Apo-II, 81 kDa), 51% lipid, and 4% carbohydrate. The two apoproteins and most of the lipid components were characteristic of the typical lipophorins that shuttle lipid through insect hemolymph, except that 11% of the INLP lipid was phosphatidyl inositol (PI). This was interesting because (1) there is evidence that the hydroxyl groups of PI can order water in an ice-like fashion via hydrogen bonding (Warner

1962), (2) PI had not been previously identified in insect lipophorins, and (3) phospholipids form a monolayer over the surface of lipophorins where they are in position to interact with water. Various manipulations of the PI (i.e., treatment with PI-specific phospholipase C), but not other phospholipids, indicated the requirement of PI integrity for ice nucleator activity. Also, delipidation of the INLP, followed by reconstitution into proteoliposomes with specific components of the native INLP demonstrated that only PI and the two apolipoproteins were sufficient and necessary to regain ice nucleator activity (Neven et al. 1989). Also, any manipulations of the PI hydroxyls, such as substitution of PI-4 monophosphate or PI-4,5 diphosphate for PI in the reconstituted proteoliposomes, produced liposomes lacking IN activity (Duman et al. 1991a, b). In contrast, manipulations of the fatty acid components of the PI had no effect on IN activity. Therefore, the surface PI of the LPIN is critical for activity. However, the two apoproteins are also essential, and while they were not sequenced, immunological evidence indicated similarities to the *Pseudomonas syringae* bacterial INP. Polyclonal antibodies raised to the *T. trivittata* INLP cross-reacted with the bacterial INP, and antibodies to the *P. syringae* INP cross-reacted with *T. trivittata* Apo-I and Apo-II, suggesting that the cranefly INLP apoproteins contain some amount of the repeat sequences similar to those that constitute ~80% of the bacterial INP (Duman et al. 1991a, b). Also, antibodies to a synthetic octapeptide representing the consensus octapeptide of the bacterial INP cross-reacted on Western blots with both INLP apoproteins. Consequently, while some regions of the apolipoproteins are buried in the interior of the INLP (as is the case with typical insect lipophorins), other regions with sequence similarity to the repeat sequences of the bacterial INP are exposed on the surface and these regions, along with PI hydroxyls, are involved in ordering water in an ice-like fashion on the surface of the INLP. This hypothesis was recently supported by a terahertz spectroscopy study of the INLP that demonstrated that the INLP had a significant effect on hydrogen bond dynamics of hydration water, and that this was directly related to the inositol hydroxyls (Baumer et al. 2016).

Multiple studies have shown that cooperation between juxtaposed INPs on the surface of neighboring ice-nucleating bacteria is required for high levels of nucleation (Southworth et al. 1988; Wolber and Warren 1989; Mueller et al. 1990). This also appears to be the case for the *T. trivittata* INLPs (Yeung et al. 1991; Duman et al. 1992, 1995). A scanning tunneling microscopy study indicated that the globular INLPs self-arrange into two side-by-side chains like pearl necklaces, perhaps positioning the water organizing regions of the INLPs to cooperate in ordering water, thereby forming a larger embryo crystal that is then capable of nucleating at a higher temperature (Yeung et al. 1991). While the two apolipoproteins are required for chain formation of INLP proteoliposomes, PI is not (recall that Apos-I and -II plus PI are required for IN activity of liposomes). Likewise, the INLP that is removed from the hemolymph of the freeze-avoiding *Ceruchus piceus* larvae in autumn also has chain-forming behavior, while the hemolymph lipophorins of larval *Manduca sexta* caterpillars and cockroach, *Periplaneta americana*, both of which lack IN ability, do not form chains (Duman et al. 1995).

6.4.2 Recrystallization Inhibition Proteins

Some freeze-tolerant insects exhibit hemolymph TH, however, as is the case with most other TH-producing freeze-tolerant organisms (most plants, freeze-tolerant frogs, etc.), the level of TH in freeze-tolerant insects is generally very low, usually only 0.2–0.6 °C (Duman 1979, 2015; Duman et al. 1982, 2004; Ramløv 2000; Walters et al. 2009c). Consequently, the responsible IBPs either are present in very low concentrations or their TH-specific activities are much lower than those of AFP-type IBPs from freeze-avoiding species in which the AFPs function as true antifreezes to prevent organismal freezing. Obviously, as freeze-tolerant species have evolved to survive extracellular freezing and many have hemolymph INPs that inhibit supercooling, the presence of more active AFPs in the hemolymph would be counterproductive. The primary function of these low TH-producing RIPs in freeze-tolerant insects, and other freeze-tolerant organisms, appears to be inhibition of recrystallization and control of ice crystal structure in the hemolymph. Consequently, RIPs is a more accurate term for them than AFPs. Two exceptions to the generally low TH present in freeze-tolerant insects were *D. canadensis* and *C. clavipes* larvae in northern Indiana. While these larvae have been freeze avoiding since the early 1980s, they were freeze tolerant when their cold tolerance was initially investigated (Duman 1979, 1980), and at this time they had the same high levels of hemolymph TH as they had after becoming freeze avoiding (Horwath and Duman 1984a; Duman 1984a).

Recrystallization is the process whereby, after initial freezing, larger ice crystals in the sample grow even larger at the expense of smaller crystals, such that after an annealing period there are fewer, but larger, crystals, even though the total volume of ice in the system has not changed. Recrystallization occurs because smaller crystals have a higher radius of curvature, and therefore a higher surface free energy, than larger crystals with a lower radius of curvature and lower surface free energy. Consequently, there is a net flux of water molecules onto larger crystals at the expense of the smaller ones (Knight et al. 1984, 1995; Knight and Duman 1986). Recrystallization can be disruptive to tissue, especially to the extracellular matrix, etc. (Mazur 1984). Consequently, the RI activity of insect IBPs (Knight et al. 1984, 1995; Knight and Duman 1986) provides protection from damage due to RI (Tursman and Duman 1995).

All TH-producing IBPs that have been investigated prevent recrystallization (Duman 2015; Olive et al. 2016), presumably by binding to the surface of ice and preventing the movement of water molecules on and off the ice surface, in the same fashion that they produce TH (Knight and Duman 1986). The equilibrium temperature difference between two ice crystals with different radii of curvature (and therefore a difference in surface free energies) is considerably smaller than the force driving seed crystal growth below the eqMP. Consequently, low-TH IBPs have RI activity, even at concentrations orders of magnitude lower than required for detection of TH. However, there is little correlation between the magnitude of the TH and RI activities produced by a given AFP or RIP, presumably resulting from differences in the crystal face to which the AFP or RIP preferentially binds (Olive et al. 2016).

While several freeze-tolerant insects exhibit low levels of hemolymph TH, to date proteins that produce this TH have not been purified and characterized from any freeze-tolerant insects. Consequently, we cannot comment on whether the low level of TH measured in these species results from the presence of IBPs with low TH-specific activities or from low concentrations of TH-producing IBPs with normal (higher) specific activities. Also, there is a possibility that the low TH found in these species is due to the presence of low levels of AFGLs in their hemolymph as is the case with *Upis ceramboides* (Walters et al. 2009c) and certain other freeze-tolerant organisms, including insects (Walters et al. 2011). Purification and determination of the specific TH activity of RIPs of freeze-tolerant insects are necessary to resolve these questions. In addition, it is possible that freeze-tolerant insects with low hemolymph TH may have cell membrane or intracellular RIPs that inhibit potentially lethal intracellular ice formation. At this time, to our knowledge, intracellular TH has not been demonstrated, much less quantified, in freeze-tolerant insects.

6.4.3 Antifreeze Glycolipids

AFGLs with TH and RI activities were initially characterized in a freeze-tolerant Alaskan Tenebrionid beetle, *Upis ceramboides* (Walters et al. 2009c), and most of the animals (other insects, frogs) and a plant later shown to have AFGLs are also freeze tolerant (Walters et al. 2011; Larsen et al. 2014). *U. ceramboides* winter as adults, and in interior Alaska are freeze tolerant to −60 °C (Miller 1982). Prior to the discovery of the AFGLs only proteins were known to produce TH. Approximately half of the mass of the AFGL is composed of repeats of a disaccharide composed of xylose residues covalently linked to mannose with a β-1-4 linkage [β-mannopyranosyl-(1→4) β-xylopyranosyl] (Fig. 6.4). The disaccharides are attached to lipid, mainly fatty acids. Additional structural studies verified this

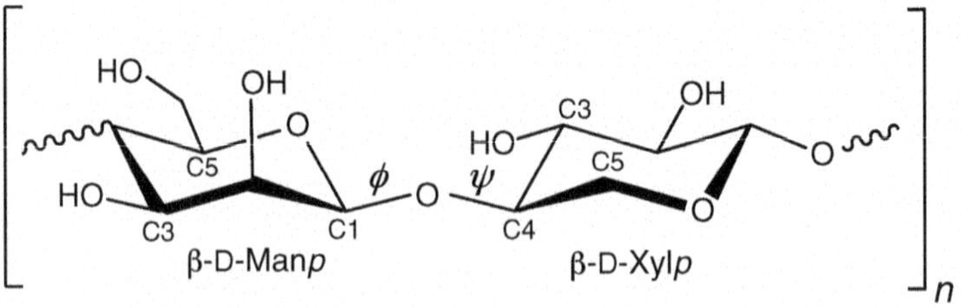

Fig. 6.4 The repeating disaccharide core structure of an antifreeze glycolipid. The β-D-Manp-(1→4)-β-D-Xylp (βManp-βXylp) of the xylomannan AFGL initially described in the Alaskan beetle *Upis ceramboides*. The saccharide, which constitutes approximately half the mass of the AFGL, is attached to lipid. Figure based on Walters et al. (2009c)

structure of the *U. ceramboides* AFGL (Crich and Rahaman 2011; Ishiwata et al. 2011; Zhang et al. 2012, 2013).

Identical and/or very similar xylomannan-based AFGLs are present in other insects, two frogs, and a plant (Walters et al. 2011; Larson et al. 2014). The ^{1}H NMR used to initially identify the nature of these AFGLs indicated minor differences in saccharide composition among some of these organisms. All these species, except some of the insects, are freeze tolerant, and in all cases the AFGL was only present in winter. An interesting freeze-tolerant species with an additional saccharide in its AFGL are nymphs of the Alaskan stonefly *Nemoura arctica*. Freshwater organisms generally are not subjected to subzero temperatures, but these nymphs were collected from the small headwaters of the Chandalar River on the south side of the Brooks Range in Arctic Alaska where the river freezes solid, top to bottom, in winter. The stream began to freeze in late September, was frozen to the bottom within 3 weeks, and temperatures at the bottom reached −13 °C in December. Nymphs collected in autumn froze when in contact with ice at −1.5 °C, but survived −15 °C (Walters et al. 2009c). The hemolymph of a few individuals collected in the autumn had measurable TH up to 1 °C, but most exhibited only RI activity. Although the original report suggested this activity was due to AFPs (Walters et al. 2009c), the TH and RI activity were later shown to result from AFGLs (Walters et al. 2011).

While purified AFGLs produce TH with specific activity equal to that of highly active insect AFPs (3.7 °C at 5 mg/ml), much of the time the insects and other organisms with AFGLs have low in vivo hemolymph TH activity, or in some situations none at all. Most individual *U. ceramboides* adults collected in the autumn, lack measurable hemolymph TH, however, a few individuals have hemolymph TH (0.4–1.4 °C) for short periods at this time. Also, cold acclimated individuals in autumn produced hemolymph TH of 0.4 °C (Duman et al. 2004). Possibly at the time of sampling these individuals the AFGL was actively being synthesized (perhaps in the fat body) and was being transported in the hemolymph to the tissues at concentrations sufficient to produce measurable TH. Winter collected individuals did not have hemolymph TH, although hemolymph RI activity was present at this time. While most of the *U. ceramboides* AFGLs are present on cell membranes, where they are positioned to inhibit lethal inoculation of the cytoplasm by extracellular ice, low levels of AFGLs are also present in the hemolymph, and they, like AFPs, have recrystallization inhibition activity. It also is possible that AFGLs may be present in the cytoplasm where they could function to prevent freezing, although this has not been demonstrated.

6.4.4 Other Freeze Tolerance Adaptations

Beyond INPs, RIPs, and AFGLs numerous additional adaptations are required to achieve freeze tolerance (Lee and Denlinger 1991; Denlinger and Lee 2010), as suggested by various "–omics" studies (i.e., Michaud and Denlinger 2010).

High polyol concentrations were the first adaptations associated with freeze tolerance (e.g., Salt 1958). Glycerol, sorbitol, mannitol, etc. are often present in molar concentrations in freeze-tolerant species, and these high levels are further magnified in the unfrozen portion of the body fluids as the hemolymph freezes and the percentage of the extracellular fluid that remains unfrozen decreases as temperature decreases (Sømme 1982; Zachariassen 1985; Lee 2010; Storey and Storey 1991). These increasing polyol, etc. concentrations increasingly lower the freezing temperature of the unfrozen fraction, acting to keep more body water, including intracellular water, unfrozen (Zachariassen 1985; Lee 2010).

These solutes undoubtedly serve additional functions, as they are likely important protections against low temperature and cell volume reduction, both potentially lethal problems for freeze-tolerant organisms (Mazur 1984). Not only polyols, but also sugars (especially trehalose) and certain amino acids (especially proline) are well known for their abilities to protect macromolecules, including membranes, from temperature extremes (see Somero et al. 2017 for a review). Also, as freezing progresses and water activity decreases, especially in the cytoplasm resulting from the outflux of intracellular water, glycerol, and other polyols become increasingly important because their preferential exclusion from the surface of proteins, etc. produces preferential hydration of proteins, membranes, etc., thereby maintaining the conformations of the macromolecules (Bolen 2004).

Proline accumulates during winter or cold acclimation in some freeze-tolerant species, including the alpine weta, *Hemideina maori*, in New Zealand (Ramløv 1999) and the Drosophilid fly *Chymomyza costata* (Rozsypal et al. 2018). Surprisingly, *Drosophila melanogaster*, a normally freeze susceptible species that cannot naturally overwinter in subzero environments, exhibits some freeze tolerance when provided with a proline augmented diet that raises proline levels of the body fluids (Kostal et al. 2012, 2016a). The specific role of proline in freeze tolerance in these cases is likely to be multifunctional (maintenance of macromolecular structure, etc.), but convincing evidence that the ability of freeze-tolerant *C. costata* to vitrify the unfrozen fraction of their body water at temperatures below −30 °C is related to elevated proline concentrations was provided recently by Kostal and collaborators (Rozsypal et al. 2018).

The importance of aquaporins in freeze tolerance of numerous organisms, including insects, has been demonstrated (Izumi et al. 2006; Philip et al. 2008; Campbell et al. 2008). In addition to allowing more rapid water flux across membranes, thereby equilibrating osmotic pressures, some aquaporins also facilitate glycerol flux. It would be interesting to investigate the aquaporins of freeze-tolerant species with widely varying supercooling points such as the two *Pytho* species, *P. deplanatus* that supercools to −56 °C before freezing and *P. americanus* with a supercooling point of just −6 °C (Ring 1982), as the rates of water flux across the cell membranes after freezing should be extremely different.

6.5 How Common Are Ice-Binding Proteins in Insects?

This is a difficult question to answer, for a number of reasons. Most insect species do not inhabit regions where freezing is a problem, so AFPs or other IBPs are not necessary in these species. Therefore, a more appropriate question is how common are IBPs in insects from areas where subzero temperatures are experienced. The answer is likely to vary with the frequency and magnitude of low-temperature periods experienced by the local insects. Also, there are a number of mechanisms for overcoming the freezing problem, and while many insects simultaneously utilize multiple mechanisms, many do not employ AFPs or other IBPs. The most common means of screening for the presence of AFPs and RIPs is to sample body fluids (usually hemolymph), and measure the melting and freezing points (the temperature (s) where the last small crystal in a frozen sample disappears and where the small crystal begins to grow, respectively) to determine the presence or absence of TH. If the MP and FP are identical the sample does not have AFPs, or RIPs. However, many insects are quite small and consequently difficult to sample, and are therefore avoided by most researchers. Also, the presence of TH in the sample does not mean that it contains AFPs or RIPs, as the TH may be due to AFGLs. In addition, there are instances where the body fluids did not have TH, but when the insect was homogenized TH was then measurable in the homogenate, suggesting that the molecules responsible for the TH are in the cuticle, membranes, and/or are intracellular. This may especially be the case when antifreeze glycolipids are responsible for the TH (Walters et al. 2009b, 2011).

With these caveats in mind, a study of insects in interior Alaska, where very low temperatures are normal, usually for prolonged periods, identified TH in the hemolymph of 18 of 75 (26%) species of insects tested (Duman et al. 2004). These included both freeze-tolerant and freeze-avoiding species. To our knowledge, this is the only published report on the frequency of TH-producing IBPs in insects. However, the TH in one of these species, the freeze-tolerant beetle *Upis ceramboides*, was later shown not to result from RIPs, but to AFGLs (Walters et al. 2009b), while other species had both AFPs/RIPs and AFGLs (Walters et al. 2011).

Based on the abovementioned study (Duman et al. 2004), it appears that while AFPs and RIPs are not universally present in insects from subzero environments they are fairly common. Thermal hysteresis has been reported in species representing numerous beetle (Coleoptera) families (including Tenebrionidae, Elateridae, Cucujidae, Pyrochroidae, Lampyridae, Coccinellidae, Scolytidae, Cerambycidae), as well as at least a few species of Plecoptera (stoneflies), Orthoptera (cockroaches, etc.), Hemiptera (true bugs), Mecoptera (scorpionflies), Lepidoptera (moths and butterflies), and Diptera (flies, midges, etc.). For a review see Duman (2001) and Duman (2015). It is interesting that, as far as the authors are aware, TH has not been reported in any Hymenoptera (bees, wasps, and ants), one of the major insect families, and one that is well represented in cold environments.

It is also difficult to gauge the prevalence of adaptive INPs in freeze-tolerant species. Ice-nucleating activity has been reported in many freeze-tolerant insects (Zachariassen 1982, 1985; Ramløv 2000; Duman et al. 2010; Zachariassen et al. 2011), and consequently benefit from the apparent advantages to freeze-tolerant insects provided by freezing at high temperatures resulting from extracellular INPs. Therefore, it is perhaps surprising to find that a number of freeze-tolerant insects lack effective INPs and some of these survive after initial freezing at extremely low temperatures. Overwintering pupae of several giant silkworm species such as *Antheraea polyphemus* undercool by approximately 20 °C prior to freezing (Duman et al. 1991b), as do several interior Alaskan insects (Miller 1982). The wheat stem sawfly *Bracon cephi* supercools about 30 °C below its −15 °C hemolymph MP (resulting from extreme polyol concentrations), yet survives this freezing (Salt 1959). The apparent champion in this regard is *Pytho deplanatus*, a beetle from the Canadian Rockies that supercools to −54 °C before freezing and surviving (Ring 1982). Interestingly, its freeze tolerant congener *Pytho americanus*, also from the northern Rockies, supercools only to −6 °C (Ring 1982).

In contrast to the freeze-tolerant insects with very low supercooling points, there are species that lack sufficiently effective INPs and die if not frozen inoculatively at high temperature by external ice in contact with the cuticle (Tanno 1977; Fields and McNeil 1986; Shimada and Riihamaa 1988; Gehrken et al. 1991). For example, if held in dry conditions the overwintering caterpillars of the moth *Cisseps fulvicollis* supercool to −12 to −17 °C before freezing, but do not survive freezing at these temperatures. However, they survive freezing if held in wet conditions where freezing is initiated by external ice at −4 °C or higher (Fields and McNeil 1986). Likewise, *Bolitophagus reticulatus* beetles in Norway freeze and die at −30 °C if kept dry, but freeze and survive at −6 °C if in contact with external ice (Gehrken et al. 1991). Linden bugs, *Pyrrhocornis apterus*, have been considered freeze avoiding as they supercool to −16 to −20 °C. However, in their normal wintering sites in the leaf litter, they are often in contact with ice, and therefore subject to inoculative freezing. When wetted and frozen they freeze inoculatively above −3 °C, but survive (Rozsypal and Košťál 2018). Consequently, it is best to determine supercooling points of insects, and their survival after freezing, in both wet conditions that subject them to extracellular ice, and dry conditions where inoculative freezing is much less likely. However, even supposed dry conditions can lead to inconsistent and misleading results. Measurements on days when relative humidity is low may produce different results than the same measurements taken on a high humidity day, as the latter may result in condensation on the cuticle leading to inoculative freezing (Tursman et al. 1994). In most cases in nature, the overwintering insect is in contact with surface ice and therefore potentially subject to inoculative freezing.

Another potential problem related to determining the presence of adaptive INPs is related to the timing and previous history of the population. Although the sub-Antarctic beetle *Hydromedion sparsutum* was freeze tolerant and initially froze at a high temperature, after thawing and subsequent refreezing the supercooling point decreased and the insects were no longer freeze tolerant (Bale et al. 2000). Likewise, wintering larvae of the hoverfly *Syrphus ribesii* initially froze and survived at approximately −7.6 °C, but subsequent thawing and refreezing of the same individuals lowered the supercooling points to as low as −28 °C, but they did not survive freezing at these lower temperatures (Brown et al. 2004).

An additional potential problem with determining the presence of adaptive hemolymph INPs is that other ice nucleators may be operative, and variations may result from population differences. While some populations of the well-studied gallfly *Eurosta solidaginis* apparently have hemolymph INPs (Sømme 1978; Lee et al. 1981; Zachariassen et al. 1982), high supercooling points in a New York population were reported to result from frass in the gall initiating nucleation on the surface of the larvae (Bale et al. 1989). Another *E. solidaginis* study reported that the hemolymph lacked effective INPs and froze at −18 °C, but that freezing was initiated at much higher temperatures by calcium phosphate crystals in the Malpighian tubules (Lee et al. 1992). Similarly, while the supercooling point of the New Zealand weta *Hemideina maori* was −4 °C (Ramløv et al. 1992), the hemolymph supercooling point was (−7.5 °C) initiated by hemolymph proteins (Wilson and Ramlov 1995).

These supercooling variations once again demonstrate the tremendous diversity in the physiology of insect cold tolerance. Many freeze-tolerant species have hemolymph INPs that severely limit supercooling, while others lack efficient INPs and consequently supercool, sometimes to extremes, prior to freezing, and others rely on inoculative freezing. Why such diversity in this regard? Zachariassen's idea that species that have more permeable cuticles evolved freeze tolerance and associated adaptive INPs to ensure freezing at high temperatures to thereby minimize water balance problems may be correct in many instances (Zachariassen 1992; Zachariassen et al. 2008, 2011). Also, some insect species may have had hemolymph proteins with limited ice-nucleating abilities due to some surface characteristics required for their original functions. The functions of these incidental INPs were perhaps essential to the insects, and therefore could not be removed, either on a seasonal or evolutionary time scale. Therefore, these incidental INPs limited the supercooling abilities of the insect, necessitating masking by antifreezes, if the insect was to be freeze avoiding. However, an alternative evolutionary strategy was for the insect to become freeze tolerant, and select for improved ice-nucleating activity in these incidental INPs resulting in the proteins becoming adaptive INPs that initiated freezing at higher temperatures more easily tolerated by the insect.

6.6 Functions of IBPs Other than Those Involved in Control of Freezing

As mentioned previously, many insects exhibiting hemolymph TH in winter have lesser magnitude TH in summer (i.e., Duman 1977b, 1980, 2015; Nickell et al. 2013). This was initially thought to result from long half-lives of the AFPs produced during cold periods. However, at least in some species, this is not always the case. Larvae of *D. canadensis* typically have hemolymph TH of 0.2–0.4 °C in the summer and continue production of mRNA encoding certain DAFP isomers throughout the summer months (Andorfer and Duman 2000). Likewise, *T. molitor* larvae have hemolymph TH (Patterson and Duman 1978) and continue to transcribe messenger RNA for some AFPs even when warm acclimated (Graham et al. 2000). This suggests that these AFPs may have functions beyond protection from freezing, and these potential functions may be critical during the entire year. While multiple function IBPs are not generally considered in insects, dual function IBPs are the rule in plants (Griffith and Yaish 2004). The following discussion concerns the possibility of multiple function IBPs in insects.

6.6.1 Inhibition of Non-ice Crystal Formation and/or Growth

D. canadensis DAFPs inhibit ice nucleation and growth of non-ice crystals known to form in the primary urine of overwintering insects (Nickell et al. 2013). This ability may be due to adsorption of DAFPs to embryo ice crystals formed on the surface of the non-ice crystal and/or by adsorption of the DAFPs directly onto the surface of the non-ice crystal. Also, DAFPs can bind to and inhibit the growth of certain nucleoside (Wang et al. 2012) and carbohydrate crystals; methyl-alpha-D-mannopyranoside (Wang et al. 2014) and trehalose (Wen et al. 2016). Since non-ice crystal formation and growth in body fluids is a potential hazard, this ability of DAFPs, and potentially other AFPs, could be of use beyond prevention of freezing.

Non-ice crystal growth inhibition may be of special benefit at low temperatures where solute solubility is decreased. This is the case with trehalose, the main hemolymph sugar of insects and a solute that often increases in concentration in winter due to its well-known abilities to protect macromolecules, membranes, etc. and in some cases to promote supercooling slightly (Wen et al. 2016). However, trehalose solubility decreases considerably at low temperature. Trehalose concentrations in *D. canadensis* hemolymph increased from 0.3 mM in summer to 90 mM in winter. When an aqueous solution of trehalose at this winter concentration was held at −5 °C for 2 h trehalose crystals formed and grew, however, winter hemolymph containing both trehalose and DAFPs at −5 or −15 °C failed to form crystals. Likewise, an aqueous 90 mM trehalose solution containing just one of the DAFPs (DAFP-1) did not form crystals at −15 °C (Wen et al. 2016). Therefore, a function of DAFPs is to prevent crystallization and growth of not only ice crystals, but also certain non-ice crystals (trehalose in hemolymph, crystals in urine) as well.

6.6.2 High Temperature Tolerance

Another possible DAFP function may be to provide protection at critically *high* environmental temperatures, both in winter and in summer (Vu and Duman 2017). This possibility was initially discovered while attempting to determine the upper lethal temperature (ULT) of *D. canadensis* larvae in winter. The study was undertaken to determine if the excessively warm temperatures that occurred during the recent record setting winter high temperatures (winter thaws, earlier springs, later winters) might be lethal. Because mean lower lethal temperatures (LLTs) of the larvae decrease from approximately 0 °C in summer to between −18 and −28 °C in winter, the winter ULTs were expected to decrease in a similar fashion. However, paradoxically the ULTs of *D. canadensis* larvae, and those of two other insects (*Tipula trivittata* and *Ceruchus piceus*), were actually higher in winter than in summer. Larval *D. canadensis* mean 24-h ULTs were 40.9 °C in winter and 36.7 °C in summer, producing a 64 °C temperature range in winter, but just a 41 °C range in summer. A number of critical physiological functions (i.e., membrane function) are likely to be negatively impacted by excessive temperatures, both high and low, and therefore multiple adaptations are potentially responsible for increased high temperature survivorship. However, the known winter adaptations to low temperature are reasonable starting points. All of the three species tested in this study have AFGLs and increased polyols in winter, while only *D. canadensis* produces AFPs. As discussed previously, polyols (glycerol and sorbitol), trehalose and proline are well known for their abilities to stabilize membrane, protein, and other macromolecular structures at both high and low temperatures (Bolen 2004; Somero et al. 2017). These microsolutes are high in winter larvae, but not summer, and therefore they may potentially be at least partially responsible for the increased winter ULTs.

AFGLs are mainly associated with membranes (Walters et al. 2009c, 2011). Might they stabilize membranes at temperature extremes? What about AFPs? While certain DAFPs continue to be expressed in summer, the number of isomers and the magnitude of DAFP concentrations are much greater in winter. Certain fish AFPs are known to stabilize membranes at low temperatures (Tomczak et al. 2002), but similar high temperature stabilization has not been investigated for any AFPs. *Microdera punctipennis*, a desert beetle from China produces AFPs similar to those of *D. canadensis*. Like *D. canadensis*, *M. punctipennis* has greater hemolymph TH in winter than summer, but summer hemolymph retains some TH and certain AFP isomers are produced only in summer (Qiu et al. 2013). When expressed in yeast or bacteria, a fusion protein of one of the two summer isomers provided increased heat tolerance in the microorganisms. Also, the protein increased lactate dehydrogenase stability at high temperatures. High temperature laboratory acclimations induced AFPs in *T. molitor* as well as *M. punctipennis* (Li et al. 2016). Also, transgenic *Drosophila melanogaster* expressing *D. canadensis* DAFP-1 or DAFP-4 isomers exhibited statistically significant increases in ULTs (Vu et al. 2019).

6.6.3 Desiccation Resistance/Water Balance

Tenebrio molitor larvae are well known for their ability to reabsorb water across the rectal epithelium using their cryptonephridial rectal complex, even to the point where they can remove water vapor from unsaturated air (down to 70% relative humidity at room temperature) taken into the rectum (Ramsay 1964). TH was greatest in the perirectal space, the region beyond the rectal epithelium into which water is reabsorbed prior to moving into the cryptonephridial Malpighian tubules (Ramsay 1964), and a later study suggested that the proteins producing the TH were at least partially responsible for the unusual water reabsorption ability (Grimstone et al. 1968). As mentioned earlier, low relative humidity induced an increase in hemolymph TH in *T. molitor*, and these larvae when subjected to low RH did not desiccate (Patterson and Duman 1978). Also, larvae that had been acclimated to either low temperature (5 °C) or short photoperiod, so as to induce increased hemolymph TH, prior to exposure to 15% RH at 22 °C, increased their time to 50% mortality to 30+ days (when the experiment was terminated), compared to 8 days to 50% mortality for unacclimated, low hemolymph TH, larvae (20 °C, 12 L/12D photoperiod, 90% RH). While these results certainly do not prove that the AFPs enhanced desiccation resistance, they strongly suggest that further study of this possibility is warranted.

6.7 Conclusions

Considerable progress has been made over the past several decades on the basic functions and biochemistry of IBPs of insects and other organisms. Basic functions of IBPs are reasonably well understood, but the biology and biochemistry of insect ice-binding proteins are far from properly understood, and there is much more to learn, especially about RIPs and IBPs. Structures of many insect AFPs are known, but there is minimal understanding of the biochemistry of RIPs or IBPs. While each new study provides clarification of certain concepts, more questions are often generated than answers provided. While this is true of IBP studies of most organisms, it is especially so for insects as they are the most abundant, evolutionarily diverse and successful of all animals. Only a small percentage of species inhabiting regions where subzero temperatures occur have been investigated, ensuring that numerous important findings are still to be made. Some suggestions for future studies follow.

At times some individuals argue against new investigations directed at "just another species" that will simply show more of the same, "stamp collecting" in their terms. However, once again, insects are so numerous and physiologically diverse that judicious searches for new and interesting species will certainly be rewarded. Understanding of insect taxonomy and evolution, plus natural history, will readily direct attention to groups that have received little, or no, attention as regard to cold tolerance. At the same time, species that have already received considerable attention to date, both freeze tolerant and freeze avoiding, should be investigated further, using new techniques and current understanding to better understand the interrelationships of their suite of numerous cold adaptations and how IBPs are incorporated. Therefore, it is important to continue to investigate other species (both freeze tolerant and freeze avoiding) for TH. Emphasis should be placed on insects that can be both studied in the field under natural conditions and reared in the laboratory.

The exact physiological function(s) of IBPs, especially RIPs, along with the magnitude of their roles in the overall cold tolerance of insects, requires further study. This is true even for the AFPs of freeze-avoiding species, in spite of their rather well-understood abilities to inhibit inoculative freezing and to promote supercooling by masking ice nucleators, as these effects have been investigated in only a small number of species. In particular, the importance of AFPs relative to other freeze-avoiding adaptations (polyols, removal of ice nucleators in winter, etc.) is virtually unknown. Certain molecular techniques such as RNA interference (RNAi) technology (Fire et al. 1998) and CRISPR (Pennisi 2013), etc. provide hope for answering these questions, but application of these techniques to AFP questions is not as straightforward as it may seem. For example, RNAi was used to inhibit AFP production in Alaskan *Cucujus clavipes* larvae during mid- to late August when AFP production begins (Sformo, Duman and Barnes, unpublished data). Although this inhibition raised larval SCPs for several days, the results were not conclusive, probably for multiple reasons. In particular, RNAi only limits transcript production for a few days, and AFP synthesis normally continues for a much longer period, through the entire winter in some species such as *D. canadensis*. Consequently, multiple treatments are required. Also, the timing of the initiation of AFP synthesis in the autumn typically does not overlap with the onset of other freeze-avoiding adaptations. For example, while AFP production is generally initiated well before actual subzero temperatures, polyol production typically does not take place until subzero temperatures occur, making comparisons of the relative values of AFP and polyol accumulation difficult at best. Also, the synergistic effects of enhancers such as polyols on AFP activity must be considered. CRISPR may be a promising technique to apply to these questions, but the multiple genes encoding AFPs and the tissue-specific expression of AFP isomers make this technique problematic as well. In particular, the function(s) of RIPs in freeze-tolerant insects is mostly conjecture. Is recrystallization inhibition in the hemolymph really their only function? Might they also be located on the cell membrane or in the cytoplasm where they could function to inhibit cytoplasmic freezing?

Determination of if and how AFPs or other IBPs have functions not related to subzero tolerance, such as desiccation or high temperature resistance, should also be amenable to application of certain molecular techniques, especially as only certain of the multiple AFP isomers are produced in summer.

The structures of only a few AFPs are known, and most of these are beetles. Determinations of structures of AFPs from other insect groups should be informative, providing important information concerning their mechanism of action, evolution, etc. Even less is known about insect RIPs and PINs as none of these have even been sequenced. This should be a priority.

Also, little is known about AFGLs, in insects or other organisms, although they appear to be common in both freeze-tolerant and freeze-avoiding species. Their function and structures should be elucidated, but these studies are not simple. Some AFGL-producing organisms have both AFPs/RIPs and AFGLs, compounding functional studies and structural understanding as the AFPs/RIPs must be separated from the AFGLs to achieve these goals (Walters et al. 2011).

These future studies of insect IBPs will not only provide important new information of import toward better understanding of the biology, physiology, biochemistry, and evolution of insect cold tolerance, but these IBPs have the potential to be of value in applications in numerous wide ranging fields: cryopreservation of biomaterials, agriculture, cryo-engineering, etc.

References

Amornwittawat N, Wang S, Duman JG, Wen X (2008) Polycarboxylates enhance beetle antifreeze protein activity. Biochim Biophys Acta 1784:1942–1948

Amornwittawat N, Wang S, Banatlao J, Chung M, Velasco E, Duman JG, Wen X (2009) Effects of polyhydroxy compounds on beetle antifreeze protein activity. Biochim Biophys Acta 1794:341–346

Andorfer CA, Duman JG (2000) Isolation and characterization of cDNA clones encoding antifreeze proteins of the Pyrochroid beetle *Dendroides canadensis*. J Insect Physiol 46:365–372

Bale JS, Hansen TN, Baust JG (1989) Nucleators and sites of nucleation in the freeze tolerant larvae of the gallfly *Eurosta solidaginis* (Fitch). J Insect Physiol 35:291–298

Bale JS, Worland MR, Block W (2000) Thermal tolerance and acclimation response of the subAntarctic beetle *Hydromediom sparsutam*. Polar Biol 23:77–84

Basu K, Graham LA, Campbell RL, Davies PL (2015) Flies expand the repertoire of protein structures that bind ice. Proc Natl Acad Sci USA 112:737–742

Baumer A, Duman JG, Havenith M (2016) Ice nucleation of an insect lipoprotein ice nucleator (LPIN) correlates with retardation of the hydrogen bond dynamics at the myo-inositol ring. Phys Chem Chem Phys 18:19318–19323

Baust JG, Zachariassen KE (1983) Seasonally active cell matrix associated ice nucleators in an insect. Cryo-Lett 4:65–71

Bennett VA, Sformo T, Walters K, Toien O, Jeannet K, Hochstrasser R, Pan Q, Serianni AS, Barnes BM, Duman JG (2005) Comparative overwintering physiology of Alaska and Indiana populations of the beetle *Cucujus clavipes* (Fabricus): roles of antifreeze proteins, polyols, dehydration, and diapause. J Exp Biol 208:4467–4477

Block W, Young SR (1979) Measurements of supercooling points in small arthropods and water droplets. CryoLetters 1:85–91

Bolen DW (2004) Effects of naturally occurring osmolytes on protein stability and solubility: issues important in protein crystallization. Methods 34:312–344

Bremdal S, Zachariassen KE (1988) Thermal hysteresis factors and supercooling of hibernating *Rhagium inquisitor* beetles. In: Sehnal F, Zabza A, Denlinger DL (eds) Proceeding from endocrinological frontiers in physiological insect physiology. Wroclaw Technical University Press, Wroclaw, pp 187–191

Brown CL, Bale JS, Walters KFA (2004) Freezing induces a loss of breeze tolerance in an overwintering insect. Proc R Soc Series B 271:1507–1511

Buch JL, Ramløv H (2017) Detecting seasonal variation of antifreeze protein distribution in *Rhagium mordax* using immunofluorescence and high resolution microscopy. Cryobiology 74:132–140

Campbell EM, Ball A, Hoppler S, Bowman A (2008) Invertebrate aquaporins: a review. J Comp Physiol B 178:935–955

Cannon RJC, Block W (1988) Cold tolerance in microarthropods. Biol Rev 63:23–77

Carrasco MA, Duman JG (2011) A cross-species compendium of proteins related to cold stress identified by bioinformatic approaches. J Insect Physiol 57:1127–1135

Carrasco MA, Buechler SA, Arnold RJ, Sformo T, Barnes BM, Duman JG (2011) Elucidating the biochemical overwintering adaptations of larval , a nonmodel organism, via high throughput proteomics. J Proteome Res 10(10):4634–4646

Carrasco MA, Buechler S, Arnold R, Sformo T, Barnes BM, Duman JG (2012) Investigating the deep supercooling ability of an Alaskan beetle, *Cucujus clavipes puniceus*, via high throughput proteomics. J Proteome 75:1220–1234

Celik Y, Graham LA, Mok Y-F, Bar M, Davies PL, Braslavsky I (2010) Superheating of ice crystals in antifreeze protein solutions. Proc Natl Acad Sci USA 107:5423–5428

Chown SL, Sinclair BJ (2010) The macrophysiology of insect cold tolerence. In: Denlinger DL, Lee RE (eds) Low temperature biology of insects. Cambridge University Press, Cambridge, pp 191–222

Chown SL, Terblanche JS (2007) Physiological diversity in insects: evolutionary and and ecological contexts. Adv Insect Physiol 33:50–152

Crich D, Rahaman MY (2011) Synthesis and verification of the xylomannan antifreeze substance from the freeze-tolerant Alaskan beetle Upis ceramboides. J Org Chem 76:8611–8620

Daley ME, Spyracopoulos L, Jia Z, Davies PL, Sykes BD (2002) Structure and dynamics of a β-helical antifreeze protein. Biochemist 4:5515–5525

Danks HV (1991) Winter habitats and ecological adaptations for winter survival. In: Lee RE, Denlinger DL (eds) Insects at low temperature. Chapman and Hall, London, pp 231–259

Davies PL (2014) Ice-binding proteins: a remarkable diversity of structures for stopping and starting ice growth. Trends Biochem Sci 39:548–555

Denlinger DL, Lee RE (eds). Low temperature biology of insects. Cambridge: Cambridge University Press; 2010. 390 pp.

DeVries AL (1971) Glycoproteins as biological antifreeze agents in Antarctic fishes. Science 172:1152–1155

DeVries AL (1986) Antifreeze glycopeptides and peptides: interactions with ice and water. Meth Enzymol 127:293–303

DeVries AL, Cheng C-HC (1992) The role of antifreeze glycopeptides and peptides in the survival of cold water fishes. In: Somero GN, Osmond CB, Bolis CL (eds) Water and life: comparative analysis of water relationships at the organismic, cellular, and molecular levels. Springer, Berlin, pp 303–315

DeVries AL, Cheng C-HC (2005) Antifreeze proteins in polar fishes. In: Farrell AP, Steffensen JF (eds) Fish physiology, vol XXII. Academic, San Diego, pp 155–201

DeVries AL, Lin Y (1977) Structure of a peptide antifreeze and mechanism of adsorption to ice. Biochem Biophys Acta 495:388–392

DeVries AL, Wohlschlag C (1969) Freezing resistance in some Antarctic fishes. Science 163:1073–1075

DeVries AL, Komatsu SK, Feeney RE (1970) Chemical and physical properties of freezing point-depressing glycoproteins from Antarctic fishes. J Biol Chem 245:2901–2908

DeVries AL, Vandenheede J, Feeney RE (1971) Primary structure of freezing point depressing glycoproteins. J Biol Chem 246:305–308

Doucet D, Tyshenko MG, Davies PL, Walker VK (2002) A family of expressed antifreeze protein genes from the moth, *Choristoneura fumiferana*. Eur J Biochem 269:38–46

Duman JG (1977a) The role of macromolecular antifreeze in the darkling beetle *Meracantha contracta*. J Comp Physiol B 115:279–286

Duman JG (1977b) Variations in macromolecular antifreeze levels in larvae of the darkling beetle *Meracantha contracta*. J Exp Zool 201:85–93
Duman JG (1977c) The effects of temperature, photoperiod, and relative humidity on antifreeze production in larvae of the darkling beetle, *Meracantha contracta*. J Exp Zool 201:333–337
Duman JG (1979) Thermal hysteresis factors in overwintering insects. J Insect Physiol 25:805–810
Duman JG (1980) Factors involved in the overwintering survival of the freeze tolerant beetle, *Dendroides canadensis*. J Comp Physiol B 136:53–59
Duman JG (1984a) Change in overwintering mechanism in the Cucujid beetle, *Cucujus clavipes*. J Insect Physiol 30:235–239
Duman JG (1984b) Thermal hysteresis antifreeze proteins in the midgut fluid of overwintering larvae of the beetle *Dendroides canadensis*. J Exp Zool 230:355–361
Duman JG (2001) Antifreeze and ice nucleator proteins in terrestrial arthropods. Annu Rev Physiol 63:327–357
Duman JG (2002) The inhibition of ice nucleators by insect antifreeze proteins is enhanced by glycerol and citrate. J Comp Physiol B 172:163–168
Duman JG (2015) Animal ice-binding (antifreeze) proteins and glycolipids: an overview with emphasis on physiological function. J Exp Biol 218:1846–1855
Duman JG, DeVries AL (1972) Freezing behavior of aqueous solutions of glycoproteins from the blood of an Antarctic fish. Cryobiology 9:469–472
Duman JG, Horwath KL (1983) The role of hemolymph proteins in the cold tolerance of terrestrial arthropods. Annu Rev Physiol 45:261–270
Duman JG, Serianni AS (2002) The role of endogenous antifreeze protein enhancers in the hemolymph thermal hysteresis activity of the beetle *Dendroides canadensis*. J Insect Physiol 48:103–111
Duman JG, Horwath KL, Tomchaney AP, Patterson JL (1982) Antifreeze agents of terrestrial arthropods. Comp Biochem Physiol 73A:545–555
Duman JG, Morris JP, Castellino FJ (1984) Purification and composition of an ice nucleating protein from queens of the hornet, *Vespula maculata*. J Comp Physiol B 154:79–83
Duman JG, Neven LG, Beals JM, Olson KR, Castellino FJ (1985) Freeze tolerance adaptations, including haemolymph protein and lipoprotein ice nucleators, in larvae of the cranefly *Tipula trivittata*. J Insect Physiol 31:1–9
Duman JG, Wu DW, Wolber PK, Mueller GM, Neven LG (1991a) Further characterization of the lipoprotein ice nucleator from freeze tolerant larvae of the cranefly *Tipula trivittata*. Comp Biochem Physiol B 99:599–607
Duman JG, Xu L, Neven LG, Tursman D, Wu DW (1991b) Hemolymph proteins involved in insect subzero temperature tolerance: ice nucleators and antifreeze proteins. In: Lee RE, Denlinger DL (eds) Insects at low temperatures. Chapman and Hall, London, pp 94–127
Duman JG, Wu DW, Yeung KL, Wolf EE (1992) Hemolymph proteins involved in the cold tolerance of terrestrial arthropods: antifreeze and ice nucleator proteins. In: Somero GN, Osmond CB (eds) Water and life. Springer, Berlin, pp 282–300
Duman JG, Wu DW, Olsen TM, Urrutia M, Tursman D (1993) Thermal hysteresis proteins. Adv Low Temp Biol 2:131–182
Duman JG, Olsen TM, Yeung KL, Jerva F (1995) The roles of ice nucleators in cold tolerant invertebrates. In: Lee RE, Warren GJ, Gusta LV (eds) Biological ice nucleation and its application. APS, St. Paul, pp 201–219
Duman JG, Parmalee D, Goetz FW, Li N, Wu DW, Benjamin T (1998) Molecular characterization and sequencing of antifreeze proteins from larvae of the beetle *Dendroides canadensis*. J Comp Physiol B 168:225–232
Duman JG, Verleye D, Li N (2002) Site specific forms of antifreeze proteins in the beetle *Dendroides canadensis*. J Comp Physiol B 172:547–552
Duman JG, Bennett V, Sformo T, Hochstrasser R, Barnes BM (2004) Antifreeze proteins in Alaskan insects and spiders. J Insect Physiol 50:259–266

Duman JG, Walters KR, Sformo T, Carrasco MA, Nickell P, Barnes BM (2010) Antifreeze and ice nucleator proteins. In: Denlinger DL, Lee RE (eds) Low temperature biology of insects. Cambridge University Press, Cambridge, pp 59–90

Ebbinghaus S, Meister K, Born B, DeVries AL, Gruebele M, Havenith M (2010) Antifreeze glycoprotein activity correlates with long-range protein–water dynamics. J Am Chem Soc 132(35):12210–12211

Ebbinghaus S, Meister K, Prigozhin MB, DeVries AL, Havenith M, Dzubiella J, Gruebele M (2012) Functional importance of short-range binding and long-range solvent interactions in helical antifreeze peptides. Biophys J 103(2):L20–L22

Elnitsky MA, Haywood SAL, Rinehart JP, Denlinger DA Lee RE (2008) Cryoprotective dehydration and resistance to inoculative freezing in the Antarctic midge *Belgica Antarctica*. J Exp Biol 211:524–530

Fall R, Wolber PK (1995) Biochemistry of bacterial ice nuclei. In: Lee RE, Warren GJ, Gusta LV (eds) Biological ice nucleation and its application. APS, St. Paul, pp 63–83

Fields PG, McNeil JN (1986) Possible duel cold-hardiness strategies in *Cisseps fulvicollis* (Lepidoptera: Arctiidae). Can Entomol 118:1309–1311

Fire A, Xu S, Montgomery MK, Kostas SA, Driver SE, Mello CC (1998) Potent and specific genetic interference by double-stranded RNA in *Caenorhabditis elegans*. Nature 391:806–811

Garnham CP, Campbell RL, Davies PL (2011a) Anchored clathrate waters bind antifreeze proteins to ice. Proc Natl Acad Sci USA 108:7363–7367

Garnham CP, Campbell RL, Walker VK, Davies PL (2011b) Novel dimeric beta-helical model of an ice nucleation protein with bridged active sites. BMC Struct Biol 11:36

Gauthier SY, Kay CM, Sykes BD, Walker VK, Davies PL (1998) Disulfide bond mapping and structural characterization of spruce budworm antifreeze protein. Eur J Biochem 258:445–453

Gehrken U (1988) Mechanisms involved in insect cold tolerance. Doctorate thesis, University of Oslo, Oslo

Gehrken U (1992) Inoculative freezing and thermal hysteresis in the adult beetles *Ips accuminatus* and *Rhagium inquisitor*. J Insect Physiol 38:519–524

Gehrken U, Stromme A, Lundheim R, Zachariassen KE (1991) Inoculative freezing in overwintering Tenebrionid beetle, *Bolitophagus reticulates* Panz. J Insect Physiol 37:683–687

Graether SP, Jia Z (2001) Modeling *Pseudomonas syringae* ice-nucleation protein as a beta-helical protein. Biophys J 80:1169–1173

Graether SP, Sykes BD (2004) Cold survival in freeze-intolerant insects. Eur J Biochem 271:3285–3296

Graether SP, Kuiper MJ, Gagné SM, Walker VK, Jia Z, Sykes BD, Davies PL (2000) β-Helix structure and ice-binding properties of a hyperactive antifreeze protein from an insect. Nature 406:325–328

Graether SP, Gagné SM, Spyracopoulos L, Jia Z, Davies PL, Sykes BD (2003) Spruce budworm antifreeze protein: changes in structure and dynamics at low temperature. J Mol Biol 327:1155–1168

Graham LA, Liou Y-C, Walker VK, Davies PL (1997) Hyperactive antifreeze proteins from beetles. Nature 188:727–728

Graham LA, Walker VK, Davies PL (2000) Developmental and environmental regulation of antifreeze proteins in the mealworm beetle *Tenebrio molitor*. FEBS J 267:6452–6458

Graham LA, Qin W, Lougheed SC, Davies PL, Walker VK (2007) Evolution of hyperactive, repetitive antifreeze proteins in beetles. J Mol Evol 64:387–398

Green RL, Warren GJ (1985) Physical and functional repetition in a bacterial ice nucleation gene. Nature 317:645–648

Griffith M, Yaish MW (2004) Antifreeze proteins in overwintering plants: a tale of two activities. Trends Plant Sci 9:399–405

Grimstone AV, Mullinger AM, Ramsay JA (1968) Further studies on the rectal complex of the mealworm, *Tenebrio molitor*. Philos Trans R Soc Lond Ser B Biol Sci 248:344–382

Guz N, Toprak U, Dageri A, Gurkan MO, Denlinger DL (2014) Identification of a putative antifreeze protein gene that is highly expressed during preparation for winter in the sunn pest, *Eurygaster maura*. J Insect Physiol 68:30–35

Hakim A, Nguyen JB, Basu K, Zhu DF, Thakral D, Davies PL, Isaacs FJ, Modis Y, Meng W (2013) Crystal structure of an insect antifreeze protein and its implications for ice binding. J Biol Chem 288:12295–12304

Han EN, Bauce E (2000) Dormancy in the life cycle of the spruce budworm: physiological mechanisms and ecological implications. Recent Res Dev Entomol 3:43–54

Holmstrup M, Sømme L (1998) Dehydration and cold hardiness in the Arctic collembolan *Onychiurus arcticus* Tullberg 1876. J Comp Physiol B 168:197–203

Holmstrup M, Westh P (1994) Dehydration of earthworm cocoons exposed to cold: a novel cold hardiness mechanism. J Comp Physiol B 164:312–315

Holmstrup M, Zachariassen KE (1996) Physiology of cold hardiness in earthworms: a review. Comp Biochem Physiol 115A:91–101

Horwath KL, Duman JG (1982) Involvement of the circadian system in photoperiodic regulation of insect antifreeze proteins. J Exp Zool 219:267–270

Horwath KL, Duman JG (1983a) Preparatory adaptations for winter survival in the cold hardy beetles, *Dendroides canadensis* and *Dendroides concolor*. J Comp Physiol B 151:225–232

Horwath K, Duman JG (1983b) Photoperiodic and thermal regulation of antifreeze protein levels in the beetle *Dendroides canadensis*. J Insect Physiol 29:907–917

Horwath KL, Duman JG (1983c) Induction of antifreeze protein production by juvenile hormone in larvae of the beetle, *Dendroides canadensis*. J Comp Physiol B 151:233–240

Horwath KL, Duman JG (1984a) Yearly variations in the overwintering mechanism of the cold hardy beetle, *Dendroides canadensis*. Physiol Zool 57:40–45

Horwath KL, Duman JG (1984b) Further studies on the involvement of the circadian system in photoperiodic control of antifreeze protein production in the beetle *Dendroides canadensis*. J Insect Physiol 30:947–955

Horwath KL, Duman JG (1986) Thermoperiodic involvement in antifreeze protein production in the cold hardy beetle *Dendroides canadensis*: implications for photoperiodic time measurement. J Insect Physiol 32:799–806

Husby JA, Zachariassen KE (1980) Antifreeze agents in the body fluids of winter active insects and spiders. Experientia 36:963–964

Ishiwata A, Sakurai A, Nishimiya Y, Tsuda S, Ito Y (2011) Synthetic study and structural analysis of the antifreeze agent xylomannan from upis ceramboides. J Am Chem Soc 133(48):19524–19535

Izumi Y, Sonoda S, Yoshida H, Danks HV, Tsumuki H (2006) Role of membrane transport of water and glycerol in the freeze tolerance of the rice stem borer, *Chilo suppressalis*, Walker (Lepidoptera: Pyralidae). J Insect Physiol 52:215–220

Jia Z, Davies PL (2002) Antifreeze proteins: an unusual receptor-ligand interaction. Trends Biochem Sci 27:101–106

Keeling CI, Henderson H, Li M, Yuen M, Clark EL, Fraser JD, Huber DP, Liao NY, RoderickDocking T, Birol I, Chan SK, Taylor GA, Palmquist D, Jones SJ, Bohlmann J (2012) Transcriptome and full-length cDNA resources for the mountain pine beetle, *Dendroctonus ponderosae* Hopkins, a major insect pest of pine forests. Insect Biochem Mol Biol 42:525–536

Knight CA (1967) The freezing of super-cooled liquids. Van Nostrand, Princeton, pp 8–48

Knight CA, DeVries AL (1989) Melting inhibition by fish antifreeze glycopeptides. Science 254:505–507

Knight CA, Duman JG (1986) Inhibition of recrystallization of ice by insect thermal hysteresis proteins: a possible cryoprotective role. Cryobiology 23:256–262

Knight CA, DeVries AL, Oolman LD (1984) Fish antifreeze protein and the freezing and recrystallization of ice. Nature 308:295–296

Knight CA, Cheng CC, DeVries AL (1991) Adsorption of alpha-helical antifreeze peptides on specific ice crystal surface planes. Biophys J 59:409–418
Knight CA, Wen D, Laursen RA (1995) Nonequilibrium antifreeze peptides and the recrystallization of ice. Cryobiology 32:23–34
Kobashigawa Y, Nishimiya Y, Miura K, Ohgiya S, Miura A, Tsuda S (2005) A part of ice nucleation protein exhibits the ice-binding ability. FEBS Lett 579:1493–1497
Koeppe JK, Fuchs M, Chen TT, Hunt L-M, Kovalick GE, Briers T (1985) The role of juvenile hormone in reproduction. In: Kerkut GA, Gilbert LI (eds) Comprehensive insect physiology, biochemistry and pharmacology, vol 8. Pergamon, Oxford, pp 165–204
Kostal V (2010) Cell structural modifications in insects at low temperatures. In: Denlinger DL, Lee RE (eds) Low temperature biology of insects. Cambridge University Press, Cambridge, pp 116–140
Kostal V, Simek P, Zahradnıckova H, Cimlova J, Stetina T (2012) Conversion of the chill susceptible fruit fly larva (*Drosophila melanogaster*) to a freeze tolerant organism. Proc Natl Acad Sci USA 109:3270–3274
Kostal V, Korbelova J, Poupardin R, Moos M, Simek P (2016a) Arginine and proline applied as food additives stimulate high freeze tolerance in larvae of *Drosophila melanogaster*. J Exp Biol 219:2358–2367
Kostal V, Mollaei M, Schottner K (2016b) Diapause induction as an interplay between seasonal token stimuli, and modifying and directly limiting factors: hibernation in *Chymomyza costata*. Physiol Entomol 41:344–357
Krieger EI, Darden T, Nabuurs SB, Finkelstein A, Vriend G (2004) Making optimal use of empirical energy functions: force-field parameterization in crystal space. Proteins 57:678–683
Kristiansen E, Pedersen S, Ramløv H, Zachariassen KE (1999) Antifreeze activity in the cerambycid beetle *Rhagium inquisitor*. J Comp Physiol B 169:55–60
Kristiansen E, Ramløv H, Højrup P, Pedersen SA, Hagen L, Zachariassen KE (2011) Structural characteristics of a novel antifreeze protein from the longhorn beetle *Rhagium inquisitor*. Insect Biochem Mol Biol 41:109–117
Kristiansen E, Wilkens C, Vincents B, Friis D, Lorentzen AB, Jenssen H, Løbner-Olesen A, Ramløv H (2012) Hyperactive antifreeze proteins from longhorn beetles: some structural insights. J Insect Physiol 58:1502–1510
Larson DJ, Middle L, Vu H, Zhang W, Serianni AS, Duman J, Barnes BM (2014) Wood frog adaptations to overwintering in Alaska: new limits to freezing tolerance. J Exp Biol 217 (12):2193–2200
Lee RE (2010) A primer on insect cold tolerance. In: Denlinger DL, Lee RE (eds) Low temperature biology of insects. Cambridge University Press, Cambridge, pp 3–34
Lee RE, Denlinger DL (1991) Insects at low temperature. Chapman and Hall, New York, p 513
Lee RE, Denlinger DL (2010) Rapid cold hardening: ecological significance and underpinning mechanisms. In: Denlinger DL, Lee RE (eds) Low temperature biology of insects. Cambridge University Press, Cambridge, pp 35–58
Lee RE, Zachariassen KE, Baust JG (1981) Effect of cryoprotectants on the activity of hemolymph nucleating agents in physical solutions. Cryobiology 18:511–514
Lee RE, Mugnano JA, Taylor RT (1992) Endogenous crystalloid spheres regulate the supercooling point of the gall fly *Eurosta solidaginis*. Cryobiology 29:750–751
Lee RE, Warren GJ, Gusta LV. Biological ice nucleation and its applications. St Paul: MN APS; 1995. 370 pp.
Lee RE, Costanzo JP, Mugnano JA (1996) Regulation of supercooling and ice nucleation in insects. Eur J Entomol 93:405–418
Li N, Andorfer CA, Duman JG (1998a) Enhancement of insect antifreeze protein activity by low molecular weight solutes. J Exp Biol 201:2243–2251
Li N, Chibber BAK, Castellino FJ, Duman JG (1998b) Mapping of disulfide bridges in antifreeze proteins from overwintering larvae of the beetle *Dendroides canadensis*. Biochemistry 37:6343–6350

Li Y, Gong H, Park HY (2000) Purification and partial characterization of thermal hysteresis proteins from overwintering larvae of the pine needle gall midge *Thecodiplosis japonesis* (Diptera: cecitomiidae). Cryo Lett 21:117–124

Li J, Ma W, Ma J (2016) Heat inducible expression of antifreeze protein genes from the beetles *Tenebrio molitor* and *Microdera punctipennis*. Cryo Lett 37:10–18

Lin F-H, Davies PL, Graham LA (2011) The thr- and ala-rich hyperactive antifreeze protein from inchworm folds as a flat silk-like β-helix. Biochemist 50:4467–4478

Lindow SE (1983) The role of bacterial ice nucleation in frost injury to plants. Rev Phytopathol 21:363–384

Lindow SE (1995) Control of epiphytic ice-nucleation-active bacteria for management of plant frost injury. In: Lee RE, Warren LGJ, Gusta LV (eds) Biological ice nucleation and its applications. APS, St. Paul, pp 239–256

Liou Y-C, Thibault P, Walker VK, Davies PL, Graham LA (1999) A complex family of highly heterogeneous and internally repetitive hyperactive antifreeze proteins from the beetle *Tenebrio molitor*. Biochemistry 38:11415–11424

Liu K, Wang C, Ji Ma J, Shi G, Xi Yao X, Fang H, Yanlin Song Y, Wang J (2016) Janus effect of antifreeze proteins on ice nucleation. Proc Natl Acad Sci USA 113:14739–14744

Lundheim R (1996) Adaptive and incidental biological ice nucleators. Doctorate thesis, Norwegian University of Science and Technology, Trondheim

Lundheim R, Zachariassen KE (1993) Water balance of overwintering beetles in relation to strategies for cold tolerance. J Comp Physiol B 163:1–4

Ma J, Wang J, Mao XF, Wang Y (2012) Differential expression of two antifreeze proteins in the desert beetle *Anatolica polita* (Coleoptera: Tenebriondae): seasonal variation and environmental effects. Cryo Lett 33:337–348

Mao XI, Liu Z, Ma J, Pang H, Zhang F (2011) Characterization of a novel β-helix antifreeze protein from the desert beetle *Anatolica polita*. Cryobiology 62:91–99

Mazur P (1984) Freezing of living cells: mechanism and implications. Am J Physiol 247:C125–C142

Meister K, Ebbinghaus Y, Xu Y, Duman JG, DeVries AL, Gruebele DM, Leitner DM, Havenith M (2013) Long-range protein-water dynamics in hyperactive insect antifreeze proteins. Proc Natl Acad Sci U S A 110:1617–1622

Meister K, Duman JG, Xu Y, DeVries AL, Leitner DM, Havenith M (2014) The role of sulfates on antifreeze protein activity. J Phys Chem B 118(28):7920–7924

Meister K, Lotze S, Olijve LLC, DeVries AL, Duman JG, Voets IK, Bakker HJ (2015) Investigation of the ice-binding site of an insect antifreeze protein using sum-frequency generation spectroscopy. J Phys Chem Lett 6(7):1162–1167

Michaud MR, Denlinger DL (2010) Genomics, proteomics and metabolomics: finding the other players in insect cold tolerance. In: Denlinger DL, Lee RE (eds) Low temperature biology of insects. Cambridge University Press, Cambridge, pp 91–115

Miller K (1982) Cold-hardiness strategies of some adult and immature insects overwintering in interior Alaska. Comp Biochem Physiol A Physiol 73(4):595–604

Miller LK, Werner R (1987) Extreme supercooling as an overwintering strategy in three species of willow gall insects from interior Alaska. Oikos 49(3):253

Mueller GM, Wolber PK, Warren GJ (1990) Clustering of ice nucleation protein correlates with ice nucleation activity. Cryobiology 27:416–422

Mugnano JA, Lee RE, Taylor RT (1996) Fat body cells and calcium phosphate sphrelules induce ice nucleation in the freeze tolerant larvae of the gall fly *Eurosta solidaginis* (Fitch). J Exp Biol 199:465–471

Neven LG, Duman JG, Beals JM, Castellino FJ (1986) Overwintering adaptations of the stag beetle, *Ceruchus piceus*: removal of ice nucleators in winter to promote supercooling. J Comp Physiol B 156:707–716

Neven LG, Duman JG, Low MG, Sehl LC, Castellino FJ (1989) Purification and characterization of an insect hemolymph lipoprotein ice nucleator: evidence for the importance of phosphatidylinositol and apolipoprotein in the ice nucleator activity. J Comp Physiol B 159:71–82

Nickell PK, Sass S, Verleye D, Blumenthall EM, Duman JG (2013) Antifreeze proteins in the primary urine of larvae of the beetle *Dendroides canadensis* (Latreille). J Exp Biol 216:1695–1703

Nicodemus J, O'Tousa JE, Duman JG (2006) Expression of a beetle, *Dendroides anadensis*, antifreeze protein in *Drosophila melanogaster*. J Insect Physiol 52:888–896

Nishimiya Y, Sato R, Miura A, Tsuda S (2006) Antifreeze protein of *Dorcus curvidens* binodulus. Submitted (JUN-2006) to EMBL/GenBank/DDBJ databases

Nutt DR, Smith JC (2008) Duel function of the hydration layer around an antifreeze protein revealed by atomistic molecular dynamics simulations. J Am Chem Soc 130:13066–13073

Olive LC, Meister K, DeVries AL, Duman JG, Guo S, Bakker HJ, Voets IK (2016) Blocking rapid ice crystal growth through non-basal plane adsorption of antifreeze proteins. Proc Natl Acad Sci USA 113:3740–3745

Olsen TM, Duman JG (1997a) Maintenance of the supercooled state in overwintering Pyrochroid beetle larvae, *Dendroides canadensis*: role of hemolymph ice nucleators and antifreeze proteins. J Comp Physiol B 167:105–113

Olsen TM, Duman JG (1997b) Maintenance of the supercooled state in the gut of overwintering Pyrochroid beetle larvae, *Dendroides canadensis*: role of gut ice nucleators and antifreeze proteins. J Comp Physiol B 167:114–122

Olsen TM, Sass SJ, Li N, Duman JG (1998) Factors contributing to seasonal increases in inoculative freezing resistance in overwintering fire-colored beetle larvae *Dendroides canadensis* (Pyrochroidae). J Exp Biol 201:1585–1594

Patterson J, Duman JG (1978) The role of thermal hysteresis producing antifreeze proteins in the low temperature tolerance and water balance of larvae of the mealworm, *Tenebrio molitor*. J Exp Biol 74:37–45

Patterson JL, Duman JG (1979) Composition of a protein antifreeze from larvae of the beetle, *Tenebrio molitor*. J Exp Zool P 210:361–367

Patterson JL, Duman JG (1982) Purification and composition of protein antifreezes with high cysteine contents from larvae of the beetle, *Tenebrio molitor*. J Exp Zool 219:381–384

Patterson JL, Kelly TJ, Duman JG (1981) Purification and composition of a thermal hysteresis producing protein from the milkweed bug, *Oncopeltus fasciatus*. J Comp Physiol B 142:539–542

Pedersen PG, Holmstrup M (2003) Freeze or dehydrate: only two options for the survival of subzero temperatures in the arctic enchytraeid *Fridericia ratzeli*. J Comp Physiol B 173:601–609

Pennisi E (2013) The CRISPR craze. Science 341:833–836

Pertaya N, Marshall CB, Celik Y, Davies PL, Braslavsky I (2008) Direct visualization of spruce budworm antifreeze protein interacting with ice: basal plane affinity confers hyperactivity. Biophys J 95:333–341

Pettersen EF, Goddard TD, Huang CC, Couch GS, Greenblatt DM, Meng EC, Ferrin TE (2004) UCSF Chimera – a visualization system for exploratory research and analysis. J Comput Chem 25:1605–1612

Philip BN, Yi S-X, Elnitsky MA, Lee RE (2008) Aquaporins play a role in desiccation and freeze tolerance in larvae of the goldenrod gall fly, *Eurosta solidaginis*. J Exp Biol 211:1114–1119

Qin W, Doucet D, Tyshenko MG, Walker VK (2007) Transcription of antifreeze protein genes in *Choristoneura fumiferana*. Insect Mol Biol 16:423–434

Qiu L, Mao X, Hou F, Ma J (2013) A novel function – thermal protective properties of an antifreeze protein from the summer desert beetle *Microdera punctipennis*. Cryobiology 66:60–68

Ramløv H (1999) Microclimate and variations in haemolymph composition in the freezing-tolerant alpine weta *Hemideina maori* Hutton (Orthoptera: Stenopelmatidae). J Comp Physiol B 169:224–235

Ramløv H (2000) Aspects of natural cold tolerance in ectothermic animals. Hum Reprod 15:26–46

Ramløv H, Bedford J, Leader JP (1992) Freezing tolerance of the New Zealand alpine weta *Hemideina maori* hutton (Orthoptera: Stenopelmatidae). J Therm Biol 17:51–54

Ramløv H, DeVries AL, Wilson PW (2005) Antifreeze glycoproteins from the antartic fish *Dissostichus mawsoni* studied by differential scanning calorimetry (DSC) in combination with nanolitre osmometry. Cryo Lett 26:73–84

Ramsay RA (1964) The rectal complex of the mealworm, *Tenebrio molitor* L. (Coleoptera, Tenebrionidae). Philos Trans R Soc Lond Ser B Biol Sci 248:279–314

Raymond JA, DeVries AL (1977) Adsorption inhibition as a mechanism of freezing resistance in polar fishes. Proc Natl Acad Sci USA 74:2589–2593

Raymond JA, Wilson PW, DeVries AL (1989) Inhibition of growth on nonbasal planes in ice by fish antifreeze. Proc Natl Acad Sci USA 86:881–885

Reid DS, Folin AT, Lem CA (1985) The effect of solutes on the temperature of heterogeneous nucleation of ice from aqueous solution. CryoLetters 6:189–198

Richardson JS, Richardson DC (2002) Natural beta-sheet proteins use negative design to avoid edge-to-edge aggregation. Proc Natl Acad Sci USA 99:2754–2759

Ring JA (1982) Freezing tolerant insects with low supercooling points. Comp Biochem Physiol 73A:605–612

Rozsypal J, Košťál V (2018) Supercooling and freezing as eco-physiological alternatives rather than mutually exclusive strategies: a case study in *Pyrrhocoris apterus*. J Insect Physiol 111:53–62

Rozsypal J, Moos M, Šimekand P, Vladimıŕ KoŠtál V (2018) Thermal analysis of ice and glass transitions in insects that do and do not survive freezing. J Exp Biol 221:jeb170464. https://doi.org/10.1242/jeb.170464

Salt RW (1936) Studies on the freezing process in insects. Tech Bull Univ Minn Agric Exp Station 116:1–41

Salt RW (1953) The influence of food on cold hardiness of insects. Can Entomol 85:261–269

Salt RW (1956) Freezing and melting points of insect tissues. Can J Zool 34:1–5

Salt RW (1958) Role of glycerol in producing abnormally low supercooling and freezing points in an insect, *Bracon cephi* (Gahan). Nature 181:1281

Salt RW (1959) Role of glycerol in the cold hardening of *Bracon cephi* (Gahan). Can J Zool 37:59–69

Salt RW (1961) Principles of insect cold hardiness. Annu Rev Entomol 6:55–74

Salt RW (1966) Factors influencing nucleation in supercooled insects. Can J Zool 44:117–133

Schneppenheim R, Theede H (1980) Isolation and characterization of freezing point depressing peptides from larvae of *Tenebrio molitor*. Comp Biochem Physiol 67:561–568

Scholander PF, Flagg W, Hock RJ, Irving L (1953) Studies on the physiology of frozen plants and animals in the arctic. J Cell Comp Physiol 42:1–56

Sformo T, Kohl F, McIntyre J, Kerr P, Duman JG, Barnes BM (2009) Simultaneous freeze tolerance and avoidance in individual fungus gnats, *Exechia nugatoria*. J Comp Physiol B 179:897–902

Sformo T, Walters K, Jeannet K, McIntyre J, Wowk B, Fahy G, Barnes BM, Duman JG (2010) Deep supercooling, vitrification, and limited survival to −100°C in larvae of the Alaskan beetle *Cucujus clavipes puniceus* (Coleoptera: Cucujuidae). J Exper Biol 213:502–509

Sformo T, McIntyre J, Walters KR, Barnes BM, Duman JG (2011) Probability of freezing in the freeze avoiding beetle larvae *Cucujus clavipes puniceus* (Coleoptera, Cucujidae) from interior Alaska. J Insect Physiol 57:1170–1177

Shier WT, Lin Y, DeVries AL (1975) Structure of the carbohydrate of antifreeze glycoproteins from an Antartic fish. FEBS Lett 54:135–138

Shimada K, Riihimaa A (1988) Cold acclimation, inoculative freezing and slow cooling: essential factors contributing to the freezing-tolerance in diapausing larvae of Chymomyza costata (Diptera: Drosophilidae). Cryo-Letters 9:5–10

Sicheri F, Yang DSC (1995) Ice-binding structure and mechanism of an antifreeze protein from winter flounder. Nature 375:427–431

Sinclair BJ (1999) Insect cold tolerance: how many kinds of frozen? Eur J Entomol 96:157–164

Somero GN, Lockwood BL, Tomanek L (2017) Biochemical adaptation: response to environmental challenges from life's origins to anthropocene. Sinauer Associates, Sunderland, p 572

Sømme L (1978) Nucleating agents in the haemolymph of third instar larvae of *Eurosta solidaginis* (Fitch) (Diptera:Tephritidae). Norw J Entomol 25:187–188

Sømme L (1982) Supercooling and winter survival in terrestrial insects. Comp Physiol Biochem 73A:519–543

Southworth MW, Wolber PK, Warren GJ (1988) Nonlinear relationship between concentration and activity of a bacterial ice nucleation protein. J Biol Chem 263:15211–15216

Storey KB, Storey JM (1991) Biochemistry of cryoprotectants. In: Lee RE, Denlinger DL (eds) Insects at low temperature. Chapman and Hall, New York, pp 64–93

Storey KB, Storey JM (2010) Oxygen: stress and adaptation in cold-hardy insects. In: Denlinger DL, Lee RE (eds) Low temperature biology of insects. Cambridge University Press, Cambridge, pp 141–165

Sun T, Lin F-H, Campbell RL, Allingham JS, Davies PL (2014) An antifreeze protein folds with an interior network of more than 400 semi-clathrate waters. Science 343:795–798

Tanno K (1977) Ecological observation and frost-resistance in overwintering pre-pupa, Sciara sp. (Sciaridae). Low Temp Sci SerB 35:63–74

Thomas MC (2002) Cucujidae Laetrille 1802. In: Arnett RH, Thomas MC, Skelley PE, Frank JH (eds) American beetles, vol 2. CRC, Boca Raton, pp 329–330

Tomchaney AP, Morris JP, Kang SH, Duman JG (1982) Purification, composition and physical properties of a thermal hysteresis "antifreeze" protein from larvae of the beetle, *Tenebrio molitor*. Biochemistry 21:716–721

Tomczak MM, Hincha DK, Estrada SD, Wolkers WF, Crowe LM, Feeney RE, Tablin F, Crowe JH (2002) A mechanism for stabilization of membranes at low temperatures by an antifreeze protein. Biophys J 82:874–881

Tursman D, Duman JG (1995) Cryoprotective effects of thermal hysteresis protein on survivorship of frozen gut cells from the freeze tolerant centipede *Lithobius forficatus*. J Exp Zool 272:249–257

Tursman D, Duman JG, Knight CA (1994) Freeze tolerance adaptations in the centipede *Lithobius forficatus*. J Exp Zool 268:347–353

Tyshenko MG, Doucet D, Davies PL, Walker VK (1997) The antifreeze potential of the spruce budworm thermal hysteresis protein. Nat Biotechnol 15:887–890

Vali G (1995) Principles of icenucleation. In: Lee RE, Warren GJ, Gusta LV (eds) Biological ice nucleation and its applications. APS, St Paul, pp 1–28

Vu HM, Duman JG (2017) Upper lethal temperatures in three cold-tolerant insects are higher in winter than in summer. J Exp Biol 220:2726–2732

Vu HM, Pennoyer JE, Ruiz KR, Portmann P, Duman JG (2019) Beetle, *Dendroides canadensis*, antifreeze proteins increased high temperature survivorship in transgenic fruit flies, *Drosophila melanogaster*. J Insect Physiol 112:68–72

Walters KR, Pan Q, Serianni AS, Duman JG (2009a) Cryoprotectant biosynthesis and selective accumulation of threitol in the freeze tolerant beetle, *Upis ceramboides*. J Biol Chem 284:16822–16831

Walters KR, Serianni AS, Sformo T, Barnes BM, Duman JG (2009b) A non-protein thermal hysteresis-producing xylomannan antifreeze in a freeze tolerant Alaskan beetle. Proc Natl Acad Sci USA 106:20210–20215

Walters KR, Sformo T, Barnes BM, Duman JG (2009c) Freeze tolerance of an Arctic Alaska stonefly. J Exp Biol 212:305–312

Walters KR, Serianni AS, Voituron Y, Sformo T, Barnes BM, Duman JG (2011) A thermal hysteresis-producing xylomannan glycolipid antifreeze associated with cold-tolerance is found in diverse taxa. J Comp Physiol B 181:631–640

Wang L, Duman JG (2005) Antifreeze proteins of the beetle *Dendroides canadensis* enhance one another's activities. Biochemist 44:10305–10312

Wang L, Duman JG (2006) A thaumatin-like protein from larvae of the beetle *Dendroides canadensis* enhances the activity of antifreeze proteins. Biochemist 215:1278–1284

Wang L, Amornwittawat N, Juwita V, Kayo Y, Duman JG, Pascal T, Goddard WA, Wen X (2009a) Arginine, a key residue for the enhancing ability of an antifreeze protein of the beetle *Dendroides canadensis*. Biochemist 48:9696–9703

Wang S, Amornwittawat N, Banatlao J, Chung M, Kao Y, Wen X (2009b) Hofmeister effects of common monovalent salts on the beetle antifreeze protein activity. J Phys Chem B 113:13891–13894

Wang S, Wen X, Nikolovski P, Juvita V, Arafin JF (2012) Expanding the molecular recognition repertoire of antifreeze polypeptides: effects on nucleoside crystal growth. Chem Commun 48:11555–11557

Wang S, Wen X, DeVries AL, Bagdagulyan Y, Morita A, Golen JA, Duman JG, Rheingold AL (2014) Molecular recognition of methyl α-D-mannopyranoside by anti-freeze (glyco) proteins. J Am Chem Soc 136:8973–8981

Warner DT (1962) Some possible relationships of carbohydrates and other biological components with the water structure at 37°C. Nature 196:1055–1058

Warren G, Corotto L (1989) The consensus sequence of ice nucleation proteins from *Erwinia herbicola, Pseudomonas fluorescens* and *Pseudomonas syringae*. Gene 85:239–242

Waterhouse A, Bertoni M, Bienert S, Studer G, Tauriello G, Gumienny R, Heer FT, de Beer TAP, Rempfer C, Bordoli L, Lepore R, Schwede T (2018) Swiss-model: homology modelling of protein structures and complexes. Nucleic Acids Res 46(W1):W296–W303

Wen X, Wang S, Duman JG, Arafin JF, Juwita V, Goddard WA, Rios A, Liu F, Kim S-K, Abrol R, DeVries AL, Henling LM (2016) Antifreeze proteins govern the precipitation of trehalose in a freeze avoiding insect at low temperature. Proc Natl Acad Sci USA 113:6683–6688

Wharton DA, Goodall G, Marshall CJ (2003) Freezing survival and cryoprotective dehydration as cold tolerance mechanisms in the Antarctic nematode *Panagroulamis davidi*. J Exp Biol 206:215–221

Wharton DA, Barrett J, Goodall G, Marshall CJ, Ramlov H (2005) Ice-active proteins from the Antarctic nematode *Panagrolaimus davidi*. Cryobiology 51:198–207

Wharton DA, Pow B, Kristensen M, Ramlov HR, Marshall CJ (2009) Ice-active proteins and cryoprotectants from the New Zealand alpine cockroach *Celatoblatta quinquemaculata*. J Insect Physiol 55:27–31

Wilkens C, Ramløv H (2009) Seasonal variations in antifreeze protein activity and haemolymph osmolality in larvae of the beetle *Rhagium mordax* (Coleoptera: Cerambycidae). CryoLetters 29:293–300

Wilson EO (1988) The current state of biological diversity. In: Wilson EO, Peter FM (eds) Biological diversity. National Academy Press, Washington, DC, pp 3–18

Wilson P, Ramlov H (1995) Hemolymph ice nucleating proteins from the New Zealand alpine weta *Hemideina maori* (Orthoptera: Stenopelmatidae). Comp Biochem Physiol B112:535–542

Wolber PK, Warren GJ (1989) Bacterial ice nucleating proteins. Trends Biochem Sci 14:179–182

Worland MR, Block W (2003) Desiccation stress at subzero temperatures in polar terrestrial arthropods. J Insect Physiol 49:193–203

Wu DW, Duman JG (1991) Activation of antifreeze proteins from the beetle *Dendroides canadensis*. J Comp Physiol B 161:279–283

Wu DW, Duman JG, Cheng C-HC Castellino FJ (1991) Purification and characterization of antifreeze proteins from larvae of the beetle *Dendroides canadensis*. J Comp Physiol B 161:271–278

Xu L, Duman JG (1991) Involvement of juvenile hormone in the induction of antifreeze protein production by fat body of larvae of the beetle *Dendroides canadensis*. J Exp Zool 258:288–293

Xu L, Neven LG, Duman JG (1990) Hormonal control of hemolymph lipoprotein ice nucleators in overwintering freeze susceptible larvae of the stag beetle *Ceruchus piceus*: adipokinetic hormone and juvenile hormone. J Comp Physiol B 160:51–59

Xu L, Duman JG, Goodman WG, Wu DW (1992) A role for juvenile hormone in the induction of antifreeze protein production by the fat body in the beetle *Tenebrio molitor*. Comp Biochem Physiol B 101:105–109
Yancey PH (2005) Organic osmolytes as compatible, metabolic, and counteracting cytoprotectants in high osmolarity and other stresses. J Exp Biol 208:2819–2830
Yancey PH, Clark ME, Hand SC, Bowlus RD, Somero GN (1982) Living with water stress: evolution of osmolyte systems. Science 217:1214–1222
Yang J, Zhang Y (2015) I-TASSER server: new development for protein structure and function predictions. Nucleic Acids Res 43:174–181
Yeung KL, Wolf EE, Duman JG (1991) A scanning tunneling microscopy study of an insect lipoprotein ice nucleator. J Vac Sci Technol B 9:1197–1201
Zachariassen KE (1982) Nucleating agents in cold hardy insects. Comp Biochem Physiol A 73:557–562
Zachariassen KE (1985) Physiology of cold tolerance in insects. Physiol Rev 65:799–832
Zachariassen KE (1992) Ice nucleating agents in cold-hardy insects. In: Somero GN, Osmond CB, Bolis CL (eds) Water and life: comparative analysis of water relationships at the organismic, cellular and molecular level. Springer, Berlin, pp 262–281
Zachariassen KE, Hammel HT (1976) Nucleating agents in the haemolymph of insects tolerant to freezing. Nature 262:285–287
Zachariassen KE, Husby JA (1982) Antifreeze effects of thermal hysteresis agents protect highly supercooled insects. Nature 298:865–867
Zachariassen KE, Baust JG, Lee RE (1982) A method for quantitative determination of ice nucleating agents in insect haemolymph. Cryobiology 19:180–184
Zachariassen KE, DeVries AL, Hunt B, Kristiansen E (2002) Effect of ice fraction and dilution factor on the antifreeze activity in the hemolymph of the cerambycid beetle *Rhagium inquisitor*. Cryobiology 44:132–141
Zachariassen KE, Kristansen E, Pedersen SA, Hammel HT (2004) Ice nucleation in solutions and freezing in insects – homogeneous or heterogeneous? Cryobiology 48:309–321
Zachariassen KE, Li NG, Laugsand AE, Kristiansen E, Pedersen SA (2008) Is the strategy for cold hardiness in insects determined by their water balance? A study on two closely related families of beetles: Cerambycidae and Chrysomelidae. J Comp Physiol B 178:977–984
Zachariassen KE, Duman JG, Kristiansen E, Pedersen S, Li N (2011) Ice nucleation and antifreeze proteins in animals. In: Graether S (ed) Biochemistry and function of antifreeze proteins. Nova Science, New York, pp 73–104
Zhang W, Oliver AV, Vu HM, Duman JG, Serianni AS (2012) Methyl 4-O-ß-D-mannopyranosyl ß-D-xylopyranoside. Acta Crystallogr C68:0502–0506
Zhang W, Oliver AG, Vu HM, Duman JG, Serianni AS (2013) Methyl 4-O-ß-D-xylopyranosyl ß-D-mannopyranoside. Acta Crystallogr C69:1047–1050

Chapter 7
Plant Antifreeze Proteins

Michael Wisniewski, Ian R. Willick, John G. Duman, David Livingston III, and Samuel S. Newton

7.1 Introduction

The onset of climate change has brought about increasing fluctuations in extreme temperatures. Early spring seeding of agronomic crops during periods of warm weather promotes rapid growth at a time when a crop is less susceptible to drought and fungal diseases but increases their potential exposure to frost. This was the case in 2015 for Canadian farmers in the provinces of Manitoba and Saskatchewan when growers experienced a complete loss of the canola (*Brassica napus*) crop at the seedling stage due to a severe frost. On a regional scale, the northwestern and eastern United States experienced an early spring warming period in April 2007 followed by a record-breaking frost that resulted in an estimated two billion dollars in crop damage (Gu et al. 2008). The high monetary return from growing frost-sensitive

M. Wisniewski (✉)
Department of Biological Sciences, Virginia Polytechnic and State University, Blacksburg, VA, USA
e-mail: wisniewski@vt.edu

I. R. Willick
Department of Plant Sciences, University of Saskatchewan, Saskatoon, SK, Canada
e-mail: ian.willick@usask.ca

J. G. Duman
Department of Biological Sciences, University of Notre Dame, Notre Dame, IN, USA
e-mail: John.G.Duman@nd.edu

D. Livingston III
United States Department of Agriculture – Agricultural Research Service (USDA-ARS), Raleigh, NC, USA
e-mail: david.livingston@ars.usda.gov

S. S. Newton
Sanford School of Medicine, University of South Dakota, Vermillion, SD, USA
e-mail: Samuel.Sathyanesan@usd.edu

H. Ramløv, D. S. Friis (eds.), *Antifreeze Proteins Volume 1*,
https://doi.org/10.1007/978-3-030-41929-5_7

crops such as corn (*Zea mays*) or soybean (*Glycine max*) has pushed their production into non-traditional Canadian and Northern European regions. Overwintering crops, such as winter wheat (*Triticum aestivum*), which has a 20–30% higher yield potential over modern spring cultivars, have been limited in northern latitudes by extreme (< −30 °C) winter temperatures (Fowler 2012). Therefore, the need to protect sensitive spring and overwintering crops from severe frosts is an ongoing management and breeding challenge.

7.2 Freezing Injury and the Acquisition of Hardiness

Any form of stress may interfere with the normal growth habit and survival of plants (Levitt 1980). Frost or freezing can be defined as a stress condition when temperatures fall below 0 °C and water in the extracellular spaces of plants freezes. Many plants have evolved mechanisms that enable them to adapt to subzero temperatures and the presence of ice in their tissues, without being injured. This ability is referred to as cold acclimation, cold hardiness, or the acquisition of hardiness.

7.2.1 Freezing Injury

All plants exhibit some ability to supercool below 0 °C to avoid freezing (see reviews by Lindow 1989; Wisniewski et al. 2014). This premise is based on the fact that while the temperature at which ice melts (0 °C) is well defined, the temperature at which water will freeze is not predetermined (Franks 1985). Pure water has the ability to supercool to temperatures approaching −40 °C (the homogeneous nucleation temperature) but freezes at much warmer, subzero temperatures due to the presence of heterogeneous nucleators that are very effective at inducing the formation of ice crystals (Franks 1985). Stated simply, heterogenous nucleators act as a template that allows water molecules to take on a crystalline arrangement. Once a core number of stable ice nuclei is present and stable, they serve as a catalyst to induce a phase transition in the surrounding water molecules. Heterogenous nucleators relevant to plant freezing can be either extrinsic or intrinsic. The presence of these nucleators can vary within and between plant species, thus making the ability to predict the temperature at which a plant will freeze problematic. The current understanding of ice nucleation and propagation in plants was reviewed by Wisniewski et al. (2014). Depending on growth conditions, the extent of supercooling and overall freezing tolerance can vary considerably between and within plant species. Field-grown plants have a higher freezing survival than plants grown in controlled environment chambers due to their exposure to additional stresses (wind, herbivory, water, UV) that induce an increase in overall stress tolerance (Gusta and Wisniewski 2013). The temperature at which ice formation is initiated in plant tissues is dependent upon the cooling rate, the source of ice nucleation, and the plant's capacity to supercool (Levitt 1980; Sakai and Larcher 1987; Gusta et al. 2009).

At the cellular level, ice nucleation initiates within the extracellular space of the apoplast. If the cooling rate is rapid, then cells cannot maintain the equilibrium between supercooled cellular water and extracellular ice (Olien 1977). The supercooled water inside cells is then susceptible to intracellular ice nucleation or ice propagation that is seeded by the extracellular ice. It is generally accepted that under natural field conditions, intracellular freezing in plants results in cell death (reviewed by Gusta et al. 2009). At slow freezing rates in cold acclimated plants, intracellular water diffuses to the growing extracellular ice crystals. The cytoplasm volume gradually decreases, which concentrates symplastic osmolytes, and depresses the freezing point (Siminovitch and Scarth 1938). In this scenario, there are multiple modes of freezing injury. Cellular dehydration becomes progressively greater with a further decrease in temperature (Gusta et al. 1975). Membrane injury occurs when freezing induced dehydration collapses the cell (cytorrhysis). Prolonged exposure to sub-zero temperatures alters membrane structure and chemical composition (Steponkus 1984; Uemura et al. 2006). During thawing, quick rehydration of cellular organelles induces further membrane injury (Arora and Palta 1988). If the cell rehydrates, freeze-induced changes, such as an alteration in the structure of the plasma membrane or the accumulation of reactive oxygen species, can further impede recovery and growth (Uemura et al. 2006; Arora 2018). Thus, protection of the plasma membrane from freeze-induced dehydration, inhibiting intracellular ice nucleation, and controlling the rate of ice propagation and crystal morphology are major determinants of freezing tolerance.

7.2.2 Cold Acclimation (Acclimatization)

To ensure survival, temperate plants continually adjust to the changing environment. Cold acclimation refers to the process that plants undergo to adapt to low, subzero temperatures. Animal researchers make a distinction between acclimation as occurring under laboratory conditions, and acclimatization that occurs under natural conditions. Plant researchers, however, do not make any such distinction. The ability of plants to survive freezing temperatures is referred to as cold hardiness, and as with supercooling, plants vary in the level of cold hardiness they can potentially achieve. The acquisition of cold hardiness is not a static entity in that a return to warmer temperatures due to the early onset of spring or temperature fluctuations in the fall or mid-winter can stop, reverse, and/or restart the acclimation process (Trischuk et al. 2014). The extent to which a plant responds to these temperature fluctuations is under genetic control (both Mendelian and epigenetic). Acclimation can be triggered by changes in photoperiod (Weiser 1970), light intensity, spectral distribution (Gray et al. 1997; Öquist and Huner 2003), or by exposure to a threshold (<14 °C) temperature (Fowler 2008). This process involves various biochemical and structural modifications regulated by the complex interactions of numerous genes and plant hormones (Thomashow 1999; Gusta et al. 2009), all of which enable plants to survive exposure to freezing and the presence of ice in their tissues.

The modification and accumulation of membrane lipids (Yoshida and Uemura 1984; Uemura et al. 1995), sugars (Uemura and Steponkus 2003), and proteins (Danyluk et al. 1998; Takahashi et al. 2016) increases the fluidity of the plasma membrane and prevents the loss of membrane function during freeze–thaw events (reviewed by Steponkus 1984; Uemura et al. 2006; Arora 2018). Cold-induced biochemical modifications (Wisniewski et al. 1991; Willick et al. 2018) increase the rigidity and reduce the porosity of the cell wall to provide a barrier to lethal penetration of ice into a cell (Wisniewski et al. 1991; Rajashekar and Burke 1996). Cold acclimation can also induce the synthesis of many different types of cryoprotective compounds, including the accumulation of soluble sugars and fructans (Livingston and Henson 1998; Livingston et al. 2005), and plant antifreeze proteins (AFPs) (Griffith et al. 1992; Duman and Olsen 1993; Duman 1994; Griffith and Yaish 2004), the latter of which is the main subject of this chapter.

Exposure to temperatures below 0 °C, further increases the level of cold hardiness in plants (Tumanov and Krasasvtsev 1959; Weiser 1970; Livingston and Henson 1998; Herman et al. 2006). Subzero acclimation induces the saccharification of fructans into sucrose, glucose, and fructose (Livingston et al. 2005; Livingston and Henson 1998) (discussed in Sect. 7.14), increases the accumulation of plant AFPs (Herman et al. 2006), reduces free mobile water (Gusta et al. 1975), and enhances the binding strength of water (Yoshida et al. 1997).

Plants have species specific strategies to survive exposure to subzero temperatures. Therefore, plants exhibit different approaches to enhancing plant cold hardiness and providing frost protection. In freezing sensitive crops, plants can enhance their freezing survival by lowering the ice nucleation temperature and inhibiting ice propagation (Wisniewski et al. 2014). In contrast, inhibiting the damage caused by ice recrystallization and extracellular freeze dehydration is an alternative approach to increasing freezing tolerance in overwintering cereals and other herbaceous crop plants (Holmberg and Bülow 1998; Pearce 2001; Griffith and Yaish 2004; Gusta et al. 2009; Duman and Wisniewski 2014). With either strategy, an increase in AFP concentration is associated with an increase in frost protection.

7.3 Ice-Binding "Antifreeze" Proteins: Criteria and Ice-Binding Mechanism

Antifreeze proteins (AFPs) have long been considered ice active if they meet the following criteria: (1) they modify ice crystal morphology and (2) inhibit ice recrystallization with either (3) high or (4) low thermal hysteresis (TH) activity. Criteria 1, 2, and 3 are characteristic of AFPs purified from species that must avoid freezing to survive, while proteins that meet criteria 1, 2, and 4 are found in species that exhibit freezing tolerance, such as cold acclimated plants (Griffith et al. 1992; Urrutia et al. 1992; Griffith and Yaish 2004). The initial definition of AFPs reflected their role as "proteins that prevented freezing." Recent advancements in our understanding of plant freezing tolerance, ice restructuring, and ice adhesion, however, have led some to argue that the original AFP definition does fully encompass our current understanding of AFP function in plants (reviewed by Davies 2014). For this reason, plant AFPs are often referred to with the more general term, ice-binding proteins (IBPs).

When discussing IBP function, authors are generally referring to the mechanism that allows IBPs to bind to ice and induce TH, recrystallization inhibition (RI), or in the case of an ice nucleating protein inhibit crystallization. In the current review, these aspects are referred to as *mechanisms*, while the term *function* is reserved to mean the *physiological function(s)* of IBPs in a plant (inhibition of freezing, RI, control of ice crystal structure, etc.). Ice-binding mechanism(s) are discussed at length elsewhere in this book, so only a brief description is provided here.

The adsorption–inhibition mechanism was initially proposed by DeVries (1971) to explain how Antarctic fish antifreeze glycoproteins (AFGPs) bind to ice and thereby lower the temperature at which the ice crystal grows (hysteretic freezing point) below the colligative equilibrium melting/freezing point determined by the solute concentration (DeVries 1971). In the context of this chapter, TH is defined as the temperature difference between the melting and freezing point of ice caused by IBP absorption to the ice surface. Experimental validation that AFGPs do indeed bind to ice was initially provided by demonstrating that AFGPs do not freeze-out of ice (Duman and DeVries 1972), and in a much more elegant fashion by later studies using ice etching, fluorescent tagging of AFPs, etc. (Raymond and DeVries 1977; Raymond et al. 1989; Knight et al. 1991; Petraya et al. 2008). Since AFPs irreversibly bind to preferred growth sites, water molecules can only be added to the crystal surface between the AFGPs in highly curved fronts with high surface free energy. Therefore, macroscopic crystal growth at the equilibrium melting/freezing point is blocked due to the Kelvin effect, and significant growth of the crystal is delayed until the temperature drops to the hysteretic freezing point where the crystal grows rapidly. In addition to the depression of the nonequilibrium freezing point, there is a smaller increase in the observed melting point, also a result of the surface adsorption of AFPs (Knight and DeVries 1989; Celik et al. 2010). Different AFPs bind predominantly to different planes of ice (basal, pyramidal, prism planes), and this can have profound effects on the relative levels of TH or RI produced (Olive et al. 2016).

The theory explaining the binding mechanism of AFPs to ice was initially based on the Antarctic fish AFGPs. These AFGPs have a very regular repeat structure consisting largely of Ala-Thr-Ala tri-peptide repeats with the disaccharide β-D-galactopyranosyl-(1 → 3)-2-acetamido-2-deoxy-α-D-galactopyranose attached to each Thr via a glycoside linkage (DeVries 1971; Shier et al. 1975). All the disaccharides extend from one side of the extended coil. Hydroxyls on the galactose residues were thought to hydrogen bond to oxygen in the ice lattice, thereby irreversibly binding to the crystal. Other fish and insect AFPs with quite different sequences also generally have regular repeat sequences that constitute the flat ice-binding site, usually with regularly spaced threonine residues with projecting hydroxyls that hydrogen bond to the crystal. Consequently, the adsorption–inhibition model based on hydrogen bonding of the Thr hydroxyls to ice appeared to fit these AFPs as well.

A second ice-binding mechanism was later proposed (reviewed by Davies 2014). This involves a highly structured ice-like hydration sphere of anchored clathrate waters around the ice-binding site. This ice-like water binds to the crystal surface thereby providing the ice-binding mechanism of the AFPs. It now appears that both ice-binding mechanisms are generally involved, with the predominant mechanism dependent on the specific AFP (Ebbinghaus et al. 2010, 2012; Meister et al. 2013). Cosolvents can extend the long-range hydration sphere and thereby increase the level of TH (Meister et al. 2014).

7.4 Initial Identification of Plants with IBPs

Although Art DeVries identified antifreeze proteins in Antarctic fish in the late 1960s (DeVries and Wohlschlag 1969; DeVries 1971), the first reports of plant TH did not appear until 20 years later (Urrutia et al. 1992; Griffith et al. 1992). This was not because TH activity is rare and difficult to identify, but rather because the possibility of TH had not been explored in plants until that time. Using size exclusion chromatography and gel electrophoresis, Marilyn Griffith and colleagues identified six major and several minor polypeptides with ice crystal modification properties and TH of up to 0.3 °C in cold acclimated winter rye (*Secale cereale*) leaves (Griffith et al. 1992). This work along with Marilyn Griffith's subsequent studies prior to her untimely death provided a major impetus to this field. For a historical review of Griffith's contributions, the readers are directed to Moffatt et al. (2006).

Urrutia et al. (1992) identified 16 dicot and monocot plant species with TH activity. Thermal hysteresis was not present in summer but only became evident after a period of cold acclimation in the fall, peaked during winter, and then declined in spring with the advent of warm weather (Duman et al. 1993). Several additional TH producing plants, representing more diverse plant groups were later identified (Duman et al. 1993; Duman and Olsen 1993). These included primitive tracheophytes, such as the gingko tree (*Ginko biloba*), ferns, horsetails (*Equisetum hiemale*), and mosses. Of the plants sampled in these studies, 40% had TH activity during the winter months. The number and types of plants identified with IBP activity were considerably increased by Doucet et al. (2000), who screened several plants and lichens from the United Kingdom, maritime Antarctica, and sub-Antarctic islands for RI activity. All the Antarctic species and 25% of the UK species had RI activity. These included several mosses and higher plants, and both monocots and dicots. A number of additional plants with IBP activity have subsequently been identified. For a recent review of plant species with known IBP activity see Gupta and Deswal (2014a).

7.5 Physiological Function

The evolution of potent AFPs with high TH activity, as well as the ability to inhibit ice nucleators and thereby enhance the level of supercooling, is an obvious advantage for freeze-avoiding species that must prevent ice formation. In contrast, freeze-tolerant species should experience significant negative selection pressure against extracellular IBPs with potent antifreeze activity (high TH and the ability to inhibit ice nucleators, thereby enhancing supercooling) as the rapid extracellular freezing that would occur following significant supercooling is likely to lead to intracellular ice formation, a lethal event in plant cells (reviewed by Wisniewski 1995). Consequently, potent extracellular AFPs with significant ice nucleator inhibition activities might be expected to be incompatible with freeze tolerance, which generally favors slow, controlled extracellular freezing after initial ice formation at high subzero temperatures (Gusta and Wisniewski 2013). The TH-producing species of plants that have been identified to date are all freeze tolerant, and the levels of TH measured in these plants are typically quite low (0.1–0.6 °C) relative to TH present in the body fluids of freeze-avoiding fish (1.0–2.5 °C) or freeze-avoiding insects (~2.0–6.0 °C) (Duman 2015). This low TH is also true of TH-producing proteins in freeze-tolerant animals (Duman 2015). However, intracellular IBPs with ice nucleator inhibition activity should be advantageous to both freeze-avoiding and freeze-tolerant species.

In fact, many freeze-tolerant animals, especially insects, produce endogenous extracellular ice nucleators to limit supercooling and induce crystallization in the extracellular fluid, thereby inhibiting intracellular ice formation (Zachariassen and Hammel 1976; Zachariassen and Kristensen 2000; Duman et al. 2010). The presence of endogenous ice nucleators in freeze-tolerant plants has been debated, however, there is general recognition that ice nucleation in plants can occur over a wide temperature range (-2.0 to ≤ -8.0 °C) and that woody, perennial plants freeze at warmer temperatures than herbaceous, annual plants (Wisniewski et al. 2014). The exception to this general statement is the deep supercooling of cells and tissues that occur in the buds and xylem tissues of some woody species (discussed in Sect. 7.16). The identification of the ice-nucleating active compounds and their role in the response of plants to freezing temperatures, however, remains an open question.

Recrystallization, even if the ice is only extracellular, can also cause lethal damage to frozen tissue (Mazur 1984; Tursman and Duman 1995), and RI along with the modification of crystal structure is generally thought to be the primary function of the low TH IBPs present in many freeze-tolerant plants and animals (Knight and Duman 1986; Griffith and Yaish 2004). Recrystallization refers to an increase in the size of larger ice crystals at the expense of smaller crystals in a frozen sample. Water molecules are constantly moving onto and off of the surface of ice crystals, but larger crystals (because of their lower radius of curvature and thus lower surface free energy) gain more water molecules than they lose at the expense of smaller crystals (with a greater radius of curvature and greater surface free energy). Even if the total percentage of ice in a sample does not change, crystal size increases over time as smaller crystals coalesce into large ice crystals. TH-producing IBPs attach to the surface of the ice and inhibit recrystallization by the same ice-binding mechanisms that produce TH (Knight et al. 1984). IBPs with high TH activity can produce RI at concentrations two-three orders of magnitude lower than required for TH (Knight and Duman 1986). Therefore, low TH IBPs such as those of plants are still potent RI proteins (Griffith and Yaish 2004).

7.6 Structural Classes of Plant IBPs

Several structurally different types of IBPs have evolved in plants. Most of these maintain their initial function, thus becoming dual function proteins. Some of these are reviewed in the following sections.

7.6.1 Pathogenesis-Related Proteins

Following the discovery of plant AFPs in cold acclimated winter rye leaves (Griffith et al. 1992), Hon et al. (1994, 1995) identified five (38, 36, 32, 29, and 26 kDa) major apoplast proteins that exhibited high ice RI activity and three (15, 13, and 10 kDa) minor proteins with low ice RI activity. No common amino acid sequence motifs were identified in the plant IBPs that had homology with known fish or insect IBPs, however, they had a 75–88% sequence similarity to pathogenesis-related (PR) proteins, including endo-β-1,3-glucanases (38 and 32 kDa), endochitinases (36 and 29 kDa) (Fig. 7.1a), osmotin/thaumatin-like proteins (26 and 15 kDa), and a lipid transfer protein (13 kDa).

Pathogenesis-related (PR) proteins that act as IBPs have been described in a number of other plants. Indeed, PR-proteins appear to be the most common type of plant IBP. Low TH chitinase-like proteins were also found in bittersweet nightshade, *Solanum dulcamara* (Duman 1994; Sathyanesan 1999) (Fig. 7.1b), brome grass, *Bromus inermis* (Nakamura et al. 2008), the late winter blooming wintersweet, *Chimonanthus praecox* (Zhang et al. 2011), and sea buckthorn, *Hippophae rhamnoides* (Gupta and Deswal 2014b). Chitinase-like IBPs have also been described in the needles of two gymnosperms, Norway spruce, *Picea abies* (Sabala et al. 1996; Jarzabek et al. 2009) and blue spruce, *Picea pungens* (Jarzabek et al. 2009), but unlike the low TH angiosperm chitinase-like IBPs, the gymnosperm IBPs have moderately high TH activity, comparable to the level of TH activity of fish AFPs (~2 °C). Thaumatin-like IBPs were described in winter wheat, *Triticum aestivum* (Chun et al. 1998; Kontogiorgos et al. 2007). Cold acclimated Japanese radish, *Raphanus sativus*, leaves and tubers have been reported to produce apoplastic β-1,3 glucanase and chitinase-like IBPs with ice structuring abilities but minimal TH activity (Kawahara et al. 2009). In addition, a small (1.32 kDa) glycosylated IBP that has a low TH activity of 0.30 °C but lacks RI activity is found in the apoplast of *R. sativus* (Kawahara et al. 2009).

Structures of chitinase-like IBPs of winter rye (*Secale cereale*) (Yeh et al. 2000; based on GenBankAF280438.1) and bittersweet nightshade (*Solanum dulcamara*) (Sathyanesan 1999; based on GenBankAY275459.1) are illustrated in Fig. 7.1 and compared to the tobacco (*Nicotiana tabacum*) chitinase (based on NCBI Reference Sequence NP_001311556.1), which lacks IBP activity. Note the similarities between the surface topologies of the two IBPs (Fig. 7.1d) and how they differ from the tobacco chitinase although there are also similarities (Fig. 7.1e). These homology

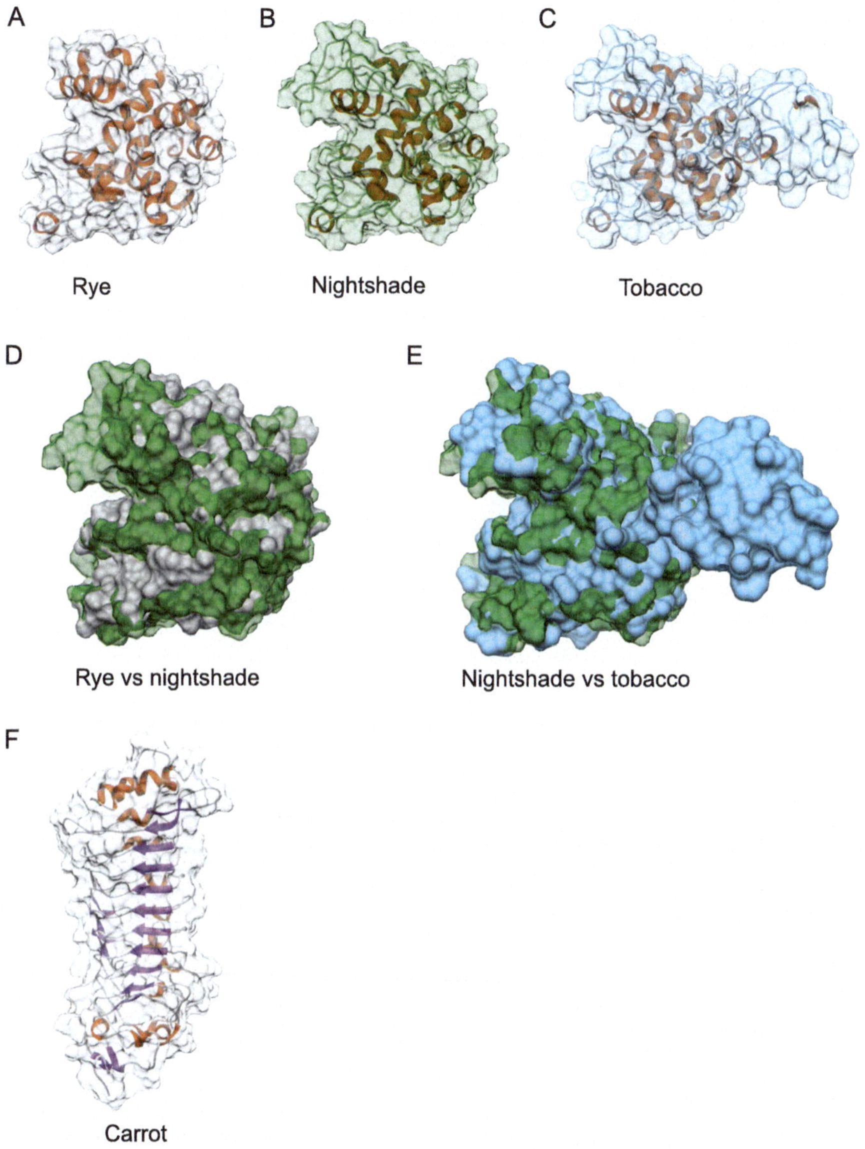

Fig. 7.1 Semitransparent molecular surface representations of homology models of various plant IBPs. (**a**) winter rye, *Secale cereale,* 24 kD endochitinase AFP (Yeh et al. 2000; GenBankAF280438.1), (**b**) bittersweet nightshade, *Solanum dulcamara,* 29 kD chitinase-like THP (Sathyanesan 1999; GenBankAY275459.1), (**c**) tobacco, *Nicotiana tabacum,* endochitinase (NCBI NP_001311556.1). (**d**) Molecular surface structures of winter rye AFP and nightshade THP are superimposed. Rye and nightshade are colored gray and green, respectively. (**e**) Molecular surface structures of nightshade THP and tobacco endochitinase are superimposed. (**f**) Carrot (*Daucus carota*) polygalacturonase inhibitor-like IBP (GenBank AFW20019.1). The present and subsequent homology models were generated using the SWISS MODEL: an automated protein homology modeling server (Schwede et al. 2003). Models were energy minimized and hydrogens were added in Yasara Structure (Krieger et al. 2004) followed by molecular dynamics simulations in water at 298°K. Molecular surface was generated in UCSF Chimera (Pettersen et al. 2004). Figures were prepared with PyMOL (Molecular Graphics System, Version 2.0 Schrodinger, LLC)

models, and the other homology models shown in subsequent figures, were prepared according to the methods described in the Fig. 7.1 legend.

The IBP present in the freeze-tolerant taproot of carrot, *Daucus carota*, has considerable sequence homology to polygalacturonase inhibitor proteins (PGIPs), another type of PR-protein (Fig. 7.1f, based on GenBankAFW20019.1) (Worrall et al. 1998; Meyer et al. 1999). Like the known PGIPs the carrot PGIP-like IBP has multiple leucine-rich repeats and has 50–65% similarity to known PGIPs, however, it lacks PGIP activity. While the ice-binding site of the carrot IBP is unknown the series of β-sheets in the center of the protein as shown in Fig. 7.1f could be a candidate. Sea buckthorn, *Hippophae rhamnoides*, also has a polygalacturonase inhibitor-like IBP (Gupta and Deswal 2012).

7.6.2 Modification of PR Proteins

The similarity and dual function of IBPs and PR-proteins raise important questions about their relationship and regulation in plant responses to abiotic and biotic stress. Do the accumulation of IBPs and PR proteins reflect a general response to stress, such that IBPs that accumulate in response to biotic stress still exhibit antifreeze activity? Conversely, do AFPs that accumulate in plants in response to low temperatures retain their antifungal activity? Additionally, if there is a difference in antifreeze activity between the IBPs that accumulate in response to cold or pathogen stress, is the difference a result of posttranslational modifications or differences associated with alternative splicing and the synthesis of isoenzymes?

To address the first question, Antikainen and Griffith (1997) characterized the various environmental conditions in which extracted PR proteins expressed antifreeze activity. In that study, 12 freeze-sensitive or -tolerant varieties of monocots and dicots were exposed to either 20 or 5 °C. High RI and ice crystal modification but low TH activity were identified in PR proteins expressed in grasses grown at 5 °C (Antikainen and Griffith 1997). Treatment of nightshade collected in November and December with citrate buffer induced a threefold increase in antifreeze activity (Huang and Duman 2001). The only samples that exhibited antifreeze activity, however, were those that were collected during those two cold months. Cold acclimated winter cereals also have greater disease resistance relative to non-acclimated plants (Hiilovaara-Teijo et al. 1999). Similar observations have been noted in Douglas fir (*Pseudotsuga menziesii*) (Zamani et al. 2003). Bark tissues inoculated with pathogens during fall and winter months had greater endochitinase activity and enhanced inhibition of ice recrystallization (Zamani et al. 2003). In general, the studies indicate that plants require low temperatures to induce the accumulation of PR proteins with TH activity.

Subsequent experiments analyzing PR proteins found that applications of abscisic or salicylic acid to winter rye resulted in an increased abundance of PR proteins but that the accumulated proteins did not exhibit any detectable RI, ice crystal modification or TH activity (Yu and Griffith 2001). Only exposure to cold, drought, or ethylene induced antifreeze activity in winter rye. Interestingly, chitinases in winter rye with RI, ice crystal modification, and TH activity induced by low temperatures had the same molecular mass as those predicted by the cDNA sequence for chitinases induced by abscisic acid (ABA) hormone treatment, indicating that the rye chitinase did not undergo posttranslational modification to either gain or lose its antifreeze ability (Yu and Griffith 2001).

Winter rye PR proteins from cold acclimated plants form oligomeric complexes (Yu and Griffith 1999). These oligomers are composed of glucanases, chitinases, thaumatin-like, and lipid transfer proteins. Formation of these complexes facilitated ice surface binding. Calcium ions (Ca^{2+}) were shown to interact and increase winter rye chitinase activity but not ice RI activity (Stressmann et al. 2004). Gupta and Deswal (2014b) confirmed that the application of Ca^{2+} to plants enhanced chitinase activity, as well as β-strand confirmation, in two class I chitinases purified from common sea buckthorn leaves.

Biochemical modifications of IBPs indicate the possibility that these oligomeric complexes are associated with oligosaccharides. Two chitinase-like proteins isolated from the rubber tree (*Hevea brasiliensis*) were analyzed using crystal structure analysis and docking experiments with oligosaccharides $(GlcNAc)_6$. The analysis provided initial evidence that PR protein oligomers are complexed with long branching sugars (Martínez-Caballero et al. 2014). A 67-kDa AFP in bittersweet nightshade loses TH activity when exposed to β-galactosidase or lectins (Duman 1994). These observations raise some interesting questions. Namely, does the accumulation of arabinoxylans and long-branching fructans, first reported by Olien and colleagues and associated with the formation of pockets of water within ice masses (Olien 1965; Kindel et al. 1989), involve the complexing of AFPs with long branching sugars? Additionally, if these putative IBP–oligosaccharide complexes are present within the apoplast, do they only form under low-temperature conditions? A more detailed discussion of fructans and freezing response is presented later in this review.

7.6.3 Other Classes of Plant IBPs

The freeze-tolerant perennial grass *Lolium perenne* produces an interesting IBP (*Lp*AFP) with low TH activity, but strong RI capability (Sidebottom et al. 2000), and the ability to inhibit bacterial ice nucleation proteins (Tomalty and Walker 2014). As seen in Fig. 7.2, *Lp*AFP contains eight 14- or 15-mer loops arranged in a left-handed β-roll structure that is reminiscent of certain high TH activity beetle AFPs (Kuiper et al. 2001). While insect AFPs typically have a flat ice-binding site on only one side of the protein, *Lp*AFP has two ice-binding sites on opposite sides of the β-roll (Kuiper et al. 2001). The reduced TH activity of *Lp*AFP is thought to be due to the smaller number of threonine residues in the ice-binding sites relative to insect AFPs (Kuiper et al. 2001; Middleton et al. 2009, 2012). The enhanced RI activity was originally thought to result from the presence of the two ice-binding sites per

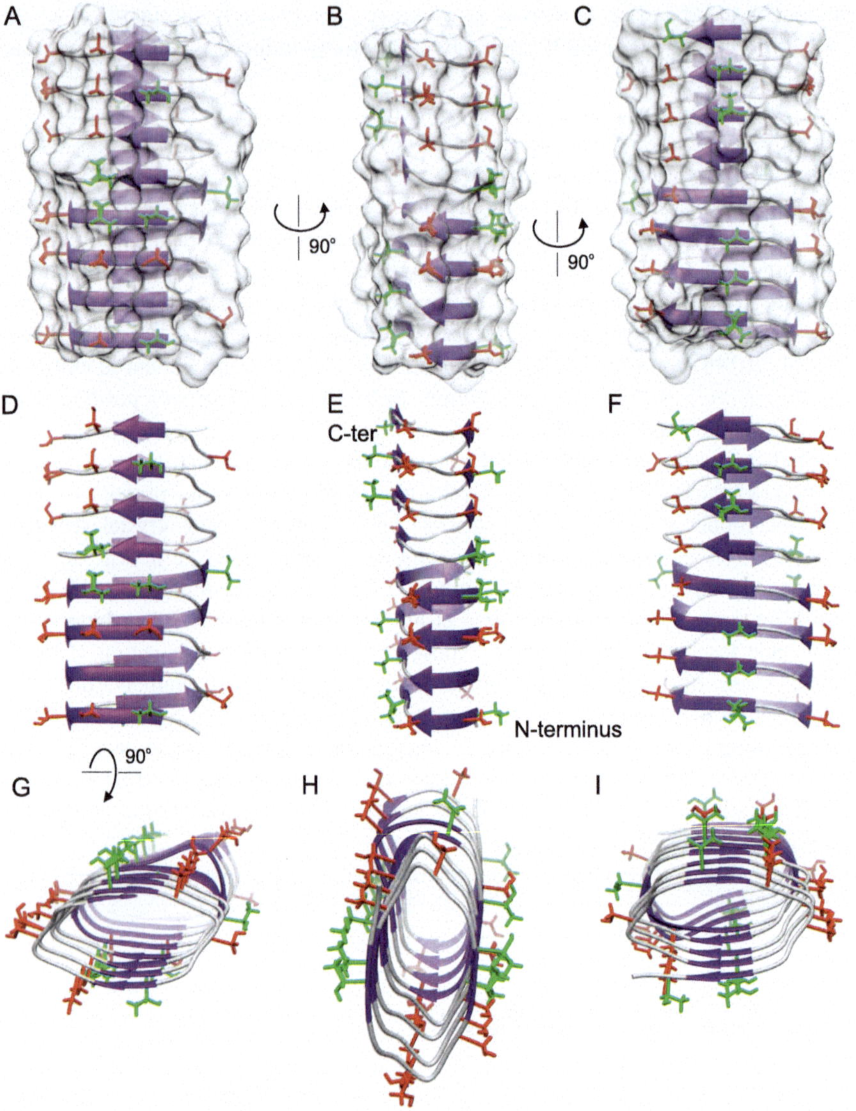

Fig. 7.2 Crystal structure of the perennial grass IBP (PDB 3ULT), *Lolium perenne*. Semitransparent molecular surface representation views are shown in the top panel, **a**, **b**, and **c**. The structures are rotated 90° to the right from **a–c**. The same views of the secondary structure are shown in the middle panel, **d**, **e**, and **f**. Beta strands are in purple and loops in gray. Threonine and serine side chains are shown in green and red, respectively. The locations of the N- and C-termini are indicated in **e**. The middle panel views are rotated 90° toward the reader in the bottom panel, **g**, **h**, and **i**. Figures were based on Middleton et al. (2012), using the methodology described in Fig. 7.1 legend

*Lp*AFP and the consequent ability to simultaneously bind to two ice crystals (Kuiper et al. 2001). Mutation experiments later demonstrated that one of the two flat ice-binding sites (shown in Fig. 7.2c) is much superior to the other (Middleton et al. 2009), however, the side with reduced ice-binding capacity may still be able to inhibit recrystallization. Middleton et al. (2012) crystallized the *Lp*AFP and generated a crystal-based structure that is the basis for Fig. 7.2.

An IBP (*Bd*IRI) with similarity to the *Lolium LP*AFP was described from the annual brome grass *Brachypodium distachyon* (Bredow et al. 2017a). Like *Lp*AFP, *Bd*IRI also has two ice-binding sites, one of which has less ice-binding capability due to a less flat surface. Although the TH activity of *Bd*IRI is low, it has high RI and crystal shaping activity. In addition, *Bd*IRI, like *LP*AFP has anti-ice nucleator activity.

As mentioned earlier, freezing following significant supercooling can lead to lethal intracellular freezing. Consequently, AFPs that promote supercooling by binding to, and inhibiting, ice nucleators should be detrimental to the survival of freeze-tolerant plants. *Lp*AFP, however, was shown to inhibit bacterial, *Pseudomonas syringae*, ice nucleation proteins, resulting in a small but statistically significant depression of the supercooling temperature (crystallization or ice nucleation temperature) of 0.9–1.9 °C (Tomalty and Walker 2014). The structurally similar *Bd*IRI had a slightly greater inhibitory effect on *P. syringae* ice nucleators, lowering the supercooling point by up to 2.26 °C (Bredow et al. 2017a). Perhaps the small ~2 °C promotion of supercooling in the apoplast is not enough to be detrimental. It should also be noted that plant TH proteins are produced in cold acclimated plants that have a general ability to survive a greater degree of supercooling prior to ice formation than non-acclimated plants (Gusta et al. 2004).

A 64-kDa IBP (STHP-64) with low TH activity and sequence homology to plant WRKY transcription factors (Fig. 7.3) is present in winter in the cytoplasm of bittersweet nightshade (Huang and Duman 2001; GenBankAF313452.1). The WRKY family of plant transcription factors is characterized by a highly conserved core WRKYGQK motif and a cysteine/histidine zinc finger region (Rushton et al. 1996; Eulgem et al. 2000). Interestingly, WRKY transcription factors are known to control the production of PR-proteins in some plants (Wang et al. 1998). STHP-64 retains its DNA-binding properties, but its unique C-terminus has 10 consecutive 13-mer repeats that contain the ice-binding motif required for TH production (Fig. 7.3b). These C-terminal repeats contain numerous lysine and histidine residues, as does the rest of the protein. STHP-64 is glycosylated, which is also required for TH activity. STHP-64 lacks a signal peptide, as would be expected for a transcription factor. Therefore, given the probable cytoplasmic location of STHP-64, its function is unlikely to involve RI, as any intracellular freezing is generally considered to be lethal. What then is the function of the TH-inducing component of the STHP-64 protein? While the promotion of extracellular supercooling followed by freezing is likely to be counterproductive in a freeze-tolerant organism, complete inhibition of intracellular ice is essential at the subzero temperatures experienced by a plant. Therefore, it is possible that STHP-64 may have an actual intracellular antifreeze function, either inhibiting cytoplasmic ice nucleators and/or preventing inoculative

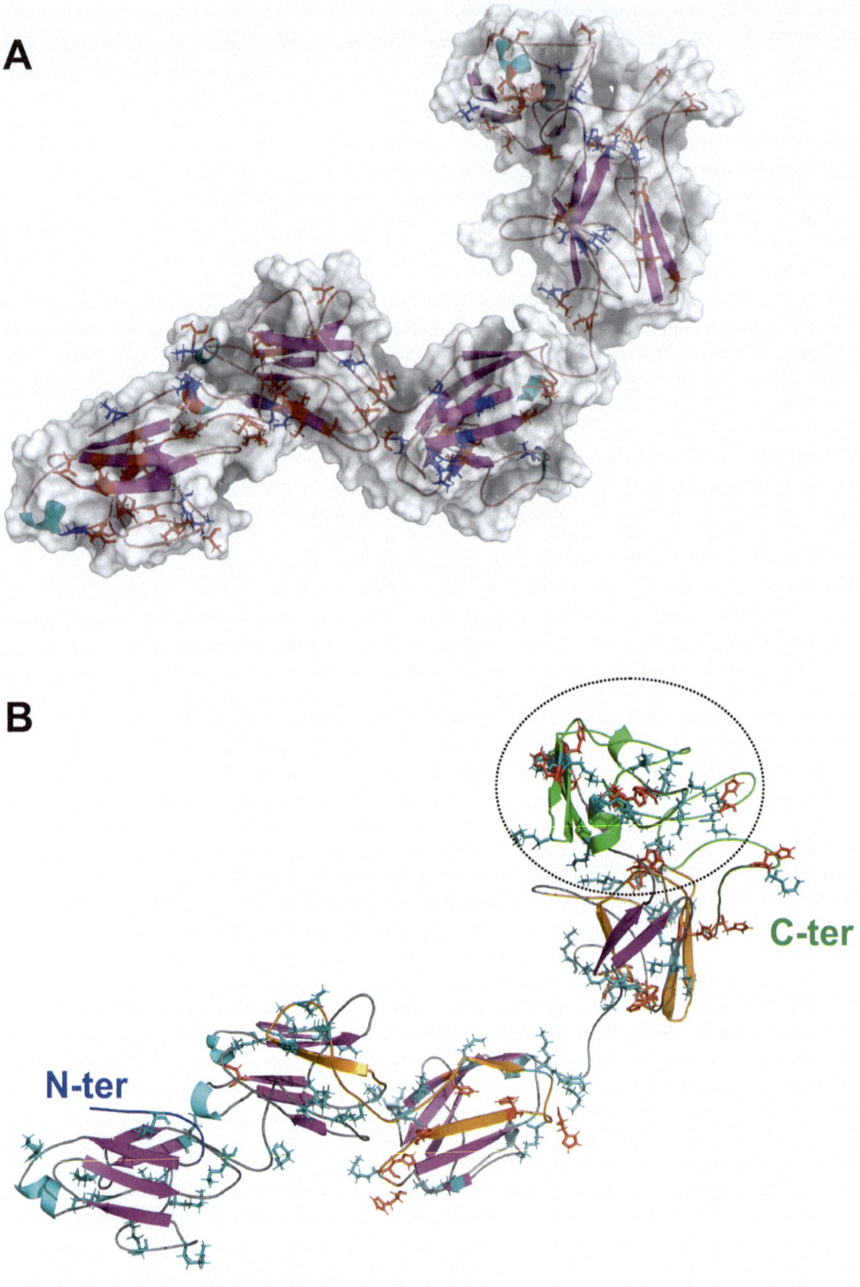

Fig. 7.3 Semitransparent molecular surface of bittersweet nightshade, *Solanum dulcamara,* 64 kDa WRKY transcription factor-like THP homology model. **(a)** Threonine and serine side chains are in blue and red, respectively. (**b**) The locations of N (blue) and C (green) termini are indicated. The beta strands comprising the two WRKY domains are in orange and the 13-mer repeat region is in light green and encircled. Lysine and histidine side chains are in teal and red, respectively. The locations of the N- and C-termini are indicated in E. Models are based on Huang and Duman (2001; GenBankAF313452.1) using the I-TASSER server (Yang and Zhang 2015). Figures were prepared with PyMOL (The PyMOL Molecular Graphics System, Version 2.0 Schrödinger, LLC)

freezing across the plasma membrane. Inhibition of ice nucleators by IBPs with low TH activities, such as STHP-64, may seem unlikely, however, recall that despite low TH activity, the IBPs from *Lolium perenne* (*Lp*IBP) and *Brachypodium distachyon* (*Bp*IRI) inhibit the ice nucleating activity of *Pseudomonas syringae* bacteria, although by only ~2 °C (Tomalty and Walker 2014; Bredow et al. 2016, 2017a). The ability, or lack thereof, of STHP-64 to inhibit ice nucleation activity and thereby promote intracellular supercooling has not been tested.

Wisniewski et al. (1999) identified a dehydrin (PCA60) in peach (*Prunus persica*) with thermal hysteresis and ice-binding activity that was localized within the cytosol, plastids, xylem nucleus, and parenchyma cells. Another dehydrin with antifreeze activity was later identified in crude extracts obtained from *Forsythia suspense* (Simpson et al. 2005).

7.7 Localization of Native Plant IBPs

The level of TH can vary with the tissue tested, however, TH is often present throughout a plant in stems, roots, tubers, and leaves (Duman et al. 1993; Duman and Olsen 1993). TH activity has even been demonstrated in flowers of the autumn blooming wood aster (*Aster cordifolius*) (Duman et al. 1993). A notable exception to widespread TH activity throughout a plant is potato (*Solanum tuberosum*), where tubers left in the ground over the winter only exhibit TH activity in axillary buds of tubers, not in the tubers themselves (Urrutia et al. 1992) Additionally, white oak (*Quercus alba*) only exhibits TH activity in acorns (Duman and Olsen 1993).

Using an immunolocalization-tissue printing method, Antikainen et al. (1996) demonstrated that IBPs were localized to intercellular spaces and within epidermal cells of cold acclimated rye leaves. Immunogold localization studies indicated that glucanase IBPs accumulated in xylem vessels, the middle lamellae, as well as cell wall junctions between the mesotome sheath cells (Pihakaski-Maunsbach et al. 1996). Pihakaski-Maunsbach et al. (1996) hypothesized that glucanase IBPs were associated with a secretory pathway and were exuded into the apoplast when plants were exposed to cold acclimating temperatures.

This premise was later confirmed in winter rye callus (Pihakaski-Maunsbach et al. 2003) and bromegrass cell-suspension cultures (Nakamura et al. 2008) that lack an organized apoplastic structure. When rye callus was grown under cold acclimation conditions, three previously identified IBPs, a 32-kDa glucanase, a 35-kDa chitinase, and a 25-kDa lipid transfer protein were identified in the growth medium but not the cells themselves. Nakamura et al. (2008) identified a significant increase in chitinase activity in the growth media but not the cells of cold acclimated bromegrass (*Bromus inermis*) suspension cultures. This indicates IBPs are readily secreted into the apoplastic space. More recent shotgun proteomic studies on apoplast fluids collected from cold acclimated *Arabidopsis* (Takahashi et al. 2016), winter oat (*Avena sativa*) and rye (*Secale cereale*) leaves (Takahashi et al. 2013), and winter wheat apical meristem and vascular crown tissues (Willick et al. 2018) identified multiple IBPs in the plant secretome. As mentioned earlier, Wisniewski et al. (1999) identified a dehydrin (PCA60) in peach (*Prunus persica*) with hysteresis and ice-binding activity that was localized within the cytosol, plastids, xylem, nucleus, and parenchyma cells. The WRKY-like transcription factor IBP (STHP-64) from bittersweet nightshade is also intracellular (Huang and Duman 2001).

Bredow et al. (2017b) reported that overexpression of *Lp*AFP in *Arabidopsis* resulted in intracellular accumulation of the protein. In all three cases, the intracellular localization may reflect a primary role in desiccation tolerance that is important to cell membranes and organelles when cells are subjected to freezing temperatures. Subsequent experiments conducted with recombinant PCA60, however, did not reveal any ability to inhibit ice recrystallization (Hughes et al. 2013). Also, as noted earlier, intracellular IBPs with ice nucleator inhibition activity, such as *LP*IBP (if it is intracellular in *Lolium*) and perhaps STHP-64 from *S. dulcamara,* would be functional in freeze-tolerant plants.

7.8 Antifreeze Glycolipids (AFGLs) in Plants

A lipopolysaccharide capable of producing TH with specific activity equal to that of highly active insect AFPs was identified in a freeze-tolerant Alaskan beetle, *Upis ceramboides* (Walters et al. 2009). Prior to this discovery, all molecules capable of producing TH were proteins. Most of the antifreeze glycolipid (AFGL) was associated with the cell membrane, and consequently, it was assumed that the AFGL might function to prevent the lethal spread of ice from the hemolymph into the cytoplasm of this freeze-tolerant insect. The AFGL also has RI activity. Approximately half of the mass of the AFGL is composed of repeats of the disaccharide β-mannopyranosyl-(1–4)β xylopyranosyl attached to a lipid, mainly fatty acids (Fig. 7.4). The xylose residues are covalently linked to mannose with a β 1–4 linkage.

Identical and/or similar xylomannan-based AFGLs were found in other insects, two frogs, and of relevance for this chapter, a plant, and bittersweet nightshade (Walters et al. 2011; Larson et al. 2014). The ^{1}H NMR used to initially identify the

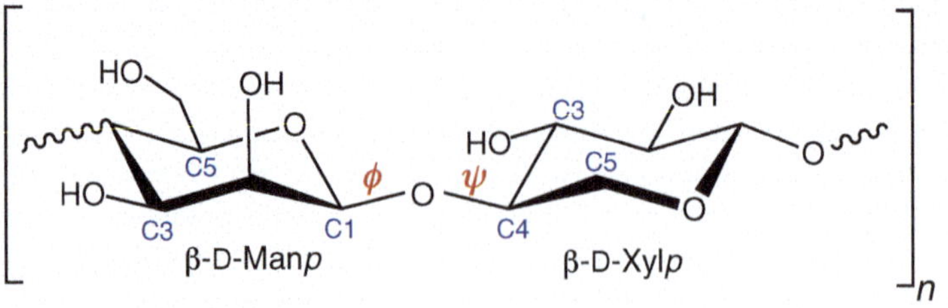

Fig. 7.4 The repeating disaccharide core structure of an antifreeze glycolipid. The β-D-Manp-(1–4)-β-D-Xylp (βManp-βXylp) of the xylomannan AFGL initially described in the Alaskan beetle (*Upis ceramboides*). The saccharide, which constitutes approximately half the mass of the AFGL, is attached to lipid. Figure based on Walters et al. (2009)

nature of these AFGLs indicated minor differences in saccharide composition among some of these organisms. All these species, except some of the insects, are freeze tolerant, and in all cases, the AFGL was only present in winter. As is the situation in the AFGL-producing animals, the bittersweet nightshade AFGL is apparently associated mainly with the cell membrane. Recall that in winter, in addition to the AFGL, bittersweet nightshade (*S. dulcamara*) has at least two types of TH-producing proteins: an apoplastic chitinase-like IBP (Sathyanesan 1999) and the cytoplasmic WRKY-transcription factor IBP (STHP-64) (Huang and Duman 2001). Therefore, bittersweet nightshade has TH-producing molecules on the plasma membrane, as well as in the apoplast and cytoplasm. Only one additional plant was checked for the presence of AFGLs. The sugar maple (*Acer saccharum*), lacked both TH activity and evidence of the AFGL. However, screening additional plants is likely to identify others with AFGLs.

7.9 Algal IBPs

Several unicellular algae (diatoms, chlorophytes, etc.) inhabit sea ice and snow. They survive, photosynthesize, and reproduce at the subzero temperatures and high levels of salinity found in the sea ice brine channels that form on the underside of ice when salts precipitate out of the frozen seawater (Priddle et al. 1986; Mock and Thomas 2005). The unfrozen brine channels in the sea ice provide a platform that positions these photosynthetic organisms near the sea surface where sunlight can penetrate when the sun returns after the long polar winter. In particular, sea ice diatoms are ecologically important, as they are often the major photosynthetic primary producers in both the Arctic and Antarctic oceans. In addition to diatoms and other algae, bacteria and various small animals inhabit the brine channels (Thomas and Dieckmann 2002). Several of these brine channel organisms are known to secrete IBPs with RI and generally low TH activities, along with the ability to influence the structure of ice and perhaps inhibit ice nucleation. These IBP activities are thought to be critical to keeping the brine channels open, thereby facilitating the channel community. In addition, the RI activity is likely responsible for the increased survival of diatoms subjected to freeze/thaw cycles (Raymond and Knight 2003). IBPs appear to be especially common in snow algae and marine sea ice diatoms from the Antarctic, but this may simply be because the Antarctic is where they have been most studied.

7.9.1 An IBP-Type Common to Different Polar Algae

IBP activity was first attributed to sea ice diatoms when interstitial water from a diatom-rich ice platelet layer in McMurdo Sound, Antarctica was shown to have ice-active substances that caused dense pitting of the basal surfaces of ice platelets at in situ concentrations (Raymond et al. 1994). The pits were similar to those produced by fish AFPs, but no TH activity was observed in water from the channels. The activity was attributed to proteins released by the diatoms because (1) activity was eliminated by protease treatment, and (2) diatom extracts had activity while diatom-free seawater did not. IBPs were partially characterized (Raymond 2000; Raymond and Knight 2003), and eventually purified and characterized from two diatom species, *Navicula glaciei* and *Fragilariopsis cylindricus* (Janech et al. 2006). Interestingly, the IBPs from both species were similar to a previously described IBP from a fungus, *Typhula ishikariensis* (47% sequence identity). The *Navicula* IBP was even more similar (76% similarity, 58% identity) to that of a bacterium, *Rhodoferax ferrireducens*, isolated from aquafer sediment. Several similar IBPs were subsequently identified from sea ice diatoms (both Arctic and Antarctic), as well as from other unicellular algae, a prasinophyte (*Pyramimonas gelidicola*), and a prymnesiophyte (*Phaeocystis antarctica*) (Bayer-Giraldi et al. 2010, 2011; Raymond and Kim 2012). Two of the diatoms (*Attheya* sp. and *Amphora* sp.) were from the Arctic, originally collected from sea ice near Baffin Island, Canada (Raymond and Kim 2012). These IBPs had (1) signal peptides indicating that they are secreted into the surrounding water of the brine channels, (2) similar sequences, and (3) appeared to have originated by horizontal gene transfer from a prokaryote source, perhaps from the IBP secreting sea ice bacterium, *Flavobacterium frigoris* that is also present in the diatom ice layer (Raymond and Kim 2012). The absence of IBP genes or IBP-homologous genes in mesophilic diatoms, along with the presence of IBPs in all sea ice diatoms investigated provided strong evidence that IBPs are an important adaptation of the diatoms and other algae, bacteria, and fungi commonly present in the sea ice (Raymond and Kim 2012). While the other Antarctic marine diatom IBPs are apparently secreted, that of *Chaetoceros neogracile*, collected from seawater rather than the sea ice environment of the other investigated diatoms, is intracellular. This IBP (*Cn*IBP) is located near the chloroplast membrane where it may function to prevent freezing and subsequent damage to the cell (Gwak et al. 2014).

In addition to the several diatoms, and the prasinophyte (*Pyramimonas gelidicola*), and the prymnesiophyte (*Phaeocystis antarctica*) (Raymond and Kim 2012) mentioned earlier, another IBP similar to those of the sea ice diatoms was described in the chlorophyte *Chlamydomonas raudensis* from ice layers in Lake Bonney, a permanently ice-covered hypersaline lake from the McMurdo Dry Valleys in the Antarctic (Raymond and Morgan-Kiss 2013). The *C. raudensis* IBP (Fig. 7.5 based on Raymond and Morgan-Kiss 2013; GenBankARM65347.1) is a close match to a hypothetical protein from a bacterium, *Stigmatella aurantiaca*, from a hypersaline pond, once again suggesting horizontal gene transfer as the origin of the *C. raudensis* IBP. Also, another chlorophyte, the Antarctic snow alga *Chloromonas brevispina*, has isoforms of more than 24 IBPs that are similar to *Chlamydomonas raudensis* IBPs from Lake Bonney and the sea ice diatoms (Raymond 2014). Therefore, all of the diatom and the two chlorophyte (*Chlamydomonas raudensis* and *Chloromonas brevispina*) IBPs are structurally similar to the bacterial and fungal (*Typhula*) IBPs. They consist of a triangular

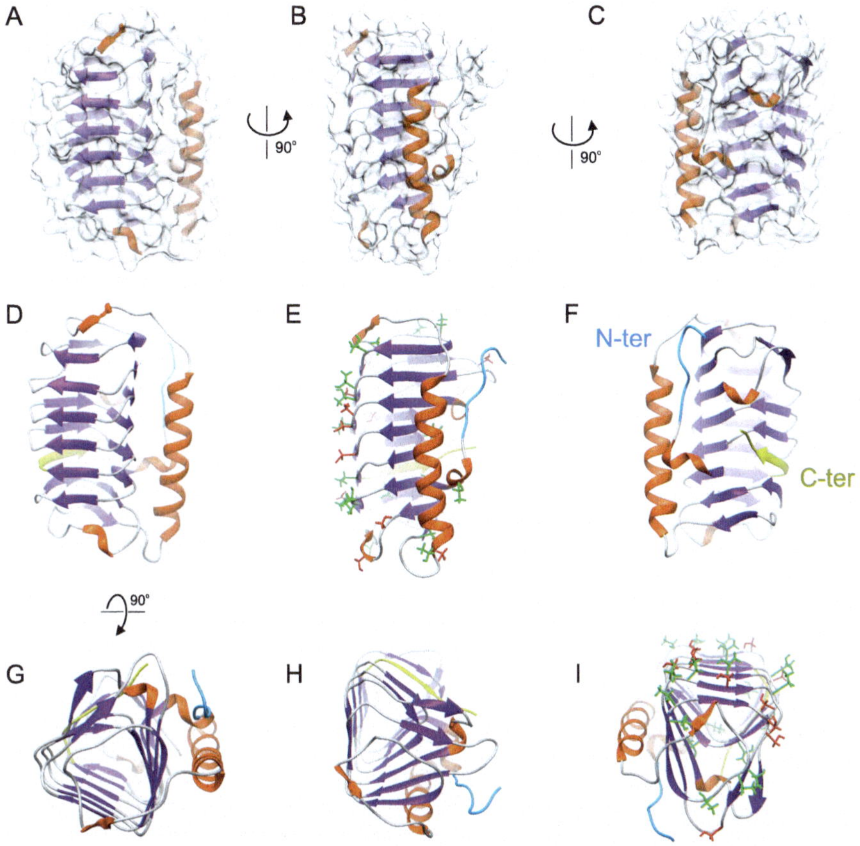

Fig. 7.5 Homology model of the *Chlamydomonas* sp. (ICE-MDV) IBP. Semitransparent molecular surface representation views are shown in the top panel, **a**, **b**, and **c**. The structures are rotated 90° to the right from **a** to **c**. The same views of the secondary structure are shown in the middle panel, **d**, **e**, and **f**. Beta strands are in purple, helices in orange and loops in gray. Threonine and serine side chains are shown in green and red, respectively in **e**. The N- and C-termini are colored blue and yellow, respectively in **f**. The middle panel views are rotated 90° toward the reader in the bottom panel, **g**, **h**, and **i**. Figures are based on Raymond and Morgan-Kiss (2017; GenBankARM65347.1) using the methodology described in Fig. 7.1 legend

β-solenoid structure reinforced by a flanking α-helix and contain the DUF3494 domain of unknown function (Fig. 7.5). The IBP genes were most likely introduced into these snow and sea ice organisms by horizontal gene transfer probably from co-inhabiting bacteria (Raymond 2014). This idea is strongly supported by the similarities of the IBP genes in spite of the incongruity between the phylogeny of the organisms (based on 18S rRNA, etc.).

7.9.2 A Second IBP-Type Present in Two Other Green Microalgae

While the *Typhula* fungal/bacterial/diatom-type IBP appears to be the most common IBP in sea ice and snow algae, including the two chlorophytes, *Chlamydomonas raudensis* and *Chloromonas brevispina*, a different IBP type has also been described in two other chlorophytes, representing both the *Chlamydomonas* and *Chloromonas* genera (Fig. 7.6). Initially, a novel IBP was identified from a *Chlamydomonas*

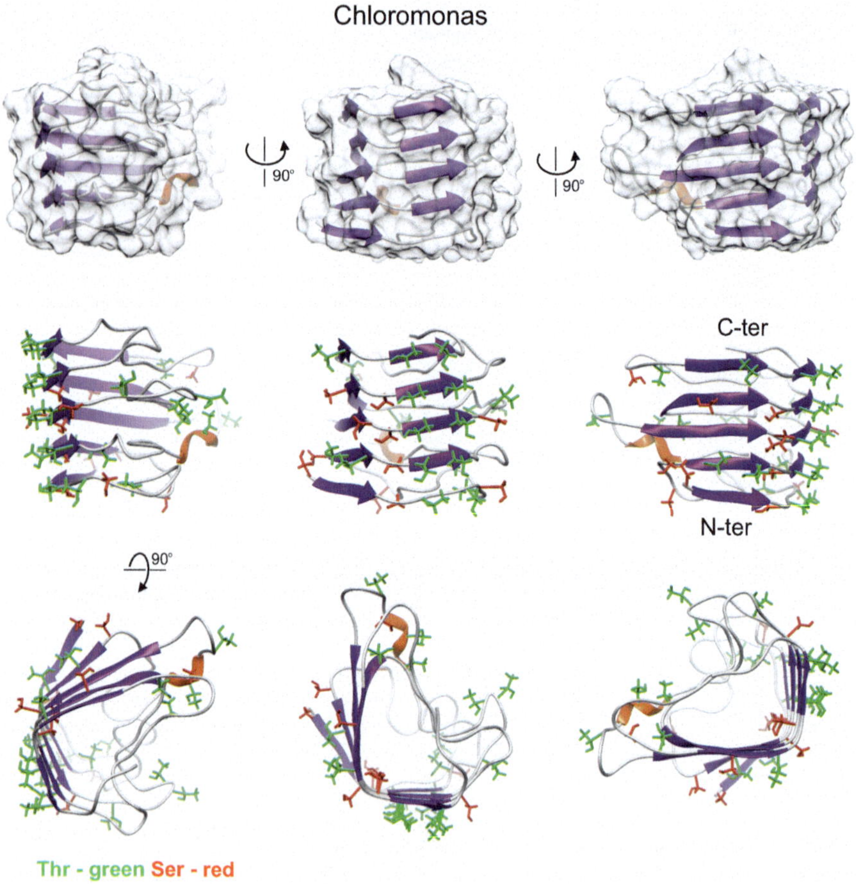

Fig. 7.6 Homology model of the freshwater *Chloromonas* sp. IBP. Semitransparent molecular surface representation views are shown in the top panel, **a**, **b**, and **c**. The structures are rotated 90° to the right from **a** to **c**. The same views of the secondary structure are shown in the middle panel, **d**, **e**, and **f**. Beta strands are in purple, helices in orange and loops in gray. Threonine and serine side chains are shown in green and red, respectively in **e**. The N- and C-termini are colored blue and yellow, respectively, in **f**. The middle panel views are rotated 90° toward the reader in the bottom panel, **g**, **h**, and **i**. Figures are based on Jung et al. (2016; GenBankHQ404890), using the methodology described in Fig. 7.1 legend

sp. (CCMP681) collected from an intertidal region in the Antarctic (Raymond et al. 2009). This alga is a unicellular euryhaline psychrophile thought to be the same as a snow alga from another Antarctic location. The IBP was present as 4+ isoforms, with signal peptides, and a presumed ice-binding site consisting of six T-F-T and one T-W-T motifs reminiscent of the T-X-T repeating units present in the ice-binding sites of certain beetle AFPs and the *Lolium* grass IBP, although this *Chlamydomonas* IBP is otherwise unique and quite different from the insect or plant IBPs. The function of this secreted IBP appears to be to maintain small pockets of unfrozen brine that resist the drainage of the brine from the ice. The secreted IBPs responsible for the RI activity in snow algae are suggested to preserve the grain boundaries between snow crystals even in glacial ice, thereby maintaining the thin liquid film that sustains microbial activity.

An Antarctic freshwater *Chloromonas* sp. collected from the ice near the King Sejong Station on King George Island produces IBPs (Fig. 7.6) similar to those of *Chlamydomonas* sp. (CCMP681), and unlike those of *Chloromonas brevispina* and the sea ice diatoms (Jung et al. 2016). This glycosylated 34 kDa cys-rich IBP (ChloroIBP) contains 15 cysteine residues, all of which are disulfide bridged except for one. The probable ice-binding site consists of a right-handed β-helix with six parallel T-X-T motifs sited on the β-2 face of the protein. Production of the protein is quickly upregulated by freezing conditions. (This *Chloromonas* sp. has high similarity to *Chloromonas* CCCryo273–06.)

Thus, four Antarctic chlorophytes, representing two different genera (*Chlamydomonas* and *Chloromonas*), have IBPs of two types, but the IBPs produced by the two genera are not common within the genera. Neither of the two types of IBPs produced by these polar diatoms and chlorophytes are similar to those found in plants.

7.10 Modification of Plants for Improved Cold Hardiness

Freeze-sensitive annual crops are routinely damaged or killed by brief late spring or early autumn frosts when temperatures are only a few degrees below 0 °C. Transgenic plants that produce IBPs with sufficient activity to inhibit freezing at these minimal subzero temperatures have the potential to have a great economic impact.

7.10.1 Transformation Studies with IBPs Confer Enhanced Freezing Tolerance in Plants

Plant IBPs, characterized by low TH activity, may not be good candidates for frost protection introduction into freeze-sensitive plants. The more active AFPs from freeze-avoiding fish, and the hyperactive insect AFPs, however, have greater potential to inhibit the propagation of ice from the exterior to the interior of the plant and/or to prevent spontaneous nucleation of water inside a plant. Even a 2 °C increase in protection would be an important improvement, and insect AFPs with this level of TH are well known. There have been several attempts to introduce fish or insect AFPs into various crop plants and Arabidopsis over the past 20 years (reviewed by Duman and Wisniewski 2014). While most of these studies generally resulted in RI and/or TH in the plants, or plant parts, unfortunately, evidence of a decrease in freezing temperature or enhanced freeze tolerance of entire plants was rarely achieved, and in many of the reports, the effect on whole plants was not tested.

In one of the earliest studies vacuum infiltration of winter flounder (*Pseudopleuronectes americanus*) AFPs into potato (*Solanum tuberosum*), canola (*Brassica napus*), and *Arabidopsis* leaves depressed the ice nucleation temperature of the leaves relative to water-infiltrated controls by 1.8 °C (Cutler et al. 1989). Additional experiments with brome grass suspension cells grown in media supplemented with flounder AFPs reduced the amount of freezable water and enhanced suspension cell survival relative to the control (Cutler et al. 1989).

Purified AFPs from beetle larvae (*Dendroides canadensis*) hemolymph were constitutively expressed in *Arabidopsis* using *Agrobacterium*-mediated transformation (Huang et al. 2002). Transformed *Arabidopsis* plants expressed the foreign AFP within apoplastic fluids and exhibited TH activity, as well as small, but statistically significantly, lowered freezing temperatures of 1–3 °C when compared to wild-type plants (Huang et al. 2002; Lin et al. 2011a, b). *Arabidopsis* transformed with an AFP from desert beetle (*Microdera punctipennis*) larvae (Wang et al. 2008) produced TH activity and increased cold tolerance relative to control plants when held at −1 °C for multiple days. Similar transformation experiments were conducted in *Arabidopsis* utilizing an AFP gene isolated from eastern spruce budworm (*Choristoneura fumiferana*) larvae (Zhu et al. 2010). The plants were placed at −20 °C for 30 min, 4 °C overnight and then held at 23 °C for varying periods. Although most of the plants died, visual inspection of the plants indicated that the transformed plants fared better than wild type.

Experiments utilizing plant IBP genes to transform *Arabidopsis* have also been conducted. The carrot (*Daucus carota*) polygalacturonase inhibitor-like IBP constitutively expressed in *Arabidopsis* conferred hexagonal ice crystal morphology after 2 h of exposure to 4 °C (Meyer et al. 1999), but cold tolerance of the plants was not tested. Interestingly, expression of perennial ryegrass (*Lolium perenne*) IBP resulted in freezing tolerance in transgenic *Arabidopsis*, especially when multiple isoforms were expressed (Bredow et al. 2017b).

7.11 Role of IBPs in Freeze-Tolerant Overwintering Crops

Chun et al. (1998) reported that IBP activity in winter rye leaves was correlated with field survival but not the LT_{50} (temperature at which half the population is unable to recover from freezing). This suggests that IBPs may not exert influence on the lowest lethal temperatures in plants but on potentially lethal processes that occur at warmer subzero temperatures. A prime example of the practical importance of plant IBPs is in the crown tissues of winter cereals. Exposure to an acute (cooling at $\leq 2\ ^{\circ}\mathrm{C\ h}^{-1}$ to a fixed subzero temperature) compared to a chronic (exposure to a fixed sublethal freezing temperature for an extended period of time) freezing stress can result in different killing temperatures in winter cereal crowns (Gusta et al. 1997), an organ that is critical to recovery and regrowth following exposure to subzero temperatures (Olien and Marchetti 1976; Chen et al. 1983; Tanino and McKersie 1985).

Snow cover provides insulation to plants under field conditions but can also affect ambient temperatures near the melting point where ice crystals recrystallize. In cold acclimated cereals, freezing first injures the vascular tissues at the base of the crown (Olien and Marchetti 1976; Tanino and McKersie 1985; Livingston et al. 2013; Willick et al. 2018). Willick et al. (2018) reported in cold acclimated "Norstar" winter wheat the preferential accumulation of apoplastic PR-AFPs in the crowns vascular transition zone.

The accumulation of proteins with the dual function of inhibiting ice recrystallization and increasing resistance to pathogens would be advantageous in winter cereals since they are subject to prolonged exposure to sub-zero temperatures over winter (Gusta et al. 1997) and are susceptible to injury from snow molds in early spring (Hiilovaara-Teijo et al. 1999). Crown vascular tissue injured by ice propagation can be further stressed by the ingress of pathogens into cavities left by melted ice crystals (Olien and Marchetti 1976; Livingston et al. 2005). This results in the formation of brown, necrotic lesions, and increased plant mortality.

7.12 Use of Plant IBPs in the Food Processing Industry

The use of plant IBPs is an attractive commercial option for the food and bioproduct industry where quality losses can be significant in frozen foods. The phase change of water and the recrystallization of ice negatively impact the quality of frozen food. Plant IBPs have potential to decrease ice crystal size and increase RI during storage to improve the stability of frozen products such as bread dough (Kontogiorgos et al. 2007; Zhang et al. 2007a, b; Xu et al. 2009) or ice cream (Regand and Goff 2006).

Reduced quality of frozen bread dough is attributed to the loss of viable yeast and the release of reducing substances from dead yeast cells (reviewed by Giannou et al. 2003). Excessive ice crystal formation will also weaken the gluten network (reviewed by Giannou et al. 2003). During freezing, water separates from the starch–gluten matrix and migrates towards the growing ice crystals. Upon thawing, this water is not incorporated into the gluten matrix. Both the reduced viability of yeast and recrystallization of ice result in longer proof times, decreased bread volume and a reduction in the desirable textural characteristics of baked bread. Frozen dough quality can be improved by using strong wheat flour, freeze-tolerant yeast or the inclusion of additives (reviewed by Giannou et al. 2003). The addition of additives can negatively impact texture properties and result in the release of volatile compounds during the baking process. One method for improving frozen dough quality that avoids the production of volatile compounds is the incorporation of plant IBPs (Zhang et al. 2007a, b). The inclusion of a 36.8-kDa carrot AFP (*Dc*AFP) in bread dough (0.06% w/w) increased RI and decreased the morality of the yeast (Zhang et al. 2007a). The addition of *Dc*AFP maintained frozen down holding loaf volume, increased dough softness, lowered the freezable water content and did not result in the release of negative volatile compounds (Zhang et al. 2007b).

A thaumatin-like IBP (21 kDa) isolated from cold acclimated winter wheat leaves was unable to protect against RI in frozen bread dough at the concentration level of 0.1% (Kontogiorgos et al. 2007). However, increased concentrations (0.3 and 0.6%) of winter wheat IBPs had a protective effect against freezing injury resulting from the duration of frozen storage or the number of freeze–thaw cycles (Xu et al. 2009). At 0.3 and 0.6% concentrations, IBPs increased water holding capacity and bread specific volume while decreasing proofing time compared to frozen dough lacking IBPs. The inclusion of cold acclimated winter wheat IBPs (at $\leq$0.13% concentration) acts as a natural ice modulator during the cold storage of ice cream (Regand and Goff 2006). The addition of cold acclimated wheat IBPs increased RI by 44%, resulting in smoother ice cream. The combination of a stabilizer (locust bean gum sugars) and the IBP had an additive effect on RI. Pasteurization did not result in a significant reduction in RI. The incorporation of similar plant IBPs may prove to be valuable ingredients with appropriate formulation testing to improved frozen food quality. A detailed, comprehensive review of the application of antifreeze compounds in the food industry is provided in Chap. 10 in Vol. II.

7.13 Sugars, Fructans, and Other Oligosaccharides Associated with Freezing Resistance in Plants

Three-dimensional reconstruction of ice formation in oats indicates that ice crystals in crown tissues can take on distinctly different forms depending on where in the crown they develop. Different cryoprotective mechanisms are likely operative in different tissues. While IBPs that inhibit recrystallization produce smaller crystals as found in leaf bases (Livingston and Tuong 2014), other mechanisms result in ice that forms vertical sheets within the crown core.

Sugars are another form of antifreeze compound and are usually considered to act in a colligative manner (Levitt 1959). Namely, their effect on freezing point depression is dependent on the number of molecules, not on their structure or size. The colligative action of sugars is the same as that of salt on roads used to lower the freezing point of water and either prevent ice from forming or melting ice already present. The end result is water with a lower freezing point. Although the colligative nature of sugars may only reduce the freezing point of water in an entire plant to a minor extent (Levitt 1959), the concentration of sugars is not homogeneous throughout all plant tissues (Canny 1995). Therefore, if sugar concentrations are higher in tissues that are crucial to the survival of meristematic regions of plants, they may indeed be cryoprotective strictly due to their colligative properties (Livingston and Henson 1998).

Livingston (2007) found that a significant amount of unfrozen water in cold acclimated oat crowns subjected to subfreezing temperatures could not be explained by colligative attributes and suggested that various non-colligative mechanisms could also be operative in winter cereal crowns. Some sugars can act in a non-colligative manner, acting as a substitute for water and incorporated within membrane bilayers increasing their stabilization (Livingston et al. 2009). In this regard, the synthetically derived sweetener Sucralose (a chlorinated sucrose molecule) has been reported to have a higher stabilizing effect on membranes than pure sucrose (Pennington et al. 2016). For other examples of cryoprotection by sugars see Sugiyama and Simura (1967), Steponkus (1968), Santerius (1973), Lineberger and Steponkus (1980).

7.14 The Significance of Sugars in Adhesive Stress

It could be supposed that every cryoprotective mechanism in plants is in opposition to a particular form of stress. One stress during freezing that has received little attention is adhesive in nature (Olien 1971, 1973, 1974; Olien and Smith 1977). It is the same phenomenon experienced when one reaches into a freezer and one's fingers adhere to an ice-cube tray. In this illustration, the fingers warm the frozen tray, melt the surface and a thin layer of water is formed. The hydrophilic nature of skin (analogous to plant cell walls and/or membranes) produces competition with the frozen ice-cube tray (analogous to ice in the apoplast of plants) for a thin intervening layer of liquid water causing strong adhesions. In plants, as temperature is reduced and dehydration becomes prominent, cells shrink in size and adhesions can tear or disrupt cells. Adhesive stress reportedly occurs during equilibrium freezing (a slow rate of freezing such that very small displacements from equilibrium occur) conditions, at temperatures between −10 and −30 °C (Olien 1977).

Visual observations of cells exposed to conditions under which adhesion stress predominates (equilibrium freezing at temperatures below −10 °C) indicate significant damage caused by adhesions (Olien and Smith 1977). The description of the damage: "…cell walls torn and plasmalemma deformed…" (Olien and Smith 1977), arguably resembles that of "fracture-jump lesions" described by Webb et al. (1994). While Olien used a light microscope to observe adhesive damage, Webb et al. used freeze–fracture techniques with scanning electron microscopy. Webb et al. describe fracture–jump lesions as regions of cells "where the fracture plane had jumped from the plasma membrane…" (Webb et al. 1994). While the exact cause of this kind of damage was not indicated, it seems possible that adhesions could be the origin.

The compliance of freezing processes to clearly defined physical laws prompted Olien to consider freezing stress using thermodynamic and kinetic principles (Olien 1971, 1973, 1974, 1977). This led to an understanding of adhesive stress from a thermodynamic perspective as well as a means by which plants could mitigate this stress (Olien and Smith 1977).

Figure 7.7a illustrates the frequency (number or amount) of molecules at their energy levels at a particular temperature. It is a graph of the Maxwell–Boltzmann distribution of the translational kinetic energy of gas molecules (Chang 1981). An assumption is made that the energy of water molecules will follow this same distribution at an ice–liquid interface. The region "A" represents ice molecules with enough energy to escape an ice lattice (melt) and "B" represents liquid

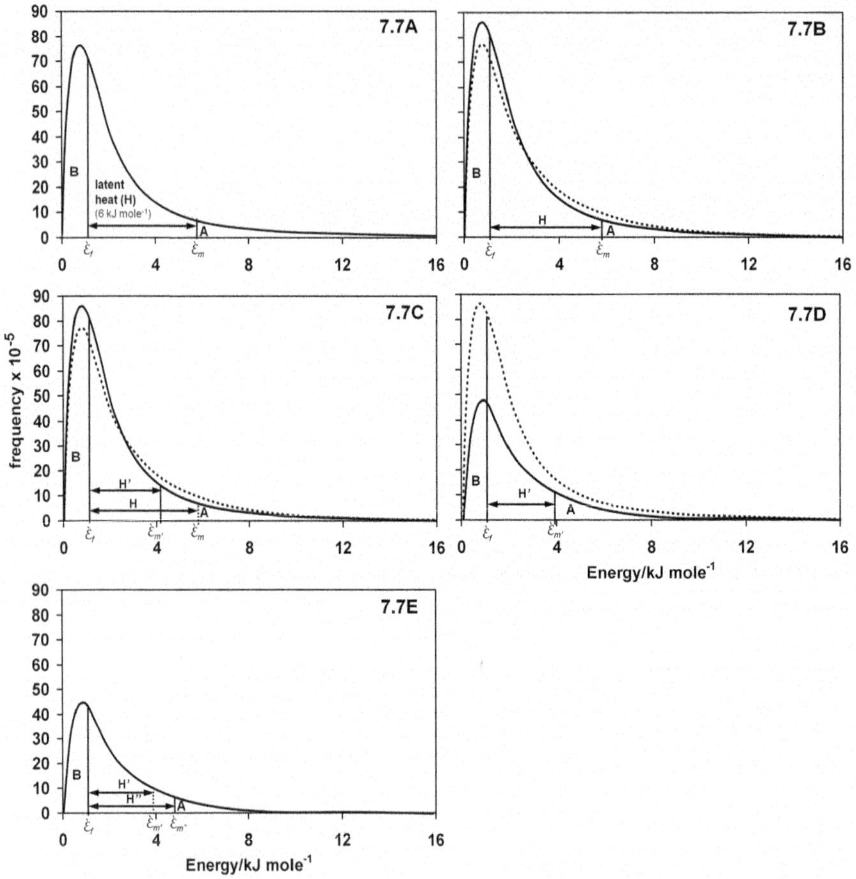

Fig. 7.7 (**a–e**) Graphical representation of adhesive stress using the Maxwell–Boltzmann distribution of molecules in a water system exposed to freezing. (**a**) Distribution of molecules in a water system at 0 °C. (**b**) Distribution of molecules in the same system after the temperature is lowered (solid line). The broken line is from **a**. (**c**) The ice–water system returning to equilibrium after the temperature shift shown in **b**. In this case equilibrium (A and B equal) was reestablished by a shift in latent heat from H to H′. (**d**) A shift in the energy distribution of water molecules as a result of solutes moving into the system. (**e**) To maintain equilibrium (equal area represented by A and B), a second shift in latent heat (from H′ to H″) occurs. Adapted from Olien (1977)

molecules with energy so low that they will add to the lattice (freeze). Because the temperature is reduced in an equilibrium manner, the value (energy) of the two integrals represented by regions A and B must be equal. The vertical line at "E_m" is the minimum energy ice molecules must have to escape the lattice (activation energy of melting) and "E_f" is the maximum energy a liquid molecule can have and still add to the lattice (activation energy of freezing). The distance between E_f and E_m is the energy that must be acquired by a molecule to escape the lattice or the amount of energy it gives up as it becomes part of the lattice (Fig. 7.7a). That energy is known as latent heat or "H" in Fig. 7.7a.

When the temperature decreases, the curve shifts, as shown by the solid line in Fig. 7.7b, reducing the number of molecules with enough energy to leave the lattice (A) and increasing the number that can freeze (B) and freezing continues. Because this is an equilibrium process (A and B are of equal energy) some modification must occur in the system to allow A and B to remain equal. One possibility is to decrease B by reducing the number of liquid molecules at the interface through dilution of interfacial water molecules with solutes. A second way for A and B to re-equilibrate is to shift the activation energy for melting downward, which increases A. If that occurs one should see a corresponding shift in latent heat (H) as in Fig. 7.7c. This shift was observed in an isothermal calorimetric analysis of a model cellulose system (Olien 1974). Olien attributed the shift in latent heat to a matric interaction of ice with hydrophilic polymers as they compete for intervening liquid (Olien 1974). This can result in highly destructive adhesions if competing polymers are part of membranes.

Adhesive stress may be relieved by releasing solutes outside the protoplast producing a fluid barrier to adhesions. This would cause a change in the number of liquid molecules in the interface without initially affecting ice. Once again A and B are out of equilibrium as shown in Fig. 7.7d. There are now a greater number of molecules with energy to escape and melting occurs. For equilibrium to be reestablished the activation energy must again change. A second shift in latent heat was observed experimentally, as shown in Fig. 7.7e. The ice lattice is now a greater distance from the protoplast with an intervening liquid of higher osmotic pressure. This establishes an osmotic gradient between the protoplast and ice crystals. As freezing progresses, ice crystals will grow at the expense of intracellular water resulting in desiccation.

After deriving a thermodynamic basis for adhesive stress, Olien hypothesized that if plants release sugars into the interface of an adhesive interaction, this form of stress could be mitigated or at least be relegated to significantly lower temperatures. This hypothesis was based partly on the discovery that fructan accumulates in crown tissue of winter cereals during cold acclimation (Livingston et al. 2009). Furthermore, when exposed to a mild freeze, fructan is hydrolyzed to its constituent components, fructose and sucrose (Trunova 1965; Tumanov et al. 1976). In an analysis of fructan concentrations in rye, Olien confirmed that fructan is hydrolyzed at below freezing temperatures (Olien 1984; Olien and Lester 1985) but also discovered that simple sugars (the products of hydrolyzed fructan) were exported into the apoplast. This finding has since been confirmed in other winter cereals and fructan, as well as fructan hydrolase and invertase, are secreted into the apoplast (Livingston and Henson 1998).

The prevention or minimization of adhesive stress when plants are frozen and sugars are released into the apoplast has not been confirmed by in vivo studies. Increases in carbohydrates during freezing were correlated with an increase in freezing tolerance as early as 1931 (cited by Trunova 1965) in wheat. This adaptation, called the "second phase of hardening," has since been termed, "subzero acclimation" (Livingston et al. 2007; Le et al. 2008; Herman et al. 2006) and was confirmed in rye and barley (*Hordeum vulgare*) (Olien and Lester 1985), oat and *Arabidopsis* (Livingston et al. 2005, 2007), and alfalfa (*Medicago sativa*) (Castonguay et al. 1993). A mild freeze (approximately −3 °C) has been shown to confer freezing tolerance beyond cold acclimation in many plant species. Whether the cause is an increase in sugars that minimize adhesions or a decrease in free water has yet to be confirmed.

7.15 Deep Supercooling of Cells and Plant Tissues: A Unique Attribute of Woody Plants

As mentioned, all plants supercool to some extent below 0 °C. This is brought about by the presence of AFPs with TH activity, sugars, and other solutes that depress the freezing temperature of cellular water, and/or the absence of nucleating agents that favor the formation of ice crystals at relatively warm, subzero temperatures. Several species of woody plants exhibit deep supercooling in which the water in cells or tissues remains unfrozen to temperatures as low as −40 °C or slightly lower, despite being surrounded by extracellular ice in the apoplast, and also do not lose water to the sites of extracellular ice (reviewed by Wisniewski 1995). This enigmatic feature requires the presence of a barrier that prevents ice propagation from outside the cell or tissue into the supercooled region where it would induce freezing, and at the same time severely inhibits or prevents water moving to the sites of extracellular ice, despite the presence of a large vapor pressure gradient. The ability to withstand a large vapor pressure gradient (approximately −400 bars at −40 °C) requires cell walls of supercooled cells to be very rigid (Burke 1979). The ability to deep supercool is present in reproductive buds and xylem parenchyma cells of several woody plant species. A full review of this topic is available in Fujikawa et al. 2009; Wisniewski 1995; Wisniewski et al. 2009; Quamme 1995). The topic was also recently reviewed by Wisniewski et al. (2014). As the focus of this chapter is on plant antifreeze proteins, only a brief summary is provided here.

The extent of deep supercooling in plants can range from −15 to −60 °C (Wisniewski 1995) and vary on a seasonal basis. Deep supercooling has been proposed to represent a limiting constraint on the distribution of species, restricting plants that exhibit this trait to below the −40 °C isotherm (George et al. 1974), however, exceptions to this limitation have been noted and largely attributed to increases in solute concentration brought about through a slow process of cellular dehydration and the concomitant increase in freezing point depression (Gusta et al. 1983). Despite the importance and prevalence of deep supercooling in numerous species of woody plants, the subject, which received a great deal of research interest in the latter part of the twentieth century (1970–1990), has not been the subject of much study in recent years.

The freezing of small clusters of cells at random throughout the xylem, as suggested by Hong and Sucoff (1980) was corroborated by infrared differential thermal analysis (IDTA) (Neuner et al. 2010), clearly indicating the presence of ice barriers between xylem parenchyma cells. Research conducted by Fujikawa, Kasuga and colleagues represent the most rigorous attempt to elucidate the mechanism of deep supercooling of xylem tissues (Kuwabara et al. 2013; Kasuga et al. 2010). The low porosity of the secondary cell wall, and specifically the pit membrane, was demonstrated to play an important role as a barrier to ice propagation and loss of water to extracellular ice in xylem parenchyma cells (Wisniewski 1995). The role of anti-nucleating substances, identified as tannin-related polyphenol and flavanol glycosides, in promoting deep supercooling has also been reported (Kuwabara et al. 2013; Kasuga et al. 2010). Importantly any mechanism purported to explain the mechanism of deep supercooling must explain the differences that exist in supercooling ability between plant genera and species, as well as the shifts in deep supercooling that occur on a seasonal basis within a species. In the case of both porosity and anti-nucleating agents, much research remains to be done before a comprehensive understanding of the mechanisms involved in deep supercooling is obtained.

7.16 Conclusions

Proteins that determine the temperature at which ice crystals will form in water-based solutions in cells and tissues and that bind to growing ice crystals, thus affecting their size, and that also impact ice recrystallization have been widely documented and studied in many plant, bacterial, fungal, insect, and animal species. The properties of these proteins have led them to be referred to as antifreeze proteins (AFPs), thermal hysteresis (TH) proteins, ice-binding proteins (IBPs), and recrystallization inhibiting (RI) proteins. In contrast to AFPs identified in insects and animals, the TH activity of plant AFPs (IBPs) is relatively weak. For this reason, their ice-binding and RI properties have been suggested to be more relevant to their role in preventing freeze injury in plants. Both ice-binding and RI activity would prevent the formation of large ice crystals that could potentially cause mechanical disruption of cells and tissues within the plant. The relatively low TH activity of plant AFPs (IBPs), however, has been solely demonstrated *in vitro* in plant or protein (purified and non-purified) extracts. This approach does not consider, however, the presence of a cell wall and the existing pore structure of the any specific cell wall. Therefore, more studies on TH activity of plant AFPs (IBPs) should be conducted in nanopore systems that simulate cell wall structure to gain a more comprehensive understanding of the potential role of the TH activity of plant AFPs in ice formation and propagation, especially since these proteins are secreted into the apoplast.

Freeze-sensitive plants are often exposed to frost conditions in which the temperature only drops a few degrees below 0 °C. In these scenarios, transgenic approaches that induce the expression of AFPs with strong TH activity in plants have the potential of providing a significant level of frost protection. Successful demonstration of this approach could save billions of dollars that are lost each year due to frost injury to crop and horticultural plants. Adequate levels of protection using this approach, however, have not yet been achieved.

In addition to AFPs (IBPs), the roles of sugars and complex carbohydrates (fructans) in improving freezing tolerance have been extensively explored. Compounds that contribute to deep supercooling, especially the anti-nucleating properties of polyphenols, flavanols, and other secondary metabolites, need to be further documented. The manipulation of these compounds, including the recent reports of glycolipids with antifreeze properties, as well as their interactions of these compounds with known AFPs remain to be explored.

References

Antikainen M, Griffith M (1997) Antifreeze protein accumulation in freezing-tolerant cereals. Physiol Plant 99(3):423–432

Antikainen M, Griffith M, Zhang J, Hon W-C, Yang DS, Pihakaski-Maunsbach K (1996) Immunolocalization of antifreeze proteins in winter rye leaves, crowns, and roots by tissue printing. Plant Physiol 110(3):845–857

Arora R (2018) Mechanism of freeze-thaw injury and recovery: a cool retrospective and warming up to new ideas. Plant Sci 270:301–313

Arora R, Palta JP (1988) In vivo perturbation of membrane associated calcium by freeze-thaw stress in onion bulb cells. Plant Physiol 87:622–628

Bayer-Giraldi M, Uhlig C, John U, Mock T, Valentin K (2010) Antifreeze proteins in polar sea ice diatoms. Environ Microbiol 12:1041–1062

Bayer-Giraldi M, Weikusat I, Besir H, Dieckmann G (2011) Characterization of an antifreeze protein from the polar diatom *Fragilariopsis cylindrus* and its relevance in sea ice. Cryobiology 63:210–219

Bredow M, Vanderbeld V, Walker VK (2016) Knockdown of ice-binding proteins in *Brachypodium distachyon* demonstrates their role in freeze protection. PLoS One 11(12): e0167941. https://doi.org/10.1371/journal.pone.0167941

Bredow M, Tomalty HE, Smith L, Walker VK (2017a) Ice and anti-nucleating activities of an ice-binding protein from the annual grass *Brachypodium distachyon*. Plant Cell Environ 41:983–992. https://doi.org/10.1111/pce.12889

Bredow M, Vanderbeld V, Walker VK (2017b) Ice-binding proteins confer freeze tolerance on *Arabidopsis thaliana*. Plant Biotechnol J 15:68–81

Burke MJ (1979) Discussion. Water in plants: the phenomenon of frost survival. In: Underwood LS, Tieszen LL, Callahan AB, Folk GE (eds) Comparative mechanisms of cold adaptations. Academic Press, New York, pp 259–281

Canny MJ (1995) Apoplastic water and solute movement: new rules for an old space. Annu Rev Plant Physiol Plant Mol Biol 46:215–236

Castonguay Y, Nadeau P, Laerge S (1993) Freezing tolerance and alteration of translatable mRNAs in alfalfa hardened at subzero temperatures. Plant and Cell Physiol 34:31–38
Celik Y, Graham LA, Mok Y-F, Bar M, Davies PL, Braslavsky L (2010) Superheating of ice crystals in antifreeze protein solutions. Proc Natl Acad Sci U S A 107:5423–5428
Chang R (1981) Physical chemistry with applications to biological systems, 2nd edn. Macmillan, New York
Chen THH, Gusta LV, Fowler DB (1983) Freezing injury and root development in winter cereals. Plant Physiol 73(3):773–777
Chun JU, Yu XM, Griffith M (1998) Genetic studies of antifreeze proteins and their correlation with winter survival in wheat. Euphytica 102:219–226
Cutler AJ, Saleem M, Kendall E, Gusta LV, Georges F, Fletcher GL (1989) Winter flounder antifreeze protein improves the cold hardiness of plant tissues. J Plant Physiol 135(3):351–354
Danyluk J, Perron A, Houde M, Limin A, Fowler B, Benhamou N, Sarhan F (1998) Accumulation of an acidic dehydrin in the vicinity of the plasma membrane during cold acclimation of wheat. Plant Cell 10(4):623–638
Davies PL (2014) Ice-binding proteins: a remarkable diversity of structures for stopping and starting ice growth. Trends Biochem Sci 39:548–555
DeVries AL (1971) Glycoproteins as biological antifreeze agents in Antarctic fishes. Science 172:1152–1155
DeVries AL, Wohlschlag C (1969) Freezing resistance in some Antarctic fishes. Science 163:1073–1075
Doucet CJ, Byass L, Elias L, Worrall D, Smallwood M, Bowles DJ (2000) Distribution and characterization of recrystallization inhibitor activity in plant and lichen species from the UK and maritime Antarctic. Cryobiology 40:218–227
Duman JG (1994) Purification and characterization of a thermal hysteresis protein from a plant, the bittersweet nightshade *Solanum dulcamara*. Biochim Biophys Acta 1206:129–135
Duman JG (2015) Anima ice-binding (antifreeze) proteins and glycolipids: an overview with emphasis on physiological function. J Exp Biol 218:1846–1855
Duman JG, DeVries AL (1972) Freezing behavior of aqueous solutions of glycoproteins from the blood of an Antarctic fish. Cryobiology 9:469–472
Duman JG, Olsen TM (1993) Thermal hysteresis protein activity in bacteria, fungi, and phylogenetically diverse plants. Cryobiology 30:322–328
Duman G, Wisniewski M (2014) The use of antifreeze proteins for frost protection in sensitive crop plants. Environ Exp Bot 106:60–69
Duman JG, Wu DW, Olsen TM, Urrutia M, Tursman D (1993) Thermal-hysteresis proteins. Adv Low Temp Biol 2:131–182
Duman JG, Walters KR, Sformo T, Carrasco MA, Nickell P, Barnes BM (2010) Antifreeze and ice nucleator proteins. In: Denlinger D, Lee RE (eds) Low temperature biology of insects. Cambridge University Press, Cambridge, pp 59–90
Ebbinghaus S, Meister K, Born B, DeVries AL, Gruebele M, Havenith M (2010) Antifreeze glycoprotein activity correlates with long-range protein−water dynamics. J Am Chem Soc 2010(132):12210–12211
Ebbinghaus S, Meister K, Prigozhin MB, DeVries AL, Havenith M, Dzubiella J, Gruebele M (2012) Functional importance of short-range and long-range solvent interactions in helical antifreeze peptides. Biophys J 103:L20–L22
Eulgem T, Rushton PJ, Robatzek S, Somssich IE (2000) The WRKY superfamily of plant transcription factors. Trends Plant Sci 5:199–206
Fowler DB (2008) Cold acclimation threshold induction temperatures in cereals. Crop Sci 48 (3):1147–1154
Fowler DB (2012) Wheat production in the high winter stress climate of the great plains of North America—an experiment in crop adaptation. Crop Sci 52(1):11–20. https://doi.org/10.2135/cropsci2011.05.0279

Franks F (1985) Biophysics and biochemistry at low temperatures. Cambridge University Press, Cambridge, p 210

Fujikawa S, Kasuga J, Takata N, Arakawa K (2009) Factors related to change of deep supercooling capability in xylem parenchyma cells of trees. In: Gusta LV, Wisniewski ME, Tanino K (eds) Plant cold hardiness: from the laboratory to the field. CABI, Oxford, pp 29–42

George MF, Burke MJ, Pellett HM, Johnson AG (1974) Low temperature exotherms and woody plant distribution. HortScience 9:519–522

Giannou V, Kessoglou V, Tzia C (2003) Quality and safety characteristics of bread made from frozen dough. Trends Food Sci Tech 14(3):99–108

Gray GR, Chauvin LP, Sarhan F, Huner NP (1997) Cold acclimation and freezing tolerance (a complex interaction of light and temperature). Plant Physiol 114(2):467–474

Griffith M, Yaish MW (2004) Antifreeze proteins in overwintering plants: a tale of two activities. Trends Plant Sci 9:399–405

Griffith M, Ala P, Yang DS, Hon WC, Moffatt BA (1992) Antifreeze protein produced endogenously in winter rye leaves. Plant Physiol 100:593–596

Gu L, Hanson PJ, Mac Post W, Kaiser DP, Yang B, Nemani R, Pallardy SG, Meyers T (2008) The 2007 eastern US spring freeze: increased cold damage in a warming world. Bioscience 58 (3):253–262

Gupta R, Deswal R (2012) Low temperature stress modulated secretome analysis and purification of antifreeze protein from *Hippophae rhamnoides*, a Himalayan wonder plant. J Proteome Res 11:2684–2696

Gupta R, Deswal R (2014a) Antifreeze proteins enable plants to survive in freezing conditions. J Biosci 39:931–944

Gupta R, Deswal R (2014b) Refolding of β-stranded class I chitinases of *Hippophae rhamnoides* enhance the antifreeze activity during cold acclimation. PLoS One 9:e91723

Gusta LV, Wisniewski ME, Nesbitt NT, Gusta ML (2004) The effect of water, sugars, and proteins on the pattern of ice nucleation and propagation in acclimated and non-acclimated canola leaves. Plant Physiol 135:1642–1653

Gusta LV, Tyler MJ, Chen THH (1983) Deep undercooling in woody plant taxa growing north of the −49 C isotherm. Plant Physiol 72:122–128

Gusta L, Wisniewski M, Trischuk R (2009) Patterns of freezing in plants: the influence of species, environment and experiential procedures. In: Gusta L, Wisniewski M, Tanino K (eds) Plant cold hardiness: from the laboratory to the field. CABI, Boston, pp 214–225

Gusta LV, Burke MJ, Kapoor AC (1975) Determination of unfrozen water in winter cereals at subfreezing temperatures. Plant Physiol 56(5):707–709

Gusta LV, Wisniewski M (2013) Understanding plant cold hardiness: an opinion. Physiol Plant 147 (1):4–14

Gusta LV, O'Connor BJ, MacHutcheon MG (1997) The selection of superior winter-hardy genotypes using a prolonged freeze test. Can J Plant Sci 77(1):15–21

Gwak Y, Jung W, Lee Y, Kim JS, Kim CG, Ju J-H, Song C, Hyun J-K, Jin ES (2014) An intracellular antifreeze protein from and Antarctic microalga that responds to various stresses. FASEB J 28:4924–4935

Herman EM, Rotter K, Premakumar R, Elwinger G, Bae R, Ehler-King L, Chen S, Livingston DP (2006) Additional freeze hardiness in wheat acquired by exposure to −3°C is associated with extensive physiological, morphological, and molecular changes. J Exp Bot 57(14):3601–3618

Hiilovaara-Teijo M, Hannukkala A, Griffith M, Yu X-M, Pihakaski-Maunsbach K (1999) Snow-mold-induced apoplastic proteins in winter rye leaves lack antifreeze activity. Plant Physiol 121 (2):665–674

Holmberg N, Bülow L (1998) Improving stress tolerance in plants by gene transfer. Trends Plant Sci 3(2):61–66

Hon WC, Griffith M, Chong P, Yang DSC (1994) Extraction and isolation of antifreeze proteins from winter rye (*Secale cereale* L.) leaves. Plant Physiol 104:971–980

Hon WC, Griffith M, Mlynarz A, Kwok YC, Yang DSC (1995) Antifreeze proteins in winter rye are similar to pathogenesis-related proteins. Plant Physiol 109:879–889

Hong S, Sucoff E (1980) Units of freezing of deep supercooled water in woody xylem. Plant Physiol 66:40–45

Huang T, Duman JG (2001) Cloning and characterization of a thermal hysteresis (antifreeze) protein with DNA-binding activity from winter bittersweet nightshade, *Solanum dulcamara*. Plant Mol Biol 48:339–350

Huang T, Nicodemus J, Zarka DG, Thomashow MF, Wisniewski M, Duman JG (2002) Expression of an insect (*Dendroides canadensis*) antifreeze protein in *Arabidopsis thaliana* results in a decrease in plant freezing temperature. Plant Mol Biol 50:333–344

Hughes SL, Schart V, Malcolmson J, Hogarth KA, Martynowicz DM, Tralman-Baker E, Patel SN, Graether SP (2013) The importance of size and disorder in the cryoprotective effects of dehydrins. Plant Physiol 163(3):1376–1386

Janech MG, Krell A, Mock T, Kang J-S, Raymond JA (2006) Ice-binding proteins from sea-ice diatoms (Bacillariophyceae). J Phycol 42:410–416

Jarzabek M, Pukacki PM, Nuc K (2009) Cold-regulated proteins with potent antifreeze and cryoprotective activities in spruces (*Picea* spp.). Cryobiology 58:268–274

Jung W, Campbell RL, Gwak Y, Kim JI, Davies PL, Jin ES (2016) New cysteine-rich ice-binding protein secreted from Antarctic microalga, *Chloromonas* sp. PLoS One 11:e0154056. https://doi.org/10.1371/journal.pone.0154056

Kawahara H, Fujii A, Inoue M, Kitao S, Fukuoka J, Obata H (2009) Antifreeze activity of Japanese radish and purification of antifreeze peptide. Cryo-Lett 30:119–131

Kasuga J, Fukushi Y, Kuwabara C, Wamg D, Nishioka A, Fujikawa E, Arakawa K, Fujikawa S (2010) Analysis of supercooling-facilitating (anti-ice nucleation) activity of flavonol glycosides. Cryobiology 60:24–243

Kindel PK, Liao S-Y, Liske MR, Olien CR (1989) Arabinoxylans from rye and wheat seed that interact with ice. Carbohydr Res 187(2):173–185. https://doi.org/10.1016/0008-6215(89)80001-1

Knight CA, DeVries AL (1989) Melting inhibition and superheating of ice by an antifreeze glycopeptides. Science 245:505–507

Knight CA, Duman JG (1986) Inhibition of recrystallization of ice by insect thermal hysteresis proteins: a possible cryoprotective role. Cryobiology 23:256–262

Knight CA, DeVries AL, Oolman LD (1984) Fish antifreeze protein and the freezing and recrystallization of ice. Nature 308:295–296

Knight CA, Cheng C-HC, DeVries L (1991) Adsorption of alpha-helical antifreeze peptides on specific ice crystal surface planes. Biophys J 59:409–418

Kontogiorgos V, Regand A, Yada RY, Goff HD (2007) Isolation and characterization of ice structuring proteins from cold acclimated winter wheat grass extract for recrystallization inhibition in frozen foods. J Food Biochem 31:139–160

Krieger EI, Darden T, Nabuurs SB, Finkelstein A, Vriend G (2004) Making optimal use of empirical energy functions: force-field parameterization in crystal space. Proteins 57:678–683

Kuiper MJ, Davies PL, Walker VK (2001) A theoretical model of a plant antifreeze protein from *Lolium perenne*. Biophys J 81:3560–3565

Kuwabara C, Wang D, Endoh K, Fukushi Y, Arakawa K, Fujikawa S (2013) Analysis of supercooling activity of tannin-related phenols. Cryobiology 67:40–49

Larson D, Middle L, Vu H, Zhang W, Serianni AS, Duman JG, Barnes BM (2014) Wood frog adaptations to overwintering in Alaska: new limits to freezing tolerance. J Exp Biol 217:2193–2200

Le MQ, Engelsberger WR, Hincha DK (2008) Natural genetic variation in acclimation capacity at sub-zero temperatures after cold acclimation at 4°C in different *Arabidopsis thaliana* accessions. Cryobiology 57:104–112

Levitt J (1959) Effects of artificial increases in sugar content on frost hardiness. Plant Physiol 34:401–402

Levitt J (1980) Responses of plants to environmental stress, Vol 1: Chilling, freezing, and high temperature stresses. Academic Press, New York

Lin X, Wisniewski ME, Duman JG (2011a) Expression of two self-enhancing antifreeze proteins from the beetle *Dendroides canadensis* in *Arabidopsis thaliana*. Plant Mol Biol Rep 29:802–813

Lindow SE (1989) Control of epiphytic ice nucleation-active bacteria for management of frost injury. In: Lee RE, Warren GJ, Gusta LV (eds) Biological ice nucleation and its applications. APS Press, Minneapolis, MN, pp 239–256

Lineberger RD, Steponkus PL (1980) Cryoprotection by glucose, sucrose and raffinose to chloroplast thylakoids. Plant Physiol 65:298–304

Lin X, Wisniewski ME, Duman JG (2011b) Expression of two self-enhancing antifreeze proteins from the beetle *Dendroides canadensis* in *Arabidopsis thaliana*. Plant Mol Biol Rep 29 (4):802–813

Livingston DP III (2007) Quantifying liquid water in frozen plant tissues by isothermal calorimetry. Thermochim Acta 459:116–120

Livingston DP III, Herman EM, Premakumar R, Tallury SP (2007) Using A*rabidopsis thaliana* as a model to study subzero acclimation in small grains. Cryobiology 54:154–163

Livingston D, Premakumar R, Tallury SP (2005) Carbohydrate concentrations in crown fractions from winter oat during hardening at sub-zero temperatures. Ann Bot 96(2):331–335. https://doi.org/10.1093/aob/mci167

Livingston DP, Henson CA (1998) Apoplastic sugars, fructans, fructan exohydrolase, and invertase in winter oat: responses to second-phase cold hardening. Plant Physiol 116(1):403–408

Livingston DP III, Tuong TD (2014) Understanding the response of winter cereals to freezing stress through freeze-fixation and 3d reconstruction of ice formation in crowns. Environ Exp Bot 106:24–33

Livingston DP III, Hincha D, Heyer AG (2009) Fructan and its relationship to abiotic stress tolerance in plants. Cell Mol Life Sci 66:2007–2023

Livingston DP, Henson CA, Tuong TD, Wise ML, Tallury SP, Duke SH (2013) Histological analysis and 3D reconstruction of winter cereal crowns recovering from freezing: a unique response in oat (*Avena sativa* L.). PLoS One 8(1):e53468

Martínez-Caballero S, Cano-Sánchez P, Mares-Mejía I, Díaz-Sánchez AG, Macías-Rubalcava ML, Hermoso JA, Rodríguez-Romero A (2014) Comparative study of two GH19 chitinase-like proteins from *Hevea brasiliensis*, one exhibiting a novel carbohydrate-binding domain. FEBS J 281(19):4535–4554

Mazur P (1984) Freezing of living cells: mechanisms and implications. Am J Phys 247:C125–C142

Meister K, Ebbinghaus Y, Xu Y, Duman JG, DeVries AL, Gruebele DM, Leitner DM, Havenith M (2013) Long-range protein-water dynamics in hyperactive insect antifreeze proteins. Proc Natl Acad Sci U S A 110:1617–1622

Meister K, Duman JG, Xu Y, DeVries AL, Leitner DM, Havenith M (2014) The role of sulfates in the enhancement of antifreeze protein activity. J Pys Chem B 118:7920–7924

Meyer K, Keil M, Naldrett MJ (1999) A leucine-rich repeat protein of carrot that exhibits antifreeze activity. FEBS Lett 447:171–178

Middleton AJ, Brown AM, Davies PL, Walker VK (2009) Identification of the ice-binding face of a plant antifreeze protein. FEBS Lett 583:815–819

Middleton AJ, Marshall CB, Faucher F, Bar-Dolev M, Braslavsky I, Campbell RL, Walker VK, Davies PL (2012) Antifreeze protein from freeze-tolerant grass has a beta-roll fold with an irregularly structured ice-binding site. J Mol Biol 416:713–724

Mock T, Thomas DN (2005) Recent advances in sea-ice microbiology. Environ Micro 7:605–619

Moffatt B, Ewart KV, Eastman M (2006) Cold comfort: plant antifreeze proteins. Physiol Plant 126:5–16

Nakamura T, Ishikawa M, Nakatani H, Oda A (2008) Characterization of cold-responsive extracellular chitinase in bromegrass cell cultures and its relationship to antifreeze activity. Plant Physiol 147:391–401

Neuner G, Xu BC, Hacker J (2010) Velocity and pattern of ice propagation and deep supercooling in woody stems of *Castanea sativa, Morus nigra* and *Quercus robur* measured by IDTA. Tree Physiol 30:1037–1045
Olien CR (1965) Interference of cereal polymers and related compounds with freezing. Cryobiology 2(2):47–54
Olien CR (1971) A comparison of desiccation and freezing as stress vectors. Cryobiology 8:244–248
Olien CR (1973) Thermodynamic components of freezing stress. J Theor Biol 39:201–210
Olien CR (1974) Energies of freezing and frost desiccation. Plant Physiol 53:764–767
Olien CR (1984) An adaptive response of rye to freezing. Crop Sci 24:51–54
Olien CR (1977) Barley: patterns of response to freezing stress. US Dep Agric Tech Bull 1558:1–8
Olien CR, Smith MN (1977) Ice adhesions in relation to freeze stress. Plant Physiol 60:499–503
Olien CR, Lester GE (1985) Freeze-induced changes in soluble carbohydrates of rye. Crop Sci 25:288–290
Olien C, Marchetti B (1976) Recovery of hardened barley from winter injuries. Crop Sci 16 (2):201–204
Olive LC, Meister K, DeVries AL, Duman JG, Guo S, Bakker HJ, Voets IK (2016) Blocking rapid ice crystal growth through non-basal plane adsorption of antifreeze proteins. Proc Natl Acad Sci U S A 113:3740–3745
Öquist G, Huner NP (2003) Photosynthesis of overwintering evergreen plants. Annu Rev Plant Biol 54(1):329–355
Pearce RS (2001) Plant freezing and damage. Ann Bot 87(4):417–424
Pennington ER, Day C, Parker JM, Barker M, Kennedy A (2016) Thermodynamics of interaction between carbohydrates and unilamellar diapalmitolyl phosphatidlcholine membranes. Evidence of dehydration and interdigitation. J Therm Anal Calorim 123:2611–2617
Pettersen EF, Goddard TD, Huang CC, Couch GS, Greenblatt DM, Meng EC, Ferrin TE (2004) UCSF chimera - a visualization system for exploratory research and analysis. J Comput Chem 25:1605–1612
Petraya N, Marshall CB, Celik Y, Davies PL, Braslavsky L (2008) Direct visualization of spruce budworm antifreeze protein interacting with ice crystals: basal plane affinity confers hyperactivity. Biophys J 95:333–341
Pihakaski-Maunsbach K, Griffith M, Antikainen M, Maunsbach AB (1996) Immunogold localization of glucanase-like antifreeze protein in cold acclimated winter rye. Protoplasma 191 (3):115–125. https://doi.org/10.1007/bf01281809
Pihakaski-Maunsbach K, Tamminen I, Pietiäinen M, Griffith M (2003) Antifreeze proteins are secreted by winter rye cells in suspension culture. Physiol Plant 118(3):390–398. https://doi.org/10.1034/j.1399-3054.2003.00110.x
Priddle J, Heywood RB, Theriot E (1986) Some environmental factors influencing phytoplankton in the Southern Ocean around South Georgia. Polar Biol 5:65–79
Quamme HA (1995) Deep supercooling in buds of woody plants. In: Lee RE, Warren GJ, Gusta LV (eds) Biological ice nucleation and its applications. APS Press, Minneapolis, MN, pp 183–199
Rajashekar C, Burke MJ (1996) Freezing characteristics of rigid plant tissues (development of cell tension during extracellular freezing). Plant Physiol 111(2):597–603
Raymond JA (2000) Distribution and partial characterization of ice-active proteins associated with sea ice diatoms. Polar Biol 23:721–729
Raymond JA (2014) The ice-binding proteins of a snow alga *Chlorospina brevispina*: probable acquisition by horizontal gene transfer. Extremophiles 6:987–994
Raymond JG, DeVries AL (1977) Adsorption inhibition as a mechanism of freezing resistance in polar fishes. Proc Natl Acad Sci U S A 86:881–885
Raymond JA, Kim KJ (2012) Possible role of horizontal gene transfer in the colonization of sea ice by algae. PLoS One 7:e35968. https://doi.org/10.1371/journal.pone.0035968.pmid:22567121
Raymond JA, Knight C (2003) Ice binding, recrystallization inhibition, and cryoprotective properties of ice-active substances associated with Antarctic Sea ice diatoms. Cryobiology 46:174–181

Raymond JA, Morgan-Kiss R (2013) Separate origins of ice-binding proteins in Antarctic *Chlamydomonas* species. PLoS One 8:e59186. https://doi.org/10.1371/journal.pone.0059186. pmid:23536869

Raymond JA, Morgan-Kiss R (2017) Multiple ICE-binding proteins of probable prokaryotic origin in an antarctic lake alga, *Chlamydomonas* sp. ICE-MDV (Chlorophyceae). J Phycol 53:848–854

Raymond JA, Wilson P, DeVries AL (1989) Inhibition of ice on nonbasal planes of ice by fish antifreezes. Proc Natl Acad Sc USA 86:881–885

Raymond JA, Janech MG, Fritsen CH (2009) Novel ice-binding proteins from a psychrophilic Antarctic alga (Chlamydomonadaceae, Chlorophyceae). J Phycol 45:130–136

Raymond JA, Sullivan CW, DeVries AL (1994) Release of an ice-active substance by sea ice diatoms. Polar Biol 14:71–75

Regand A, Goff HD (2006) Ice recrystallization inhibition in ice cream as affected by ice structuring proteins from winter wheat grass. J Dairy Sci 89(1):49–57

Rushton PJ, Torres JT, Parniske M, Wernert P, Hahlbrock K, Somssich IE (1996) Interaction of elicitor-induced DNA-binding proteins with elicitor response elements in the promoters of parsley PR1 genes. EMBO J 15:5690–5700

Sabala I, Egertsdotter U, Fircks HV, Arnold SV (1996) Abscisic acid-induced secretion of an antifreeze-like protein in embryogenic cell lines of *Picea abies*. J Plant Physiol 149:163–170

Sakai A, Larcher W (1987) Frost survival of plants. Ecological studies 62. Springer, Berlin

Santerius KA (1973) The protective effect of sugars on chloroplast membranes during temperature and water stress and its relationship to frost, desiccation and heat resistance. Planta 113:105–114

Sathyanesan SN (1999) Purification and identification of thermal hysteresis proteins and other proteins in the bittersweet nightshade (*Solanum dulcamara*). PhD Thesis, Department of Biological Sciences, University of Notre Dame, p. 157

Schwede T, Kopp J, Guex N, Peitsch M (2003) SWISS-MODEL: an automated protein homology-modeling server. Nucleic Acids Res 31:3381–3385

Shier WT, Lin Y, DeVries AL (1975) Structure of the carbohydrate of antifreeze glycoproteins from an Antarctic fish. FEBS Lett 54:135–138

Sidebottom CS, Buckley P, Pudney S, Twigg C, Jarman C, Holt J, Telford A, McArthur D (2000) Heat-stable antifreeze protein from grass. Nature 406:256

Siminovitch D, Scarth GW (1938) A study of the mechanism of frost injury to plants. Can J Res 16 (11):467–481

Simpson DJ, Smallwood M, Twigg S, Doucet CJ, Ross J, Bowles DJ (2005) Purification and characterisation of an antifreeze protein from *Forsythia suspensa* (L.). Cryobiology 51 (2):230–234

Steponkus PL (1968) The relationship of carbohydrates to cold acclimation of *Hedera helix* L. cv. Thorndale. Physiol Plant 21:777–791

Steponkus PL (1984) Role of the plasma membrane in freezing injury and cold acclimation. Annu Rev Plant Physiol 35(1):543–584

Stressmann M, Kitao S, Griffith M, Moresoli C, Bravo LA, Marangoni AG (2004) Calcium interacts with antifreeze proteins and chitinase from cold-acclimated winter rye. Plant Physiol 135(1):364–376

Sugiyama N, Simura T (1967) Studies on the varietal differentiation of frost resistance of the tea plant. IV. The effects of sugar level combined with protein in chloroplasts on the frost resistance. Jpn J Breed 17:292–296

Takahashi D, Kawamura Y, Uemura M (2013) Changes of detergent-resistant plasma membrane proteins in oat and rye during cold acclimation: association with differential freezing tolerance. J Proteome Res 12(11):4998–5011

Takahashi D, Kawamura Y, Uemura M (2016) Cold acclimation is accompanied by complex responses of glycosylphosphatidylinositol (GPI)-anchored proteins in Arabidopsis. J Exp Bot 67(17):5203–5215

Tanino KK, McKersie BD (1985) Injury within the crown of winter wheat seedlings after freezing and icing stress. Can J Bot 63(3):432–436

Thomas DN, Dieckmann GS (2002) Antarctic Sea ice - a habitat for extremophiles. Science 295:641–644

Thomashow MF (1999) Plant cold acclimation: freezing tolerance genes and regulatory mechanisms. Annu Rev Plant Biol 50(1):571–599

Tomalty HE, Walker VK (2014) Perturbation of bacterial ice nucleation activity by a grass antifreeze protein. Biophys Biochem Res Commun 452:636–641

Trischuk RG, Schilling BS, Low NH, Gray GR, Gusta LV (2014) Cold acclimation, de-acclimation and re-acclimation of spring canola, winter canola and winter wheat: the role of carbohydrates, cold-induced stress proteins and vernalization. Environ Exper Bot 106:156–163

Trunova TL (1965) Light and temperature systems in the hardening of winter wheat and the significance of oligosaccharides for frost resistance. Fiziol Rast 12:70–77

Tumanov II, Trunova TI, Smirnova NA, Zvereva GN (1976) Role of light in development of frost resistance of plants. Fiziol Rast 23:109–114

Tumanov I, Krasasvtsev O (1959) Hardening of northern woody plants in temperatures below zero. Sov Plant Physiol 6:654–657

Tursman D, Duman JG (1995) Cryoprotective effects of thermal hysteresis protein on survivorship of frozen gut cells from the freeze tolerant centipede *Lithobius forficatus*. J Exp Zool 272:249–257

Uemura M, Joseph RA, Steponkus PL (1995) Cold acclimation of *Arabidopsis thaliana* (effect on plasma membrane lipid composition and freeze-induced lesions). Plant Physiol 109(1):15–30

Uemura M, Steponkus P (2003) Modification of the intracellular sugar content alters the incidence of freeze-induced membrane lesions of protoplasts isolated from *Arabidopsis thaliana* leaves. Plant Cell Environ 26(7):1083–1096

Uemura M, Tominaga Y, Nakagawara C, Shigematsu S, Minami A, Kawamura Y (2006) Responses of the plasma membrane to low temperatures. Physiol Plant 126(1):81–89

Urrutia M, Duman JG, Knight CA (1992) Plant thermal hysteresis proteins. Biochim Biophys Acta 1121:199–206

Walters KR, Serianni A, Sformo T, Barnes BM, Duman JG (2009) A nonprotein thermal hysteresis-producing xylomannan antifreeze in the freeze-tolerant Alaskan beetle *Upis ceramboides*. Proc Natl Acad Sci U S A 106:20210–20215

Walters KR, Serianni AS, Voituron Y, Sformo T, Barnes BM, Duman JG (2011) A thermal hysteresis-producing xylomannan glycolipid antifreeze associated with cold tolerance is found in diverse taxa. J Comp Physiol B 181:631–640

Wang Y, Qiu L, Dai C, Wang J, Luo J, Zhang F, Ma J (2008) Expression of insect (*Microdera puntipennis dzungarica*) antifreeze protein MpAFP149 confers the cold tolerance to transgenic tobacco. Plant Cell Rep 27(8):1349–1358

Wang ZP, Yang PZ, Fan BF, Chen Z (1998) An oligo selection procedure for identification of sequence-specific DNA-binding activities associated with the plant defense response. Plant J 16:515–522

Webb MS, Uemura M, Steponkus PL (1994) A comparison of freezing injury in oat and rye: two cereals at the extremes of freezing tolerance. Plant Physiol 104:467–478

Weiser C (1970) Cold resistance and injury in woody plants. Science 169(3952):1269–1278

Willick IR, Takahashi D, Fowler DB, Uemura M, Tanino KK (2018) Tissue-specific changes in apoplastic proteins and cell wall structure during cold acclimation of winter wheat crowns. J Exp Bot 69:1221–1234. https://doi.org/10.1093/jxb/erx450

Wisniewski ME, Gusta LV, Fuller MP, Karlson D (2009) Ice nucleation, propagation and deep supercooling: the lost tribes of freezing studies. In: Gusta LV, Wisniewski ME, Tanino K (eds) Plant cold hardiness: from the laboratory to the field. CAB International, Cambridge, pp 1–11

Wisniewski M (1995) Deep supercooling in woody plants and the role of cell wall structure. In: Lee RE, Warren GJ, Gusta LV (eds) Biological ice nucleation and its applications. APS Press, Minneapolis, MN, pp 163–181

Wisniewski M, Davis G, Arora R (1991) Effect of macerase, oxalic acid, and EGTA on deep supercooling and pit membrane structure of xylem parenchyma of peach. Plant Physiol 96 (4):1354–1359

Wisniewski M, Gusta L, Neuner G (2014) Adaptive mechanisms of freeze avoidance in plants: a brief update. Environ Exp Bot 99:133–140

Wisniewski M, Webb R, Balsamo R, Close TJ, Yu X-M, Griffith M (1999) Purification, immunolocalization, cryoprotective, and antifreeze activity of PCA60: a dehydrin from peach (*Prunus persica*). Physiol Plant 105(4):600–608. https://doi.org/10.1034/j.1399-3054.1999.105402.x

Worrall D, Elias E, Ashford D, Smallwood M, Sidebottom C, Lillford P, Telford J, Holt C (1998) A carrot leucine-rich repeat protein that inhibits ice recrystallization. Science 282:115–117

Xu H-N, Huang W, Jia C, Kim Y, Liu H (2009) Evaluation of water holding capacity and breadmaking properties for frozen dough containing ice structuring proteins from winter wheat. J Cereal Sci 49(2):250–253

Yang J, Zhang Y (2015) I-TASSER server: new development for protein structure and function predictions. Nucleic Acids Res 43:174–181

Yeh S, Moffatt BA, Griffith M, Xiong F, Yang DSC (2000) Chitinase genes responsive to cold encode antifreeze genes in winter cereals. Plant Physiol 124:1251–1264

Yoshida M, Abe J, Moriyama M, Shimokawa S, Nakamura Y (1997) Seasonal changes in the physical state of crown water associated with freezing tolerance in winter wheat. Physiol Plant 99(3):363–370

Yoshida S, Uemura M (1984) Protein and lipid compositions of isolated plasma membranes from orchard grass (*Dactylis glomerata* L.) and changes during cold acclimation. Plant Physiol 75 (1):31–37

Yu XM, Griffith M (1999) Antifreeze proteins in winter rye leaves form oligomeric complexes. Plant Physiol 119:1361–1369

Yu XM, Griffith M (2001) Winter rye antifreeze activity increases in response to cold and drought, but not abscisic acid. Physiol Plant 112(1):78–86

Zachariassen KE, Kristensen E (2000) Ice nucleation and antinucleation in nature. Crybiology 41:257–279

Zachariassen KE, Hammel HT (1976) Nucleating agents in the haemolymph of insects tolerant to freezing. Nature 262:285–287

Zamani A, Sturrock R, Ekramoddoullah A, Wiseman S, Griffith M (2003) Endochitinase activity in the apoplastic fluid of *Phellinus weirii*-infected Douglas-fir and its association with over wintering and antifreeze activity. For Pathol 33(5):299–316

Zhang C, Zhang H, Wang L (2007a) Effect of carrot (*Daucus carota*) antifreeze proteins on the fermentation capacity of frozen dough. Food Res Int 40:763–769

Zhang C, Zhang H, Wang L, Gao H, Guo XN, Yao HY (2007b) Improvement of texture properties and flavor of frozen dough by carrot (*Daucus carota*) antifreeze protein supplementation. J Agric Food Chem 55:9620–9626

Zhang SH, Wei YL, Liu J, Yu HM, Yin JH, Pan HY, Baldwin TC (2011) An apoplastic chitinase CpCHT1 isolated from the corolla of wintersweet exhibits both antifreeze and antifungal activities. Biol Plant 55:141–148

Zhu B, Xiong A-S, Peng R-H, Xu J, Jin X-F, Meng X-R, Yao Q-H (2010) Over-expression of ThpI from *Choristoneura fumiferana* enhances tolerance to cold in Arabidopsis. Mol Biol Rep 37 (2):961–966

Chapter 8
Antifreeze Proteins in Other Species

John G. Duman and Samuel S. Newton

8.1 Introduction

After antifreeze glycoproteins (AFGPs) were first discovered in Antarctic fish (DeVries and Wohlschlag 1969; DeVries 1971) antifreeze proteins (AFPs, lacking carbohydrate), with many different structures were found in numerous marine fish from cold oceans (see Chap. 5). The majority of subsequent non-fish AFP studies over the next 20 years were done on insects (Chap. 6). The unique ability of the AF (G)Ps to bind to the surface of ice crystals, thereby inhibiting their growth by preventing adsorption of additional water molecules to the crystal, lowers the nonequilibrium (hysteretic) freezing point (hFP) of water below the equilibrium freezing/melting point (eqFP/MP) by ~1 °C to 2 °C in fish and by ~2 °C to 6 °C in insects. This produces their characteristic thermal hysteresis (TH), the difference between the hysteretic hFP and the eqFP/MP. AF(G)Ps also have a small effect on the eqMP also, raising the hMP slightly. These AF(G)Ps function to prevent lethal freezing in these nonfreeze-tolerant (freeze avoiding) organisms, often far beyond the experimentally determined hFP by (1) inhibiting inoculative freezing across the body surface by contact with external ice and, (2) in freeze-avoiding insects, promoting supercooling by inhibition of ice nucleators. The advantage of AF(G)Ps over colligative low molecular mass antifreezes, such as glycerol and other polyols, is that they lower the freezing temperature from the eqFP to the hFP, and promote supercooling while having only limited effect in raising osmotic pressure.

J. G. Duman (✉)
Department of Biological Sciences, University of Notre Dame, Notre Dame, IN, USA
e-mail: duman.1@nd.edu

S. S. Newton
Sanford School of Medicine, University of South Dakota, Vermillion, SD, USA
e-mail: Samuel.Sathyanesan@usd.edu

H. Ramløv, D. S. Friis (eds.), *Antifreeze Proteins Volume 1*,
https://doi.org/10.1007/978-3-030-41929-5_8

AFPs with very limited TH-producing abilities (a few tenths of a degree) were identified in the hemolymph of many freeze-tolerant insects, and later in the 1990s in the extracellular fluid of many freeze-tolerant plants (Chap. 7). It appeared unsuitable to label these proteins AFPs due to their limited antifreeze ability, and their presence in freeze-tolerant organisms that were obviously evolved to survive extracellular freezing. A probable function for these proteins was presented when AFPs were demonstrated to have recrystallization inhibition activity (Knight et al. 1984; Knight and Duman 1986). Recrystallization occurs when water molecules preferentially migrate from smaller crystals (with higher radius of curvature and surface free energy) to larger ones (with lower radius of curvature and surface free energy). This process eventually results in the change from an initial population of many small ice crystals to a population of fewer but larger crystals, even though the total volume of ice does not change. All known TH-producing AFPs also have RI activity, generally at a much (sometimes orders of magnitude) lower concentration than that required for TH production, but there is not a direct correlation between the magnitude of TH and RI activity (Olive et al. 2016). Recrystallization can be damaging to the extracellular matrix, etc. Consequently, the function of the low TH-producing AFPs in freeze-tolerant species appears to be recrystallization inhibition. Therefore, they are now properly called recrystallization inhibition proteins (RIPs), although it is also possible that they help to prevent lethal cytoplasmic freezing.

True AFPs and RIPs both bind to ice by the same mehanism (Part III, Volume 2 of this book) and consequently both are ice-binding proteins (IBPs). Although it may appear strange, ice nucleator proteins (INPs) are a third type of IBP, even though they have a function opposite to that of AFPs. INPs organize water in an ice-like embryo crystal on their surfaces, limiting supercooling when the embryo crystal becomes sufficiently large to induce ice nucleation at a temperature above that where freezing would otherwise occur. "Adaptive" INPs are common in freeze-tolerant insects where they function to nucleate the extracellular fluid at high subzero temperatures, thereby permitting slower and more manageable ice formation and growth that excludes lethal intracellular freezing largely by permitting outflux of water down the osmotic gradient created by the extracellular ice. "Incidental" INs are proteins and other surfaces that may be endogenous or ingested (bacteria, etc.) that can limit supercooling and cause lethal ice formation in freeze-avoiding species at temperatures above where ice would form in their absence. Consequently, these incidental ice nucleators must be seasonally eliminated, or masked by AFPs. INPs were found in numerous bacteria, many of which are important plant pathogens that cause frost damage to crops. This, along with their demonstrated importance in atmospheric freezing events important in initiating precipitation, drew considerable attention to structure/function studies of bacterial nucleation (reviewed by Lee et al. 1995).

Antifreeze glycolipids (AFGLs) with thermal hysteresis similar to those of the most active insect AFPs have also been described in insects and one plant (Walters et al. 2009, 2011).

The AFPs and RIPs discovered in fish, insects, and plants, although quite diverse, all have the common feature of binding to ice to achieve their physiological functions. Given the diversity of IBP structures and physiological functions in fish, insects, and plants, it should not come as a surprise that numerous "other" organisms (other than fish, insects, and plants) have also evolved IBPs and AFGLs of various functional and structural types. The ever-expanding list of these "others" includes bacteria, archaea, fungi (including yeasts), nematode worms, cestodes (fish parasites), a nemertean (proboscis worm), various noninsect arthropods (an Antarctic sea ice calanoid copepod, a centipede, spiders, a tick, collembola), mollusks, a sponge, frogs, and even mammals and birds. The IBPs, especially the physiological functions, of these "others" are the topic of this chapter. They will be discussed without regard to phylogenetic order.

8.2 Animals

IBPs are important components of the suites of adaptations evolved by many animals to survive subzero temperatures. However, there are numerous and varied additional adaptations (polyol accumulation, removal of incidental ice nucleators in freeze-avoiding species, behavioral adaptations such as microhabitat selection, etc.) that receive only minor mention in this chapter. Although they pertain mostly to insects, the following reviews cover these topics and are also applicable to many of the animals addressed in this chapter (Danks 1991; Ramløv 2000; Lee 2010; Lee and Denlinger 2010; Denlinger and Lee 2010; Storey and Storey 2010; Michaud and Denlinger 2010; Kostal 2010; Carrasco and Duman 2011; Duman and Newton, Chap. 6).

8.2.1 Arthropods

In addition to the insects, other arthropod groups are abundant in cold regions, and some have members known to have IBPs: arachnids (spiders, mites, ticks), centipedes, a crustacean, and collembola. For those not familiar with arthropod phylogeny Table 8.1 is provided, identifying those arthropod groups known to have species that demonstrate TH and/or RI activity. Many of the basics concerning AFPs and other IBPs in insects are also applicable to these other arthropods. Consequently, it may be useful to the reader to become familiar with Chap. 6.

8.2.1.1 Arachnids (Spiders, Mites, and Ticks)

The arachnids are composed of numerous groups, and three of the major groups contain species that have TH-producing factors: spiders, mites, and ticks (Table 8.1).

Table 8.1 Arthropod subphyla and classes

Subphyla	Class
Trilobitomorpha	*Trilobita* (Trilobites, extinct)
Chelicerata	*Arachnida* (Spiders[a], scorpions, mites[a], ticks[a]) *Merostomata* (Horseshoe crabs, *Euryptids* (extinct), etc.) *Pycnogonida* (Sea spiders)
Myriopoda	*Chilopoda* (Centipedes[a]) *Diplopoda* (Millipedes) *Pauropoda* *Symphyla*
Crustacea	*Branchiopoda* (Brine shrimp, tadpole shrimp, etc.) *Remipedia* (Blind crustaceans) *Cephalocarida* (Horseshoe shrimp) *Maxillopoda* (Copepods[a], barnacles, fish lice, etc.) *Ostracoda* (Seed shrimp) *Malocostraca* (Shrimps, crabs, lobsters)
Hexapoda	*Insecta* (Insects[a]) *Entognatha* (Collembola[a], etc.)

[a]Groups in which at least one species is known to have thermal hysteresis and/or recrystallization inhibition producing proteins (AFPs and/or RIPs)

All Arachnids that have been investigated have been shown to be freeze avoiding, except for certain scorpions (Whitmore et al. 1985), and the red velvet mite, *Trombidium holosericeum* (Anthony and Sinclair 2019), neither of which is known to produce IBPs.

Spiders (Araneae)

After insects, the first arthropod group in which thermal hysteresis (TH) was identified was the spiders (Duman 1979; Husby and Zachariassen 1980). Numerous European spiders had previously been shown to be freeze avoiding (see review by Kirchner 1973). The presence of TH was not examined in these species, but when polyols were measured glycerol was sometimes present (Kirchner and Kestler 1969). Two freeze-avoiding North American spider species, a crab spider (*Philodromus* sp.) from northern Indiana (42°N latitude) and a sac spider (*Clubiona* sp.) from Pennsylvania (40° 29′), had hemolymph TH (Duman 1979). *Philodromus* in winter had a mean supercooling point (SCP) of −26.2 °C and a hemolymph TH of 2.44 °C, but lacked TH in summer. The winter MP was −2.74 °C, due mainly to glycerol accumulation (Duman 1979). The *Clubiona* sp. had hemolymph TH of 1.88 °C in January and the MP was −2.87 °C, the latter also due primarily to glycerol. Glycerol presence is important not only for its colligative antifreeze activity, but it also enhances the activity of AFPs of the beetle *Dendroides canadensis* (Li et al. 1998), and perhaps does so in these spiders as well. The SCP of *Clubiona* was −15.4 °C. Both species winter in thermally exposed sites under the loose bark of standing dead trees where they are somewhat protected from predation by birds, etc.

but are not much thermally buffered from low air temperatures, although the silk sac in which *Clubiona* winter may protect them from contact with water and ice. Neither spider could be identified to species, as such identification requires adults (morphology of external reproductive organs is required) and both species apparently winter only as juveniles.

A winter active Norwegian spider, (*Bolyphantes index*) was identified with high hemolymph TH (5.2 °C) and a mean whole body SCP of −15.2 °C (Husby and Zachariassen 1980). The combination of a lowered hemolymph MP and the TH depressed the hFP to −6.9 °C, an important point since contact with ice in this winter active species could render them especially susceptible to inoculative freezing. The authors also provided indirect evidence that the hemolymph AFPs were sufficient to inhibit inoculative freezing and lower the SCP by demonstrating that the measured TH was inversely dependent upon the size of the ice crystal used in the measurement—smaller crystals yielded higher TH values. They extrapolated their TH measurements to the small size of the cuticular pores through which surface ice would need to penetrate to inoculate the hemolymph and found that the AFPs provided sufficient protection. This also assumes that the spider, and the insects they included in the study, either did not have efficient incidental internal ice nucleators or the AFPs were able to mask such ice nucleators.

A later study of Alaskan insects and spiders (Duman et al. 2004) identified three additional spider species with hemolymph TH: a *Philodromus* sp., a *Pardosa* sp., and a *Gnaphosa* sp. Once again only juveniles were found, so identification to species was not possible. Also, only a few individuals were collected, and consequently, SCPs were not determined. A few *Philodromus* were collected in April (still a cold period in interior Alaska) when the hemolymph TH was just 1.39 °C and the MP was a somewhat depressed −1.74 °C, the latter indicating the presence of small unknown antifreezes. The following September additional *Philodromus* were collected and then cold acclimated, resulting in approximately the same TH (1.46 °C), but a lower MP (−3.47 °C), indicating more significant low molecular mass antifreeze levels. The responsible low mass antifreezes were not determined. The two other spider species were also collected in September and cold acclimated, both showing low TH and MPs (TH = 0.25 °C, MP = −3.40 °C in *Pardosa*; TH = 0.85 °C, MP = −3.90 °C in *Gnaphosa*). It is unwise to place too much emphasis on the low TH as just one individual of each species was cold acclimated. However, even these low values demonstrate the presence of AFPs, or possibly AFGLs or RIPs (unlikely as no spiders are known to be freeze tolerant).

Unfortunately, spider AFPs have not been purified and sequences are not known. This should be rectified. While a few species of spiders are known to have TH-producing factors, many do not. This absence of TH is not often reported, partially due to the difficulty of having species identified. Spiders are quite common in cold regions, even in far northern latitudes such as Arctic tundra (Danks 1981). Perhaps surprisingly, even though some of these tundra species have been investigated for the presence of TH, few have been found to have it. Wolf spiders (Lycosidae) are especially abundant in many of these tundra sites, yet those that have been studied lack hemolymph TH (Duman and Kukal personal observations).

Mites (Acari)

Mites are especially well represented in cold climes, especially in the Arctic, both in species numbers and biomass (Danks 1981). All previously investigated Acari were shown to be freeze avoiding, except for the recently studied red velvet mite, *Trombidium holosericeum* (Anthony and Sinclair 2019).

TH was first discovered in mites in a maritime Antarctic oribatid species, *Alaskozetes antarcticus*, collected from Signy Island in the South Orkney Islands (60° 43″S) (Block and Duman 1989). *A. antarcticus* had been much studied previously, primarily by researchers from the British Antarctic Survey (Block 1977, 1980; Cannon and Block 1988). This mite lives under rocks and on the soil surface, is cold hardy year around, and winters as both adults and nymphs. It is especially hardy in winter with SCPs of −30 °C due to gut clearing, high concentrations of glycerol and several other polyols, desiccation during low relative humidity periods (Young and Block 1980; Cannon and Block 1988), and AFPs (Block and Duman 1989). Hemolymph TH was present in both adults (1.20 °C) and nymphs (1.86 °C) that had been collected during the Antarctic summer and cultured at 4 °C prior to hemolymph sampling for TH measurements. It is likely that TH levels would be higher in hemolymph taken directly from winter-collected individuals (Block and Duman 1989).

Another oribatid mite from the Antarctic, *Maudheimia wilsoni*, also has TH, which lowers the hFP to between −4.7 °C and −6.1 °C (Sømme et al. 1993). A third oribatid with hemolymph TH was studied in the cold mountainous Finse region of Norway (Sjursen and Sømme 2000). These alpine mites have limited TH in summer, but TH increases in winter when, along with high polyol concentrations, AFPs assist in lowering SCPs to −35 °C.

The factors responsible for the TH in the three species of mites mentioned above are not known. However, putative AFPs were identified in the two-spotted spider mite, *Tetranychus urticae*, in a genome-wide expression analysis that investigated factors up- and downregulated during facultative reproductive diapause (Bryon et al. 2013). The biology of this species has been heavily studied, including various aspects of its diapause, because it a serious pest of agricultural crops (reviewed in Bryon et al. 2017). Diapause occurs in the adults, but diapause induction in *T. urticae* is cued mainly by photoperiod during the nymphal stages that initiate significant signal transduction pathways. Diapause involves developmental and reproductive arrest, decreased metabolic rate, improved stress tolerance to cold and desiccation (Khodayari et al. 2012), and metabolic pathway changes that include accumulation of various low molecular mass antifreezes, especially mannitol and inositol (Khodayari et al. 2013). SCPs are lower in diapausing mites, especially when they are held at low temperatures (0 °C and 5 °C) (Khodayari et al. 2012). Most importantly for this chapter, the microarray study of diapausing *T. urticae* identified 20+ genes that coded for proteins with significant similarities to known beetle AFPs (Bryon et al. 2013). The putative mite AFPs exhibit multiple continuous 12-mer repeat sequences consisting of Asn-Cys-Thr-X-Cys-X-X-Cys-X-Asn-Cys-X. Like several beetle AFPs (Davies 2014; Duman 2015; Chap. 7), the putative spider mite AFPs have numerous cysteines at least every six residues. In these beetle AFPs, the cysteines (involved in intra-repeat disulfide bridges) are usually flanked by threonines, producing the T-C-T sequence that comprises much of the ice-binding site of certain beetle AFPs (i.e., *Tenebrio molitor, Dendroides canadensis*). In the proposed spider mite AFPs, the cysteines are flanked by asparagine, aspartate, or threonines. Most of these putative AFPs are upregulated by diapause, but some can be induced in nondiapausing adults by cold acclimation (Bryon et al. 2013). The function and TH-producing abilities of these putative AFPs are not known at this time.

Ticks (Acari)

Ticks are a third major arachnid group known to have AFPs, although at this time only two species have been shown to do so.

Supercooling points of 14 species of ticks from multiple geographic areas were determined to range from −17 °C to −23 °C (Dantel and Knulle 1996). Some species were kept in outdoor enclosures over winter while others were held under various laboratory conditions. Season, photoperiod, diapause, and other environmental conditions had little or no effect on SCPs. Note that all SCPs were fairly low, suggesting some level of protection from freezing. This is an important point as the species tested were not freeze tolerant, and no freeze-tolerant ticks are known. One central European species, *Dermocentor marginatus*, with a mean SCP of −21.5 °C was shown to have hemolymph thermal hysteresis ranging from 0.27 °C to 0.56 °C. This species winters in the litter as both nymphs and adults, but only adults were tested for TH. The factor responsible for the hemolymph TH was not investigated.

Ixodes scapularis, the black-legged tick, vectors multiple bacterial diseases (of which Lyme disease is the best known) when infected ticks blood feed on vertebrate hosts. *I. scapularis* has recently spread through the northeastern and midwestern United States where it winters as larvae, nymphs, and adults. It can be active during warmer periods in winter, actively seeking out and feeding on hosts. Somewhat surprisingly, *I. scapularis* produces antifreeze glycoproteins (AFGPs) (Neelakanta et al. 2010) similar to the well-known AFGPs of polar marine fish (DeVries 1971; Raymond et al. 1975; VanVoorhies et al. 1978; Chap. 5). While numerous arthropods have evolved AFPs to improve subzero survival, this is the only known instance of an arthropod with these AFGPs. In fish the AFGPs are present in multiple isomers with sizes ranging from 3.5 to 32 kDa, consisting mainly of tripeptide repeats of Ala-Thr-Ala with the disaccharide β-D-galactopyranosyl-(1→3)-2-acetamido-2-deoxy-α-D-galactopyranose, that is O-glycoside linked to every threonine (DeVries 1971; Shier and DeVries 1975). Some of the fish AFGPs also have tripeptide repeats containing proline residues (Pro-Thr-Ala) (Lin et al. 1972). The *I. scapularis* AFGPs (IAFGP) has approximately 70% identity to known fish AFGPs, consisting of Ala-Ala-Thr and Pro-Ala-Thr repeats interspersed between a 6 amino acid spacer (Pro-Ala-Arg-Lys-Ala-Arg), approximately every 21 amino acids. The IAFGP was identified from an expressed sequence tag (EST) from the *I. scapularis* database. Consequently, the carbohydrate component was not verified, but it is assumed that a saccharide is attached to each threonine. This is likely to be the case, and also the same disaccharide known from the fish AFGPs is perhaps present as any manipulation of the disaccharide of the fish AFGPs results in inactivation of the TH ability of the AFGPs. For example, borate treatment, which affects the 3–4 *cis*-hydroxyls of the galactose residues renders the fish AFGP completely without TH (DeVries 1971). The *iafgp* gene contains a signal peptide, indicating that the IAFGP is secreted.

Using QRT-PCR the authors demonstrated expression of IAFGP in nymphs, larvae, and especially in adults (Neelakanta et al. 2010). Not surprisingly, relative to high temperatures of 23 °C or 10 °C, expression of IAFGP was increased fourfold if the ticks were cold acclimated (4 °C or 0 °C).

Interestingly, ticks infected with *Anaplasma phagocytophilum* bacteria (the cause of human granulocytic anaplasmosis) survived −20 °C for a longer time than uninfected ticks. Also, *A. phagocytophilum* infection increased IAFGP expression threefold at all temperatures tested, while infection by other bacteria did not. Ticks were also treated with interference RNA (RNAi) to downregulate IAFGP and the resulting IAFGP deficient ticks showed depressed survival and mobility at −20 °C compared to wild type. In addition, cells from an *I. scapulara* cell line that harbor the *A. phagocytophilum* infection had increased IAFGP expression and survival, as well as decreased membrane damage at −20 °C relative to uninfected controls (Heisig et al. 2014).

This appears to be a symbiotic relationship between the tick and bacteria. Enhanced low-temperature survival of *I. scapularis* is not the only function of IAFGP as it also benefits the *A. phagocytophilum* bacteria as well. Infection induces increased expression of IAFGP in the tick gut as well as throughout the rest of the body, and IAFG has anti-virulence properties against several bacteria, but not against *A. phagocytophilum*, apparently by inhibiting the formation of microbial biofilms (Heisig et al. 2014). Also, transgenic mice and *Drosophila melanogaster* expressing IAFGP were resistant to infection by various bacteria. Not only does IAFGP inhibit bacterial competitors, but it also assists *A. phagocytophilum* infection of the tick by decreasing the thickness of the peritrophic matrix when the ticks take blood containing *A. phagocytophilum* from an infected host. The peritrophic matrix lies just outside the gut epithelium and functions to hinder bacterial and fungal access to the gut epithelial cells. IAFG attaches to certain peptides in the glycoprotein rich matrix, thereby inhibiting its thickness and allowing easier access of *A. phagocytophilum* into the epithelium (Abraham et al. 2017).

The presence of the antifreeze glycoproteins in this tick extends the convergent evolution of the AFGPs beyond their previously described presence in unrelated Antarctic notothenioid and northern Gadid (cod family) fishes (Chen et al. 1997) to arthropods. In addition, AFGPs had previously been described in intertidal mussels, *Mytilus edulis* (Theede et al. 1976).

8.2.1.2 Centipedes (Chilopoda)

The subphylum Myriapoda consists of four classes (Table 8.1), but the two major groups are centipedes (Chilopoda) and millipedes (Diplopoda). There has been little cold tolerance work done on them, and among these groups just one centipede species, *Lithobius forficatus*, is known to exhibit TH (Tursman et al. 1994). Previously, a centipede from a mountainous region of New Mexico, *Scolopendra polymorpha*, was shown to have rather high SCPs (−7 °C) in winter, and many individuals survived if held at that temperature until frozen (Crawford and Riddle 1974; Crawford et al. 1975). Although the authors labeled these individuals as "moderately freeze avoiding," with the benefit of hindsight today we would probably call them moderately or slightly freeze tolerant. *S. polymorha* did not accumulate polyols or sugars in winter, and TH activity was not tested. With lower lethal temperatures just slightly below −7 °C, the winter microhabitat of these centipedes, under large rocks where they are somewhat thermally buffered from air temperatures, is important.

L. forficatus is native to temperate deciduous woods in northeastern North America, where they are common in and under decomposing logs. In late autumn they move to the underside of the logs where they encounter less extreme temperatures. This insulation is critical as, although *L. forficatus* is freeze tolerant, they are only slightly so, with 24 h lower lethal temperatures of only −6 °C (Tursman et al. 1994). When placed in a temperature gradient containing substrate in the laboratory the centipedes move just ahead of the freeze line. In the field in winter they likewise are found mainly in the loose bark and leaf litter on the underside of the logs generally just below frost line. Hemolymph SCPs ranged from −7 °C to −9 °C, indicating the likely presence of hemolymph ice-nucleating proteins. However, the centipedes died if freezing was not initiated by inoculative freezing across the cuticle by surface ice at temperatures above the SCPs. Centipede whole body SCPs had to be measured in the presence of desiccant in the freezing chamber, otherwise SCP values were inconsistent, with lower SCPs of approximately −3 °C to −4 °C predominating on low humidity days, but high SCPs near −1 °C on high humidity days (when condensation formed on the centipedes as temperature was decreased, and this subsequently froze and then inoculated the centipedes). If frozen by contact with ice above −2 °C survival of winter centipedes was high, but low if frozen at lower temperatures. Centipedes lack the cuticular wax coating of insects and consequently are more susceptible to both desiccation and inoculative freezing. *L. forficatus* does not accumulate low molecular mass cryoprotectants (polyols, etc.) in winter. Only occasionally were centipedes collected from the field (generally in late autumn, November) that had measurable hemolymph TH, ranging from 0.33 to 1.35, however, recrystallization inhibition in the hemolymph, indicating RIPs, was present in winter, but not summer. All known TH-producing proteins and AFGLs whether from insects, fish, plants, fungi, etc. inhibit recrystallization, and do so at concentrations much lower than those required for thermal hysteresis (Knight et al. 1984; Knight and Duman 1986; Duman 2015).

Numerous freeze-tolerant organisms (some insects, plants, fungi, etc.) exhibit TH, but at much lower levels than in freeze-avoiding species (Duman 2015). For example, while most freeze-avoiding insects have 2 °C–6 °C of hemolymph TH, and some as much as 13 °C, freeze-tolerant insects usually have just 0.2 °C–0.6 °C (Duman 2015). Consequently, the primary function of the TH-producing proteins in these species is generally thought to be recrystallization inhibition (Griffith and Yaish 2004; Duman 2015). This was shown to be the case with *L. forficatus* (Tursman and Duman 1995). Isolated *L. forficatus* midgut cells, both summer and winter, frozen under conditions designed to promote recrystallization had increased survivorship if purified AFP from the beetle *Dendroides canadensis* (DAFP) was added to the bathing medium at a very low concentration (0.02 mg/ml). DAFP was used in these experiments because anti-DAFP antibody cross-reacted with *L. forficatus* hemolymph on immunoblots, indicating that the centipede AFP/RIP has at least some epitopes in common with DAFP. In addition, fluorescence tagged anti-DAFP antibody cross-reacted with winter cells (primarily the cell membrane), but not summer cells. Also, the LT_{50} (temperature at which there was 50% mortality) of the winter cells was lowered from −12.1 °C to −15.1 °C by addition of DAFP, and the LT_{50} of summer cells decreased from −8.2 °C without DAFP to −15 °C (the winter value) with the DAFP added. In addition, when summer cells were incubated with DAFP in the medium and then washed repeatedly with fresh medium lacking DAFP (to washout the DAFP), the fluorescent anti-DAFP antibody showed DAFP remained associated with the cell membrane, and the LT_{50} was lowered from −8.2 °C to −14.5 °C, a value comparable to winter cells. This indicates that the membrane-associated centipede RIP may inhibit inoculation of the cytoplasm by extracellular ice. In summary, these experiments showed that the centipede RIP (1) has some structural similarity with the *D. canadensis* beetle DAFP, (2) is present in the hemolymph and on the cell membrane, and (3) functions to (a) prevent damaging recrystallization in the extracellular fluid and (b) inhibit lethal intracellular freezing.

8.2.1.3 Crustacea (A Calanoid Copepod)

Most marine invertebrates, because they are isosmotic or slightly hyperosmotic to seawater, do not generally have a problem with freezing unless they are positioned in a site, such as the intertidal region at low tide, where they are exposed to temperatures below the freezing point of the seawater (approximately −1.86 °C). The Antarctic calanoid copepod *Stephos longipes* occupies brine channels in the surface sea ice in the Antarctic Ocean where temperatures can reach well below freezing in the hypersaline water of the channels (Schünemann and Werner 2005). It is the only known crustacean to have hemolymph TH (Kiko 2008, 2010). The TH is produced by at least two IBP isoforms (*St*AFPs) (Fig. 8.1a–c) with considerable sequence homology to known IBPs from various unicellular algae (diatoms and chlorophytes, See Chap. 7), bacteria such as *Colwellia* strain SLW05 and a snow mold, *Typhula ishikariensis*, IBP (*Tis*AFP) that will be discussed later in this chapter. Compare the

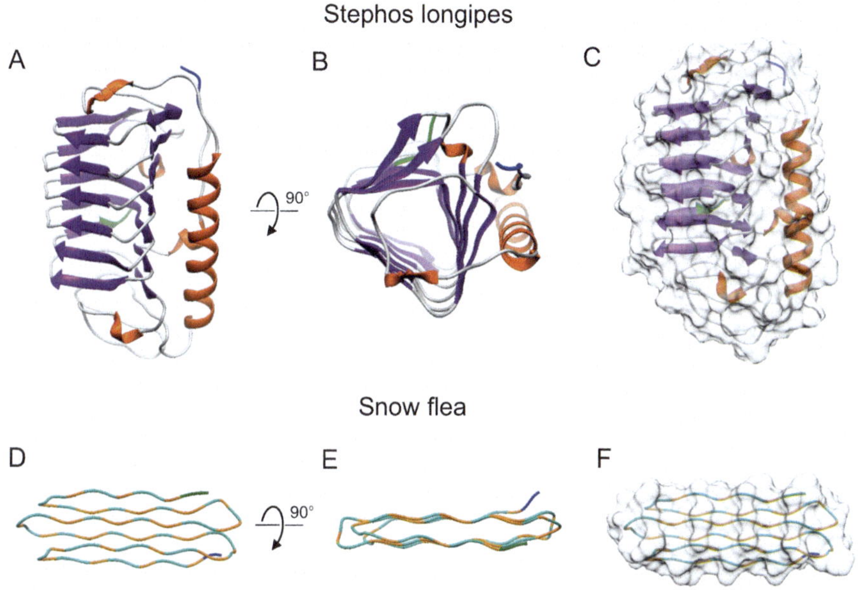

Fig. 8.1 The structures of AFPs of two arthropod species are shown. (**a–c**) Homology model of the crustacean, *Stephos longipes*, antifreeze protein, *St*AFP (GenBank FJ483935.1). Model was generated using the SWISS-MODEL server. (**a**) Secondary structure is shown with helices in orange, beta strands in purple and loops in gray. N-terminus is colored blue and C-terminus is in green. (**b**) View in A is rotated 90° toward the reader. (**c**) Semitransparent molecular surface representation of the view in (**a**). (**d–f**) Crystal structure of the snow flea antifreeze protein, sfAFP, (PDB 3BOG). Polyproline helices are in cyan and glycine residues are colored orange. The width of the orange bands indicates single or double glycine residues. The N- and C-termini are colored blue and green, respectively. (**e**) View in (**d**) is rotated 90° toward the reader. (**f**) Semitransparent molecular surface representation of the view in (**d**)

*St*AFP structure in Fig. 8.1a–c with those of the snow mold (*Tis*AFP, Fig. 8.3a–c) and *Colwellia* (*Col*AFP, Fig. 8.4d–f). Some of these organisms inhabit the same sea ice brine channels with *S. longipes* and are thought to have received their AFPs via horizontal gene transfer, perhaps from co-inhabiting bacteria and/or fungi (Raymond and Kim 2012).

The brine channels inhabited by *S. longipes* and these other organisms form as salts freeze out of seawater ice, concentrating the salts in unfrozen water in the channels, thereby increasing the salinity and keeping them from freezing (Priddle et al. 1986; Mock and Thomas 2005). Numerous organisms (including unicellular algae, bacteria, fungi, and some metazoans) depend on this environment, especially the photosynthetic organisms as the ice provides a mechanism to keep them at the surface where they are exposed to the sun (Thomas and Dieckmann 2002). However, organisms inhabiting the channels must be able to cope with the hypersalinity and low temperatures. Some of these brine channel organisms produce and secrete TH-producing IBPs into the channels where they are thought to assist in keeping the channels open and to inhibit freezing of the organisms (Raymond and Knight

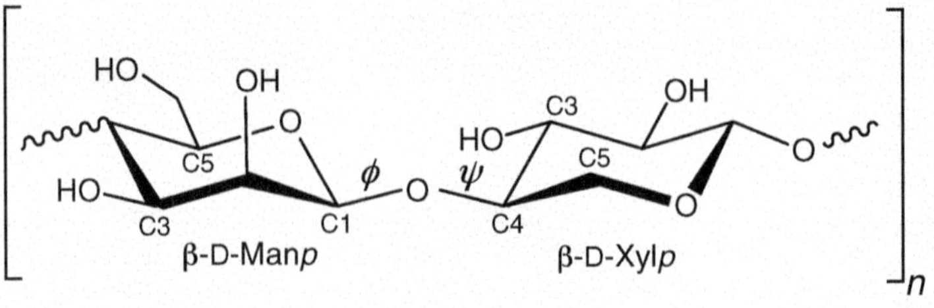

Fig. 8.2 The repeating disaccharide core structure of an antifreeze glycolipid is depicted. The β-D-Manp-(1→4)-β-D-Xylp (βManp-βXylp) of the xylomannan AFGL initially described in the Alaskan beetle *Upis ceramboides*. The saccharide, which constitutes approximately half the mass of the AFGL, is attached to lipid. Figure based on Walters et al. (2009)

2003; Raymond and Kim 2012; Wisniewski et al., Chap. 7). Certain juvenile stages of *S. longipes* winter in the water column, and in late winter move to the undersurface and bottom layers of the ice where they develop into adults and reproduce. Their nauplius and early juvenile stages move through brine channels throughout the ice, including the surface layers where the temperatures are lowest. The nauplius stages remain in the channels over the winter (Kiko 2008, 2010). When *S. longipes* were held at 4 °C they had hemolymph TH ranging from 0 °C to 3.8 °C and remained isosmotic over a range of salinities. A probe based on an IBP from the sea ice diatom *Navicula glaciei* identified two isoforms in an *S. longipes* cDNA library. Blast searches showed considerable sequence homology to IBPs from *Navicula glaciei*, as well as to a sea ice bacterium (*Colwellia* sp.), and the snow mold *T. ishikariensis*, but no homology to any known metazoan IBPs. The high sequence homology of the *S. longipes* AFPs with those of these other brine channel organisms, along with the absence of homologs in any other known metazoans suggests that the *S. longipes* AFPs were obtained via horizontal gene transfer. The two isoforms originally designated as *Stephos*-Ice binding proteins (St-IBPa and b) had high sequence homology to one another (98% at the amino acid level), but they had different signal peptides, suggesting secretion to different sites. Expressed recombinant *Stephos*-IBPs in *E. coli* were active, producing RI and TH. In situ hybridization suggested that the IBPs were produced in all cells throughout the body, including eggs. As mentioned previously, the high level of relatedness of the IBPs produced by phylogenetically unrelated brine channel organisms suggests that horizontal gene transfer is common in this environment (Raymond and Kim 2012; Kiko 2008, 2010).

8.2.1.4 Hexapoda: Insects and Collembola (Springtails, Snow Fleas)

Numerous insects produce AFPs and other IBPs (Duman 2001, 2015; Duman et al. 2010; Duman and Newton, Chap. 6). Collembola (springtails) were long considered by most entomologists to be insects. However, more recently, while still placed in the subphylum Hexapoda with the insects, they are located in the class Entognatha, along with other noninsect hexapods (Table 8.1). Other phylogenies place the

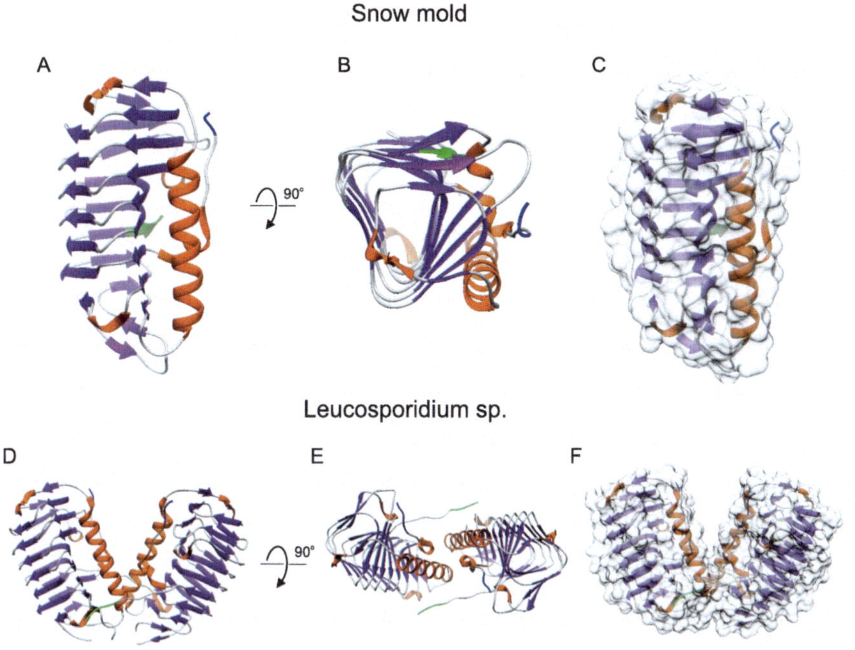

Fig. 8.3 Two fungal AFP structures are presented. (**a–c**) Crystal structure of the snow mold *Typhula ishikariensis* AFP, *Tis*AFP, (PDB 5B5H). (**a**) Secondary structure is shown with helices in orange, beta strands in purple and loops in gray. N- and C-termini are colored blue and green, respectively. (**b**) View in A is rotated 90° toward the reader. (**c**) Semitransparent molecular surface representation of the view in (**a**). (**d–f**) Crystal structure of the arctic yeast, *Leucosproridium* sp. AFP, *Le*AFP, (PDB 3UYU). (**d**) Secondary structure is shown with helices in orange, beta strands in purple and loops in gray. N- and C-termini are in blue and green. (**e**) View in (**d**) is rotated 90° toward the reader. (**f**) Semitransparent molecular surface representation of the view in (**d**)

noninsect hexapoda, collembola, and others, into separate unrelated lineages that are equally distant from one another. However, these phylogenetic controversies do not affect our discussion of this interesting group.

Collembola are very small, wingless, and can be quite numerous, in some areas reaching densities of 100,000 per m^2. As a group they have a worldwide distribution, inhabiting most terrestrial environments, including the most extreme polar regions (Danks 1981; Hopkin 1997). While most species of collembola are found near the soil surface (detritus, leaf litter, moss, decayed logs, etc.) some are arboreal and therefore are not insulated from extremes of air temperatures. Certain species are winter active, even moving upward through the snow on warm days to form large aggregations on the snow surface (e.g., the dark purple *Hypogastrura nivicola*).

All of the various species of collembola that have been studied from the standpoint of cold tolerance are freeze avoiding. Some accumulate low molecular mass antifreezes (polyols and/or sugars), including glycerol, in winter (Sømme and Conradi-Larsen 1977; Block and Zettel 1980; Block 1982; Schnenker 1983). Also, a few springtails, especially polar species, inhibit freezing by cryoprotective dehydration (Holmstrup and Sømme 1998; Worland et al. 1998; Worland and Block 2003). Their small size and cuticular respiration can make them especially susceptible to dehydration. However, some species are well known for their abilities to tolerate dry conditions that result in their desiccation, and some species even become anhydrobiotic. When in the anhydrobiotic state *Folsomides angularis* survives −180 °C (Poinsot-Balaguer and Barra 1983). Ice in the environment causes a decreased atmospheric vapor pressure that results in water flux out of the body of the springtails. This rapid equilibration results in cryoprotective dehydration as such small porous animals quickly concentrate body fluid solutes such that the eqMP/FP of the organism stays lower than the surrounding temperature and the animal does not freeze (Lundheim and Zachariassen 1993; Worland and Block 2003). This process can be taken to the extreme, resulting in anhydrobiosis when body water is reduced to 10% or less.

Several species of collembola are also known to produce hemolymph TH, presumably resulting from AFPs that assist freeze avoidance. Hemolymph TH was first identified in a detailed study of alpine collembola in Switzerland (Zettel 1984). Sixteen species of Collembola were collected from six different habitats of varying winter extremes ranging from hygrogaphic (near or in contact with water) and terrestrial lowlands where winter temperatures, while often subzero, are not extreme, to species active on the snow surface in winter (some at temperatures as low as −3 °C), and to a glacier dwelling species and a tree-dwelling species that winter in a quiescent state. Hemolymph osmolality was measured along with the presence or absence of TH, both winter and summer, and species were compared seasonally and to one another. Air and microhabitat temperatures were also monitored. The general trend was that those species with minimal subzero temperature exposure exhibited no or very minor (0.04 °C–0.06 °C) hemolymph TH and very little increase in hemolymph osmolality (indicating only minor accumulations of low molecular mass solutes) in winter relative to summer. These species move vertically downward in the litter ahead of the frost line, and thereby avoid freezing mainly by behavioral means. In contrast, species with the most exposure to potential freezing conditions (those active on the snow surface at temperatures down to −3 °C, and the arboreal species exposed to temperatures as low as −15 °C for long periods) had hemolymph TH mean values ranging from 1.60 °C to 2.00 °C. Supercooling points were not determined in this study, but the SCP of the tree-dwelling *Entomobrya nivalis* was previously shown to be −18 °C in winter (Zettel and von Allmen 1982). The hemolymph osmolality increase in winter was also greatest in these more exposed species, suggesting winter polyol/sugar concentrations of approximately 100–300 mM. The greatest TH was found in the glacier-dwelling species (*Isotoma* sp.) in the summer (2.20 °C). This species could not be collected in winter, but its habitat temperature in the glacier is quite stable at temperatures between +1 °C and −5 °C throughout the year, so its winter TH was thought to be similar to the measured summer values. In contrast to the species active on the snow surface at subzero temperatures, the species active on the snow surface only at higher temperatures had less hemolymph TH (0.21 °C–0.31 °C) and lower osmolality, and these species remained in the litter at subzero temperatures. This study showed that those collembola exposed to the lowest temperatures had significant TH to protect them from inoculative freezing when in contact with ice and to lower their SCPs.

In regard to inoculative freezing, the following observations from this study are informative. The cold hardiness of one of the species active on the snow surface at low subzero temperatures, *Isotoma hiemalis*, demonstrates cyclomorphosis (a seasonal change in phenotype). However, even in January part of the population exists in the summer morph, and these individuals stay in the leaf litter and do not move into the snow along with the winter morph. The summer morph has much lower hemolymph TH (0.02 °C–0.12 °C) in January compared to the winter morph (1.18 °C–1.34 °C). When placed on ice at −2 °C (in January) the summer morph froze and died, even though their SCPs were −7 °C, while the winter morph did not. However, both morphs survived contact with plaster (no ice) at −2 °C. These results indicate that inoculative freezing across the cuticle initiated by contact with external ice is problematic in the summer morph, but inoculative freezing is inhibited in the winter morph, probably aided by AFPs (Zettel 1984).

A subsequent study of the arboreal *Entomobrya nivalis* (Meier and Zettel 1997) found winter hemolymph TH as high as 3.5 °C. Also, at this time osmolality increased by 300–400 mOsm relative to summer, mainly due to the accumulation of ribitol and lesser amounts of arabitol and threitol. TH increases were triggered in the autumn by low temperature and short photoperiod.

The Antarctic springtail *Gomphiocephalus hodgsoni* also produces hemolymph TH and has high concentrations of polyols that assist in depressing its SCPs to values as low as −38 °C (Sinclair and Sjursen 2001). This study was done on Ross Island, McMurdo Sound (77°S), but *G. hodgsoni* is found further south in even colder environments including the snow-free (therefore uninsulated) areas in the Antarctic Dry Valleys. Because of difficult working conditions during the Antarctic winter, the study was conducted between mid- to late-October, when conditions are transitioning from winter to spring (but with long periods of −20 °C and lower air temperatures persisting), and into early summer (December). *G. hodgsoni* were found under rocks, often in areas with minimal snow cover. SCPs varied greatly, especially during the warmer periods, with SCPs of −13.2 °C to −38.0 °C on October 21 and −4.3 °C to −32.9 °C on December 24. These SCP variations were thought to result from initiation of feeding when ice nucleators are potentially introduced into the gut. Hemolymph osmolality (1502–1755 mOsm) and glycerol concentrations were very high in November, but transitioned quickly to low levels by late December. Likewise, hemolymph TH was 1.1 °C in late November, but disappeared by late December, although recrystallization inhibition activity of whole-body homogenates was still present. The possibility that antifreeze levels were higher in winter, and that *G. hodgsoni* employs cryoprotective dehydration during the long and severe Antarctic winter were discussed, but this is still unknown. Three species of collembola from sub-Antarctic Marion Island were found not to have thermal hysteresis (Sinclair and Chown 2002).

The only collembolan antifreeze protein that has been purified and well-characterized comes from the common Nearctic collembolan *Hypogastrura harveyi*, collected from snowbanks in Ontario, Canada (Graham and Davies 2005). Two isoforms (6.5 and 15.7 kDa) of this unique glycine-rich snow flea AFP (sfAFP) were described. The amino acid compositions of the two isoforms are similar, but have some differences. Both are composed of approximately 46 mol% glycine and 15 mol % alanine, with limited cysteine residues that form intramolecular disulfide bonds (two disulfides in the smaller and one in the larger isoform). The smaller isoform is composed primarily of a tripeptide repeat -Gly-X-X-, where the second position is a small side-chain amino acid (often another glycine) and the third position is variable but often has charged or hydrophilic residues or proline. The sfAFP is structurally different from any other known AFPs, but its TH activity is comparable to those of the very active insect AFPs.

The small size of *G. harveyi*, along with the inability to collect them in sufficient numbers to purify reasonable amounts of sfAFP has hampered structural analyses of the AFP. In addition, their unusual structure creates folding problems for expressed or synthetic sfAFPs. Also, this unique structure means that even somewhat similar known high glycine proteins, such as collagen peptides, may not be satisfactory models for determining sfAFP structure. However, in spite of these drawbacks reasonable theoretical structures for sfAFPs have been proposed (Fig. 8.1d–f). The smaller, 81 residue, sfAFP is suggested to form six short polyproline type II helices (similar to those in certain collagens) that combine into a bundle with prolines at the ends of five of the six helices. The prolines provide a "kink" that permits the helices to turn back 180°, to form two sheets of three antiparallel helices each. The central helices are rich in glycine that, because they lack side chains, allow hydrogen bonding to four neighbor helices via carbonyl-amide hydrogen bonding. The disulfide bridges provide further stabilization. The modeled protein is amphipathic and has a hydrophobic binding surface on one side (Lin et al. 2007; Pentelute et al. 2008). The smaller of the two sfAFPs is more abundant, but has much less specific activity than the larger isoform.

The larger 203 residue isoform has a low level of sequence identity (<50%) with the smaller, however, the presence of similar tripeptide repeats and turns allowed modeling of the larger isoform, that indicated the larger isoform consists of 13 polyproline II helices (rather than six in the smaller isoform) connected by proline-containing loops that form two flat antiparallel sheets (Mok et al. 2010). The larger isoform, like the smaller, is amphipathic, with the hydrophobic side forming the ice-binding surface that is approximately twice the surface area of that of the smaller isoform. This larger ice-binding surface is likely responsible for the greater TH activity. Both isoforms, while binding primarily to the basal plane of ice, also bind to additional surfaces (Todde et al. 2014).

AFPs were isolated, but not described, from the Antarctic collembolan *Gressitacantha terranova* (Hawes et al. 2011). In addition, AFPs were purified from the previously mentioned Antarctic springtail, *G. hodgsoni*, and the amino acid composition of one, a 9-kDa isoform, was provided (Hawes et al. 2014). Insufficient material was available for further description, however, the amino acid composition is revealing. The three most prominent amino acids were cysteine (13.8 mol%), histidine (11.5 mol%), and glycine (11.5 mol%), suggesting that this AFP is different from other known AFPs, including those of *G. harveyi*. Indications are that all 12 cysteines in this AFP form disulfide bridges, similar to the situations in several beetle AFPs. Also, the glycine content, although high, is not nearly as prevalent as in the *H. harveyi* sfAFP (46 mol%) mentioned above. The histidine content is unique, and taken together these points indicate a new AFP in *G. hodgsoni*.

8.2.2 Mollusks

An antifreeze glycoprotein (AFGP) like that present in many Antarctic Notothenioid fishes (DeVries 1971) was reported to be responsible for the 0.38 °C TH present in the blood of the freeze-tolerant intertidal mussel *Mytilus edulis* from the coast of Europe (Theede et al. 1976). The presence of these AFGPs in four unrelated taxa [*M. edulis* mussels, Antarctic Notothenioid fishes (DeVries 1971), Arctic Gadid fishes (Raymond et al. 1975; VanVoorhies et al. 1978), and *Ixodes scapularis* ticks (Neelakanta et al. 2010)] makes the apparent convergent evolution (Chen et al. 1997) of these AFGPs, especially significant since in this case horizontal gene transfer seems very unlikely. *M. edulis* are freeze tolerant, unlike the other freeze-avoiding AFGP-producing animals. During the normal twice per day low tides, the mussels are exposed to air temperatures and potential freezing. Consequently, the function of the AFGPs in the mussels is probably recrystallization inhibition, rather than as an antifreeze to prevent freezing. TH was not present in *M. edulis* collected in late December from Cape Cod Massachusetts (Duman unpublished data), so there may be population variation in this species.

TH has also been measured in another mollusk, a gastropod, the nudibranch *Tergipes antarcticus*, found in brine channels in the Antarctic sea ice (Kiko 2008). Blood from one of the two adults collected had TH of 1.49 °C. Egg clutches collected from the ice also had TH, albeit at low levels (0.16 °C–0.19 °C in three samples). The eggs and adults were not freeze tolerant, but the SCPs of both were below the lowest temperatures measured in the ice during this austral spring cruise (−10.9 °C). The origin of the AFPs responsible for the measured blood TH in the adults was undetermined, although the author speculated that food containing AFPs could be the source since TH-producing sea ice diatoms and bacteria inhabit the brine channels and are consumed by the nudibranchs. Horizontal gene transfer is also a possibility.

When the Antarctic limpet, *Patinigera polaris*, becomes frozen into the anchor ice near the shoreline, where the temperature can reach well below the −2 °C freezing point of the limpet, they secrete an enveloping mucus that protects from inoculative freezing initiated by the ice down to approximately −10 °C. It was reported that the mucus has thermal hysteresis activity as part of its protective activity (Hargens and Shabica 1973). A recent study of another Antarctic limpet *Nacella concilla* was unable to confidently identify the mucus as being a feature in the inhibition of inoculative freezing, but the physical inhibition of ice propagation due to the viscosity of the slime was suggested to perhaps play a role (Hawes et al. 2010).

8.2.3 Nemerteans (Ribbon Worms, Also Known as Proboscis Worms)

An intertidal Antarctic Nemertean, *Antarctomertes validium*, was shown to have hemolymph TH of 1.4 °C (Waller et al. 2006). This study of intertidal invertebrate fauna near the Rothera Research Station on Adelaide Island in the Antarctic Peninsula (64°34′S, 68°07′W) during the Antarctic summer (December and January) identified the presence of a surprisingly large number of invertebrate animals and studied their ecophysiological adaptations. Nine mobile and three sessile (encrusting) species representing multiple classes remain in the intertidal throughout the winter. Because marine invertebrates are typically isosmotic to seawater unless they are exposed to temperatures below the freezing point of the seawater (approximately −1.9 °C) they are not in danger of freezing. Of course, this does not hold for intertidal species from cold environs as they are exposed to colder air temperatures at low tide, generally twice per day. Also, the scouring action of ice as the tides change can cause physical disruption that eliminates many species. In addition, even in the subtidal zone in the Antarctic the formation of "anchor ice" can surround and entrap sessile or slow-moving organisms as it forms down to about 33 m depth (Dayton et al. 1969). The winter minimum temperature was −15 °C in the high intertidal at the study site and was a fairly constant −5 °C in the low intertidal. During the summer study of the various species only the nemertan, *A. validium,* demonstrated TH activity (1.4 °C). At this time, its SCP of approximately −10 °C, most likely assisted by AFPs, protected it from freezing as the minimum temperature recorded in situ was −5 °C.

8.2.4 Cestodes (Leeches: Annelid, Hirudinea)

AFPs in two species of fish leeches, *Cryobdella antarctica* and *Cryobdella levigata*, present on Antarctic Notothenioid fishes in McMurdo Sound has been reported (Kolb 2013). These Piscicolida leeches of the order Rhyncobdellidae are obligate ectoparasites typically attaching to, and sucking blood from, the gills of host fishes. As these host fishes have antifreeze glycoproteins (AFGPs) in their blood (DeVries 1971; DeVries and Cheng 2005), the presence of AFGPs in the digestive fluids of the leeches is expected. Indeed, digestive fluid from *C. levigata* had a mean TH of 2.84 °C. Also, the MP of this fluid was 2.46 °C (lower than that of the seawater), and the hFP of the digestive fluid was −5.30 °C, well below the FP of the seawater. The small size and flattened shape of the leeches made the acquisition of samples (other than digestive fluid) sufficient for TH determinations difficult. Consequently, body fluid TH determinations could not be made, and additional measurements were done only on the whole body and tissue homogenates or on tiny body fluid samples that required dilution prior to measurement. However, all samples had some level of TH, ranging from a few hundredths to a few tenths of a degree C, as well as the typical bipyramidal crystal growth of ice formed in the presence of AFGPs. Immunohistochemistry with AFGP antibodies was used to identify AFGPs in protein vesicles in digestive system epithelial cells, indicating endocytosis of AFGPs. The TH of body fluid was presumed to result from gut uptake of at least some amount of functional AFGPs originating from the fish blood. This mechanism would be analogous to the situation in the fish themselves, where most of the AFGPs are produced in the pancreas, secreted via the pancreatic duct into the small intestine, taken up across the gut, and then distributed throughout the body by the vascular system (Evans et al. 2012).

Tandem time-of-flight mass spectroscopy and protein sequencing determined that the AFGPs in the internal body fluids of the leeches were undigested AFGP-7 and -8. These are the smaller AFGPs produced by the fish hosts, with lesser specific TH activity. Along with the characteristic -Ala-Ala-Thr- tripeptide repeats that compose the fish AFGP-7 and -8, some repeats have proline substituted for certain alanines (Lin et al. 1972).

It appears that these two leeches have co-opted the fish host AFGPs for their own use to inhibit lethal freezing. However, mRNA investigation indicated that in one of the species, *C. levigata*, horizontal gene transfer, perhaps from host to parasite, introduced the AFGP genes into the leech genome and that this species produces some level of its own AFGPs, in addition to using those of the host.

8.2.5 Sponges

TH was identified in a homogenate prepared from the Antarctic sponge *Homaxinella balfourensis* collected in McMurdo Sound (Wilkins et al. 2002). Supernatant was applied to HPLC, and activity based on bipyramidal crystal morphology and low-level TH was used to select active HPLC fractions for further analysis, eventually purifying a 2457-Da peptide with a partial sequence of X-Pro-His-Gln-Ser-Arg-Gly-Ala-Gln-Arg, with X representing an unknown amino acid. Insufficient material was available for further analysis. While this species is found at depths down to 500 m, it is also present in more shallow areas, including sites where it can be overgrown by anchor ice. Like most sponges, it is a filter feeder, so there is a possibility that the TH-active peptide originates from a source other than the sponge itself. The possible TH/RI activity of *H. balfourensis* deserves additional study, as from an evolutionary standpoint this is an important addition to the list of animals with such IBPs.

8.2.6 Nematodes

Nematodes are an ubiquitous and major component of the soil fauna. They are especially resistant to dry conditions, sometimes undergoing anhydrobiosis when desiccating conditions are severe. In addition, although typically insulated from the extremes of air temperatures by their soil habitat, they are often exposed to subzero temperatures, especially in polar and alpine environments (Treonis et al. 2000). Consequently, as a group, they demonstrate considerable adaptation to cold (Wharton 1995, 2003). Because of their small size (high surface to volume ratio) and the permeability of their cuticles, many have evolved cryoprotective dehydration as a subzero survival mechanism (Wharton et al. 2003). In addition, some nematode species are freeze tolerant, even surviving intracellular freezing, a normally lethal occurrence for most animals (Wharton 1995, 2003; Wharton and Ferns 1995; Wharton et al. 2005b). Most relative to this review is the finding that recrystallization inhibition and ice restructuring have been described in some nematode species, suggesting the presence of RIPs, although the responsible proteins have not been identified.

The Antarctic nematode *Panagrolaimus davidii* (now renamed *Panagrolaimus* sp. DAW1) is the best studied nematode from the standpoint of cold tolerance and RIPs. Freeze tolerance, cryoprotective dehydration to avoid freezing, and RIPs are important mechanisms of the subzero tolerance of this species (Wharton 2003), however, transcriptomics has identified numerous additional components (Thorne et al. 2014, 2017). In the laboratory, sufficiently slow cooling rates that allow equilibration of the unfrozen worms with the surrounding lower vapor pressure in the frozen soil cause outflux of water permitting cryoprotective dehydration that keeps the worms from freezing. However, faster cooling rates that do not allow for vapor pressure equilibrium result in extracellular freezing that the worms typically survive (Wharton et al. 2005a). At even faster cooling rates intracellular freezing occurs, but many worms are even able to survive this usually lethal event (Wharton and Ferns 1995; Wharton et al. 2005b). Recrystallization inhibition has been reported multiple times in *P. davidii* (Wharton and Ferns 1995; Ramløv et al. 1996; Marshall 2003; Wharton et al. 2005a, b), and this activity is likely of value in mitigation of both extracellular and intracellular ice formation by minimizing the size and controlling the shape of ice crystals. The responsible proteins have not been identified, although there have been multiple attempts, including the use of modern molecular techniques (Thorne et al. 2014, 2017). The latter suggests a new type of IBP.

Recrystallization inhibition was also reported in additional nematodes from both Antarctic and temperate environs. Homogenates of two freeze-tolerant Antarctic nematodes, *Scottnema lindsayae*, and *Plectus murrayi*, exhibited low-level thermal hysteresis activity, although the homogenates were very concentrated (Wharton and Raymond 2015). A desiccation induced transcriptome of *P. murrayi* from the very cold and desiccating Antarctic Dry Valleys had previously identified a transcript in *P. murrayi* that had sequence homology to a type II fish AFP similar to that from

Atlantic herring, *Clupea harengus*, but whether the putative AFP has TH or RI activity was not determined (Adhikari et al. 2009). Also, *Steinernema feltiae* is a temperate zone nematode that in its juvenile stages infects insects. These juveniles are freeze tolerant, even surviving intracellular freezing, and they have RI activity, but do not exhibit TH (Ali and Wharton 2016).

In summary, it appears that multiple species of nematodes exhibit RI, and sometimes low levels of TH, that is, a functional component of their freeze tolerance. However, the responsible RIPs have not been identified. The possibility that the RI might be at least partially induced by antifreeze glycolipids has not been investigated.

8.2.7 Vertebrates

Because marine teleost (boney) fishes are hypoosmotic to seawater they are in danger of freezing in ice-laden seawater where the water temperature is lower than the eqMP/FP of their body fluids. Therefore, as mentioned previously (and treated in Chap. 5), AFPs have evolved multiple times in these teleosts as a mechanism to avoid freezing without raising their osmotic concentration and consequently losing their hypoosmotic condition and thereby disrupting macromolecular structure/function (DeVries 1971; Chap. 5; DeVries and Cheng 2005). Put simply, it is apparently easier to evolve AFPs than to reevolve multiple macromolecules that remain functional at higher osmotic concentrations.

In contrast to teleosts, Elasmobranch fishes osmoconform to their normal seawater habitat, and consequently the few species that enter subzero ice-laden seas would seem to be protected from freezing. Slight increases in their inorganic ion, urea, and/or trimethylamine oxide concentrations would render them unfreezable without causing major osmotic problems. Still, it would be interesting to test for TH in the blood of Greenland sharks, *Somniosus microcephalus,* that commonly venture into the Arctic Ocean. Peptides with TH have been isolated from skin collagen digestions of sharks (Wang et al. 2014), chickens, and mammals. These will be discussed in a later Sect. 8.2.7.2.

Although no fishes are known to be freeze tolerant, a few amphibians and reptiles have adaptations producing varying levels of freeze tolerance (Costanzo and Lee 2013; Storey and Storey 2017). No reptiles, except perhaps for birds (see below) have been shown to have IBPs or AFGLs, however, this could change with additional studies. Consequently, we will not discuss reptiles, other than birds, further in this review. However, two species of freeze-tolerant frogs have IBPs and/or AFGLs that are likely involved in their freeze tolerance.

8.2.7.1 Frogs

In regions where subzero air temperatures occur, many amphibians overwinter in microhabitats where they are not exposed to freezing—at the bottom of ponds and streams, or in the soil below the frost line (Storey and Storey 2004). However, there are various species of terrestrial frogs that winter in shallow leaf litter above the frost line. While the temperatures in these sites are thermally buffered from air temperatures, especially if covered by snow, freezing temperatures do occur there, and these species typically survive freezing (Schmid 1982; Storey and Storey 1984, 2017; Costanzo and Lee 2013). Among these, the freeze tolerance adaptations of the North American wood frog, *Lithobates* (previously *Rana*) *sylvatica*, are the best understood. The distribution of *L. sylvatica* in North America is extensive, ranging in the east from the southern Appalachian Mountains north into Canada and west across much of Canada into Alaska above the Brooks Range (~69°N) (Martof and Humphries 1959). While wood frogs are freeze tolerant to approximately −7 °C in the more southerly portions of their range such as Ohio (Layne and Lee 1987), as might be expected they are considerably more cold tolerant in Interior Alaska where they were monitored in the field and shown to be frozen for approximately six months at temperatures as low as −18 °C (Larson et al. 2014). The permeable skin of frogs provides little protection from inoculative freezing, so they freeze at a temperature only slightly below the Mp/Fp of their body fluids (Layne 1991). Freezing then proceeds slowly through the body and can take 1–2 days to come to equilibrium, until as much as 65% of the body water is frozen (Layne and Lee 1987). Wood frogs accumulate glucose as a cryoprotectant (Storey and Storey 2004), while others such as the tree frogs *Hyla versicolor* and *H. chrysosceelis* concentrate glycerol (Zimmerman et al. 2007). In the autumn, prior to freezing wood frogs also slowly accumulate urea as a cryoprotectant (Costanzo and Lee 2013), and then when cued by the initiation of the freezing process they quickly build up glucose from stored glycogen (Storey and Storey 2004). The metabolic pathways, cues, etc. are well understood (Storey and Storey 1984, 2017). Redistribution of water takes place during freezing as water preferentially is moved and frozen in coelomic and lymphatic regions, leaving less water to freeze in organs (Lee and Costanzo 1998).

Although an IBP that produces TH and/or inhibits recrystallization has not been definitively identified in any amphibians, a possible IBP with these capabilities has been proposed for *L. sylvatica* (Biggar et al. 2013; Sullivan et al. 2015).

Also, AFGLs have been identified in *L. sylvatica* (Larson et al. 2014) and the somewhat freeze-tolerant European frog *Rana lessonae* (Walters et al. 2011).

Production of two proteins, Fr10 and Li16, is induced by cold in *L. sylvatica* and these proteins provided protection from freeze damage to an insect cell line, suggesting that they do so in the frogs as well (Biggar et al. 2013). Fr10 is secreted while Li16 is intracellular. When frogs are subjected to anoxia, dehydration, or freezing the gene encoding the 12-kDa Fr10 is upregulated in various tissues and organs in a tissue-specific manner (Sullivan et al. 2015). Modeling studies indicated that although Fr10 does not have sequence homology to any known AFP, it has structural similarities to a fish Type IV AFP from the sculpin *Myoxocephalus octodecimspinosus* (Sullivan et al. 2015). However, its abilities to produce TH or recrystallization inhibition have not been tested.

AFGLs, similar to those initially described in the freeze-tolerant Alaskan beetle *Upis ceramboides* (Walters et al. 2009) have been identified in several additional insects, plants, and the freeze-tolerant European frog *Rana lessonae* (Walters et al. 2011), as well as in wood frogs from interior Alaska (Larson et al. 2014). The AFGL backbone is composed of repeating units of a disaccharide, β-mannose, and β-xylose residues in 1→4-linkage, plus lipid made-up primarily of fatty acid (Fig. 8.2). The TH activity of the AFGLs is comparable to those of the hyperactive insect AFPs, but they are primarily found on cell membranes. The AFGLs were purified from skeletal muscle and skin of *R. lessonae*, but the blood lacked TH or recrystallization activity. Likewise, Alaskan *L. sylvatica* lacked blood recrystallization inhibition, while AFGLs were found mainly on cell membranes in skeletal muscle and various internal organs, with lesser amounts in skin. This suggests that the primary function of the AFGLs in frogs is to inhibit lethal inoculation of the cytoplasm across the cell membrane by extracellular ice. Recent reviews of these and other freeze tolerance adaptations of amphibians may be consulted by the interested reader (Storey and Storey 2017; Costanzo 2019).

8.2.7.2 Endothermic Vertebrates (Birds and Mammals)

Although mammals and birds, due to their endothermic homeothermy and resulting high core body temperature, have not traditionally been thought to need AFPs or RIPs this may not always be the case. As core body temperature in these animals begins to drop below the hypothalamic set point, in order conserve heat to maintain core temperature they typically decrease blood flow to the skin and extremities via general peripheral vasoconstriction, and often employ heat exchangers in the extremities, allowing the temperature of the periphery to decrease well below normal core set point temperatures. If environmental temperatures are especially cold this could result in skin temperature dropping to near subzero levels or less, especially in extremities, and particularly if these are poorly insulated. Consider the bare legs and feet of birds, the sparsely furred or bare tails, etc. of many mammals such as beaver and other rodents, many of which are small. The bare ears, nostrils and tail of the northward range expanding American opossum (*Didelphis virginia*) in the northern areas of its range (northeastern USA and southern Canada) are often visibly freeze damaged (Walsh and Tucker 2017). Consequently, limited antifreeze or RI activity in some endotherms may be useful. The skin of baleen whales (Balaena mysticetus) collected from individuals legally killed by native Eskimo hunters near Barrow Alaska had thermal hysteresis ranging from 0.75 °C to 2.97 °C, although not all individuals exhibited TH (Sformo, Vu, George, Champion, Serianni, Duman unpublished data). These whales routinely swim in ice laden water and are exposed to subzero air temperatures when surfacing, suggesting that the TH may have a protective function. The TH active material was a protein, although proteomic studies failed to identify the AFP. Recent research has shown that collagen, or perhaps more accurately certain of the multiple types of peptides that compose collagen, in many endotherms have TH and/or RIP properties (Damodaran 2007; Wang and Damodoran 2009; Wang et al. 2014; Du and Betti 2016; Cao et al. 2016). Might these functions to inhibit inoculative freezing across the skin or mitigate freeze damage in some mammals and birds that we typically do not recognize as having these problems?

Collagen is composed of an extremely common, but heterogeneous and complicated family of proteins that, in spite of being heavily studied for many years, is far from completely understood. Collagen accounts for approximately one-third of all the proteins in mammals, and is an important structural component of skin (especially dermis), cartilage, extracellular matrix, etc. In vertebrates, collagen types I and II are the most common forms, but 28 different collagens composed of some 50+ different peptides are recognized, and collagen-like domains are present in numerous other proteins (Shoulders and Raines 2009). A common structural motif of collagens is three tightly packed parallel polypeptide strands in a left-handed polyproline II-type (PPII) helical conformation that coil around one another with a one residue stagger that forms a right-handed triple helix. Because of the tight packing of the triple helix every third amino acid is a glycine resulting in a tripeptide repeat of (G-X-Z) where X is often proline or hydroxyproline and Z may be any amino acid. The glycine is often buried on the inside of the coil.

Recall that, as described earlier (Fig. 8.1), the *S. harveyi* snowflea AFPs (sfAFP) form short polyproline II helices (Graham and Davies 2005). Given the similarities between the sfAFP and predominant features of collagen, perhaps it is not surprising that peptides with TH and/or RI activities have been derived from proteolytic treatments of collagen from the skin of mammals (cows and pigs) and chickens. These studies stemmed from efforts to identify uses for the skin and other byproducts of these animals that are leftover from meat processing.

Peptides with RI activity were initially produced by papain or alcalase digestion of bovine gelatin (Damodaran 2007; Wang and Damodoran 2009). Small peptides were most active. Likewise, a PPII collagen peptide derived from chicken collagen with a sequence of G-S-P-G-A-D-G-P-I-G-A-Hyp-G-T-Hyp-G-P-Q-G-I-A-G-Q-R has RI activity (Du and Betti 2016).

Alkaline hydrolysis of porcine collagen and subsequent chromatographic procedures purified a 1163 Da peptide with high TH (5.28 °C, measured using differential scanning calorimetry) and RI activity (Cao et al. 2016). This is interesting as pigs have very sparse hair on the skin, and therefore lack protective insulation. In addition, the peptide decreased the glass transition (vitrification) temperature of ice cream to −17.6 °C, and raised the melting temperature of ice cream as well. [Recall that a small portion of the TH activity of AFPs results from a rise in the hMP above the eqMP, as well as the hFP decrease (Knight and DeVries 1989)]. This AP-3 ice-binding collagen peptide has the characteristic collagen tripeptide repeat sequence of GLLGPLGPRGLL, although there are multiple leucine residues.

Similar collagen peptides were also isolated from sharks (Wang et al. 2014). (The species was not given, but the source was identified as "the local supermarket.") An

11-residue peptide (SsC-AFP) with the PPII sequence (G-I-G-P-A-G-P-L-G-P) provided protection to *Lactobacillus bulgaricus* at −20 °C over a 24-h period.

At this time it is not known if these collagen-derived peptides actually function as AFPs or RIPs in the skin, etc. in their natural state as components of collagen. However, this possibility begs further inquiry. Also, as collagens are important components of extracellular structures such as basement membranes, and extracellular matrix perhaps these PPII peptide components contribute RI activity along with soluble RIPs when they are present in any organism.

8.3 Fungi

Given their importance and abundance in cold areas of the world it is surprising that more cold tolerance studies have not been done on fungi, which, of course, include the single-celled fungi—the yeasts. A few cold-adapted fungi concentrate polyols and sugars (Robinson 2001; Tereshina and Memorskaya 2005) and several species have IBPs that probably function as RIPs, as the TH is typically low and the fungi freeze tolerant. The first report of fungal TH was based on TH in homogenates of fruiting bodies from four of five species collected in late winter from northern Indiana, USA (~42°N latitude)—the oyster mushroom *Pleurotus ostreatus*, the winter mushroom *Flammulina velupites*, and two bracket fungi *Coriolis versicolor* and a *Stereum* sp. (Duman and Olsen 1993). The *Stereum* sp. is an ascomycete and the others are basidiomycetes. All were collected in early March following 2 days with subzero temperatures and snow. The fruiting bodies of *P. ostreatus* and *F. velupites* emerged during a brief thaw prior to this return of cold weather. The common name, winter mushroom, of *F. velupites* derives from its habit of sprouting during winter thaws, which indicates its likely cold tolerance. All were frozen when collected. This, along with their low level of TH (0.30–0.35), suggests they were freeze tolerant. Samples collected from the same species in summer lacked TH. All samples lost TH after treatment with bacterial protease, indicating that a protein produced their TH activity. However, a xylomannan AFGL with low TH (0.10 °C) and RI activity has been described in fruiting bodies and mycelia of commercial *F. velupites* from Japan (Kawahara et al. 2016). This AFGL is composed of mannose and xylose in a 2/1 ratio, which is somewhat different from that of the previously described xylomannan AFGLs from insects, frogs, and plants with a 1.3/1 ratio of the two saccharides shown in Fig. 8.2 (Walters et al. 2009, 2011). The purified *F. velupites* AFGL protected hamster CHO cells from freeze damage at −20 °C for 2 days.

The first fungal TH producing IBP described was that of the snow mold *Typhula ishikariensis* (Hoshino et al. 2003b). Snow molds are psychrophilic and psychrotrophic, growing mycelia and reproducing in winter under the snow. Consequently, these phytopathogenic fungi are able to attack plants when they are in a vulnerable dormant state. The snow molds have representatives in the two major fungal groups, ascomycetes, and basidiomycetes, as well as the oomycota.

T. ishikariensis is freeze tolerant (Hoshino et al. 1998), and produces multiple isoforms of a potent IBP that is secreted into the extracellular fluid and probably functions as a RIP (Hoshino et al. 2003a, b). Culture filtrates of several species of snow mold fungi, including multiple strains of *T. ishikariensis* and other *Typhula* species were tested for IBP activity by means of the ability, or lack thereof, to produce TH and to structure ice (Hoshino et al. 2003a, b). Only basidiomycetes, four *Typhula* species and *Coprinus psychromorbidus*, had IBP activity consisting of very low TH of 0.015 °C–0.033 °C and the ability to structure ice. Multiple isoforms of IBPs from both *T. ishikariensis* and *C. psychromorbidus* were purified with molecular masses of approximately 22 kDa and 25 kDa, respectively. The N-terminal sequences were similar and *T. ishikariensis* genes were cloned, providing the cDNA sequence of one of the *T. ishikariensis* IBPs. The encoded proteins were expressed, and unsuccessfully tested for the ability to protect cultured *T. ishikariensis* from freeze damage. However, the authors suggested that the IBPs function as RIPs.

An isoform of a *T. ishikariensis* IBP (*Tis*AFP8) was further described and compared to an IBP from a psychrophilic ascomycete *Antarctomyces psychrotrophicus* (Xiao et al. 2010). Although the two had somewhat similar N-terminal amino acid sequences they differed sufficiently that the authors stated that they evolved from different genes. The *A. psychrotropicus* IBP (*Anp*AFP) was somewhat larger than the *Tis*AFP and contained a significant carbohydrate component. Also, Asx was the most abundant amino acid of the *Anp*AFP (17.4%) while Asx composes only 3.4% of *Tis*AFP. The crystal structure of a different *Tis*AFP isoform (*Tis*AFP6) provided the conformation and ice-binding site information of this interesting IBP (Kondo et al. 2012). It forms a β-helical solenoid that is triangular in cross section. The solenoid barrel is composed of six right-handed β-helical loops, each made up of 18–27 amino acids each, as well as a reinforcing α-helix that lies parallel to the axis of the barrel and perpendicular to the coils of the loops (Fig. 8.3a–c). Surprisingly, the first β-helical loop (originating near the N-terminus) lies next to the sixth loop (from near the C-terminus) such that the loops are arranged in an unusual side-by-side pattern, 1-6-2-3-4-5. The flat ice-binding site is composed of 5–6 amino acids in the middle of the loops, and located on the same side of the six β-helical loops, which are superimposable. The residues in positions 1, 2, 4, and 6 of each loop project their side chains out toward the solvent while those in positions 3 and 5 project into the inside of the β-helix, forming a regular pattern that permits the ice-binding site to attach to the surface of ice both by means of hydrogen bonding and by the formation of ordered ice-like waters on its surface (22 of the 29 amino acids composing the ice-binding site have hydrophobic side chains). The crystal structure of the larger *Tis*AFP8, with considerably greater TH activity than *Tis*AFP6, has also been described (Cheng et al. 2016). *Tis*AFP8 also folds into a right-handed β-helix barrel with a flat IBS, and the parallel α-helix. This hydrophobic ice-binding site of *Tis*AFP8 also structures waters in an ice-like fashion, and the IBS is both larger and more complementary to the ice surface, allowing improved attachment of the protein to ice. The ice-like waters are organized into two groupings that can separately mediate binding of *Tis*AFP8 to ice, thereby increasing its antifreeze activity relative to *Tis*AFP6.

IBPs with the typical pit forming capabilities of AFPs on the surface of ice were identified in multiple additional fungi (enoli, *F. populacola*; shitaki, *Lentinula elodes*; king oyster, *Pleurotus, eryngii*; shimeji, *Lyophyllum shimeji*), plus the sequences of IBPs from two of these (enoki, *F. populicola*; and shitoki, *L. edodes*) were determined (Raymond and Janech 2009). The amino acid sequences of these probable RIPs are very similar to one another and to the *Typhula* snow mold IBP (*Tis*AFP) with 50%–55% identity and 68%–70% similarity to the *Typhula* IBP. Both have signal peptides, and their predicted molecular masses are 24.9 kDa (enoki) and 28 kDa (shitoki). In addition to their sequence similarity to *Typhula* IBPs, all three have significant similarity to IBPs of some sea ice bacteria and diatoms, once again indicating possible horizontal gene transfer of IBP genes from bacteria to these eukaryotes. An interesting phylogenetic tree based on sequences of IBPs of fungi, diatoms, and bacteria, illustrating likely relationships, was presented. The authors also noted that many bacterial and archaeal sequences of unknown function encoding complete IBP sequences are in the genomic databases.

Many yeasts are cold adapted and well represented in polar and other less extreme cold areas (Buzzini and Margesin 2014), and some are known to produce IBPs (Kim et al. 2014). A secreted, glycosylated dimer-forming IBP (*Le*IBP) with an amino acid sequence similar to *Tis*AFP (56% identity) was described from the Arctic psychrophilic yeast, *Leucosporidium* sp. AY30 isolated in a freshwater ice core from Svalbard Island, Norway (Lee et al. 2010). A subsequent X-ray crystallographic study of expressed *Le*IBP, both the native glycosylated (expressed in *Pichia pastoris*) and the non-glycosylated (expressed in *E. coli*) forms, showed that *Le*IBP is a homodimer (Park et al. 2011; Lee et al. 2012) (Fig. 8.3d–f). Each monomer is a right-handed β-solenoid structure with an associated α-helix, similar to that of the *Tis*AFP. The dimers are connected at the ends of the α-helices and the C-terminal loops that extend out from each of the two β-helices, forming a V structure. The conformation of the natural glycosylated *Le*IBP is quite similar to that of the non-glycosylated form. The glycosylation site is Axn^{185} and four sugars were identified—two *N*-acetyl-β-D-glucosamines, one β-D-mannopyranose, and one α-D-mannopyranose. Because the TH activities of the glycosylated and non-glycosylated forms are identical and the carbohydrates are oriented away from the IBS the authors suggest that the sugars are not involved in ice binding, but indicated that they are likely to be important in the folding and secretion of *Le*IBP.

A psychrophilic Antarctic yeast *Glaciozyma antarctica* PI12, initially collected from sea ice, produced and secreted IBPs into the culture medium (Hashim et al. 2013). (Note that *G. antarctica* was originally described as *Leucosporidium antarctica*.) A sequence survey of the genome identified a novel IBP with just 30% similarity to *Tis*IBP. This protein (AFP1) was described from cDNA as a 177 residue IBP. Expressed AFP1 had RI activity and low TH (just 0.08 °C). A second IBP (AFP4) was also described from *G. antarctica* PI12 (Hashim et al. 2014). The 25-kDa (261 amino acid) AFP4 has 93% similarity to *Le*IBP and folds into a β-helix with three distinct faces, one of which is the ice-binding surface. Recombinant AFP4 produced TH of 0.80 °C at a concentration of 5 mg/ml. Another *G. antarctica* IBP (18 kDa) with very low sequence similarity to any known

AFPs, is composed of four α-helical and three β-strand regions, and is predicted to have a globular shape (Shah et al. 2012). This study primarily looked at α-helical peptide regions of the IBP, suggested to have AFP activity, and noted their similarity to the α-helical Type-I winter flounder AFPs. Designed peptide fragments based on the native α-helices had antifreeze activity.

Lichens are extremely well represented in cold environments, including polar regions. Crude homogenates of several lichens were shown to have IBPs capable of RI in a large screen (120 species) of plants and ten species of lichens (Doucet et al. 2000). Nine of the ten lichens tested positive for RI—all five species from South Georgia Island in the maritime Antarctic and four of five species tested from Great Britain. Whether the fungal or the photosynthetic (a unicellular alga, a cyanobacterium, or both) component of the symbiotic, mutualistic association (or both) produces the RIPs is unknown, but the sharing of RIPs is another example of the benefits of this association.

Fungi are also known to produce INPs (reviewed by Ashworth and Kieth 1995). These include both free-living and lichen fungi. They appear to be functional in these freeze-tolerant organisms, inducing ice nucleation at high subzero temperatures to allow better control of ice growth. Also, it is been suggested that the lichen fungi INPs could function to increase uptake of atmospheric water vapor.

8.4 Bacteria

Thermal hysteresis and recrystallization inhibition activity have been identified in several bacteria from multiple locations and environments—soils (temperate and polar), ice (sea, freshwater, glacial), etc. Many of the known TH/RI IBP-producing bacteria, especially polar bacteria, are psychrophilic (optimal growth rates <15 °C), but others from more temperate regions are psychrotolerant (optimal growth rates >15 °C, but survive lower temperatures). While many bacteria are generally considered to be freeze tolerant, it is not always obvious if these single-celled organisms are freeze tolerant or freeze avoiding. Do they survive freezing of the periplasmic fluid (between the cell membrane and the cell wall), intracellular freezing? The functions, and even the ultimate locations (cytoplasmic, periplasmic fluid, the water immediately outside the cell wall surrounding the organism) of the IBPs, are not always clear. Some IBPs have N-terminal signal peptides that target them for secretion. But, are they secreted into the periplasmic space, beyond into the surrounding environs, both? What are the in situ TH and RI activities in these spaces? Although the reported activity for a given IBP is sometimes variable and low, this could be due to a number of reasons such as the purity of the sample, structural accuracy and stability of the purified or expressed proteins, the presence or absence of enhancers, etc.

As we will see, many of the TH and RI producing bacteria that have been identified are sea ice bacteria. These and the numerous other IBP-producing organisms (fungi, diatoms, chlorophytes, the copepod *Stephos longipes*, etc.) that constitute this interesting sea ice community are mainly found in hypersaline brine channels in the often meters thick ice. It is important to the inhabitants that these channels remain unfrozen (Priddle et al. 1986; Mock and Thomas 2005), and the TH/RI-producing IBPs produced by the inhabitants are often secreted into the surrounding water of the ice channels where they assist in keeping the hypersaline ice channels open, even when the temperatures in the channels dip as low as −20 °C. The same could be the case for bacteria inhabiting glacial or soil ice and snow as RI keeps the size of ice crystals small. For more on the sea ice and its inhabitants see Chap. 7 by Wisniewski et al. However, the TH-producing IBPs could also function to inhibit RI in the periplasmic space and/or prevent cytoplasmic freezing.

Bacterial ice-nucleating proteins (INPs) have been well studied and reviewed multiple times since they were initially discovered. Consequently, they will be discussed here only briefly. For a comprehensive review see Lee et al. (1995).

The first evidence of TH in bacteria was based on homogenates of cultures (grown on agar at 3 °C) of the common soil bacterium *Rhodococcus erythropolis* (isolated from the gut of larvae of the AFP producing beetle *Dendroides canadensis*) and of the cryophilic *Micrococcus cryophilus*, but the responsible proteins were not identified (Duman and Olsen 1993).

The growth-promoting rhizobacterium *Pseudomonas putida* GR12-2, originally isolated from soil in the Canadian High Arctic, produces and secretes TH-producing IBPs that probably function as RIPs (Sun et al. 1995). The initial impetus of this study was to determine if *P. putida* could survive and be used to promote the growth of crop plants at low soil temperatures in colder regions of Canada. Plant growth-promoting rhizobacteria are critical to plant growth, performing numerous services for the plant such as production of antibiotics, phytohormones, and siderophores; enhancement of nitrogen fixation; phosphate uptake, etc. *P. putida* was indeed shown to accelerate plant growth, even at 5 °C. *P. putida* survived freezing at -50 °C, and part of this ability was attributed to the production and secretion of IBPs with likely RI function. Growth medium of *P. putida* grown at 5 °C, but not 25 °C, demonstrated the hexagonal ice crystal shapes that are common in the presence of low activities of TH-producing IBPs. When the growth medium was concentrated a low TH (0.11 °C) was measured. While the periplasmic space and cytoplasm had ice structuring activity it was less than in the growth medium, suggesting that *P. putida* secretes most of the IBPs into the medium. Various manipulations of the IBP samples indicated that the 32–34 kDa IBP was indeed a protein and that it lacked disulfide bridges. However, a subsequent study (Xu et al. 1998) by the same group isolated and described a large 164 kDa IBP from *P. putida* culture filtrate that not only had TH activity, but also had some ice-nucleating activity (initiating nucleation at approximately −10 °C). This study stated that the earlier described 32–34 kDa IBP was probably a breakdown product of this larger IBP. The 164-kDa protein contained lipid and a large carbohydrate component, while the amino acid composition was more comparable to those of previously described bacterial INPs than any known AFPs. The unusual combination of TH and IN activity of the large IBP suggested to the authors that this protein combined the functional properties of RIPs and INPs found in many freeze-tolerant organisms into one protein that could (1) induce ice nucleation at a higher temperature, resulting in slower and more controlled ice growth, and (2) produce RI to keep the numerous small ice crystals from coalescing into potentially damaging larger crystals. It is now known that AFPs, RIPs, and INPs all organize water on their surfaces in a similar ice-like manner, so this duplication of functions into one protein is not as unlikely as it might seem (Kobashigawa et al. 2005; Davies 2014). The intimate contact of this rhizobacterium with plant roots suggests that the secreted *P. putida* IBPs may benefit the plant as well as the rhizobacterium.

There have been numerous studies that identified Antarctic bacteria with varying levels and types of TH and RI producing IBPs. The first of these discovered six Ross Island bacteria with antifreeze activity (Yamashita et al. 2002). One of these, a species of *Moraxella*, secreted a 52-kDa lipoprotein with antifreeze activity into the media when cultured at cold temperatures.

An Antarctic strain of *Pseudomonas fluorescens* KUAF-68 produced two functionally different types of IBPs—an INP and an antifreeze protein with weak activity, probably a RIP (Kawahara et al. 2004). *P. fluorescens* was previously known to have INPs (Warren and Corotto 1989; Graether and Jia 2001), but not TH-producing IBPs.

Eleven strains of bacteria with RI activity were isolated from brackish Antarctic lakes (water FP approximately −1.0 °C) in the Vestfold Hills region of eastern Antarctica that had been connected to the ocean at an earlier time (Gilbert et al. 2004). The most active of these bacteria was initially identified as *Marinomonas protea*, but it was later shown to be a strain of *Marinomonas primoryensis* that produced a novel Ca^{++}-dependent IBP (*Mp*AFP) with TH activity of 0.80 °C at 11 mg/ml (Gilbert et al. 2005). Antifreeze activity was present only in cell lysates, not secreted into the culture medium. This, along with the high TH, suggested to the authors that the *Mp*AFP could function as an AFP, perhaps in the periplasmic space, to keep *M. protea* from freezing. A subsequent study of the *Mp*AFP showed it was indeed hyperactive, Ca^{++}-dependent, and very large (>1 MDa) (Garnham et al. 2008). A 322 amino acid portion of the protein that enclosed the 122 residue TH active ice-binding region was characterized, and the expressed protein showed TH of approximately 2 °C at 0.5 mg/ml. A model of the antifreeze region indicated a novel β-helical fold that includes the ice-binding site (Fig. 8.4a–c). A row of bound internal Ca^{++} ions occurs along one side of the protein, with a hydrophobic core down the opposite side and parallel repeats containing Thr and Asp/Asn located along the Ca^{++} binding site of the protein. The Ca^{++} ions are required to maintain the structure of the protein, and they are themselves coordinated by the carboxyl side chains of aspartates and the carbonyl groups of glycines in the protein backbone. The *Mp*AFP is now understood to be just one of multiple domains making up a large adhesin protein that attaches *M. putida* to the surrounding ice. The ice-binding active domain functions to secure *M. putida*, an obligate aerobe, to the undersurface of ice where it is exposed to O_2, nutrients, etc. The bulk (90%) of the massive adhesin protein contains some 120 tandem repeats (each made up of 104 residues) that serve to extend the ice-binding site domain out from the base domain of the protein that functions to bind the adhesin to the cell membrane (Guo et al. 2012).

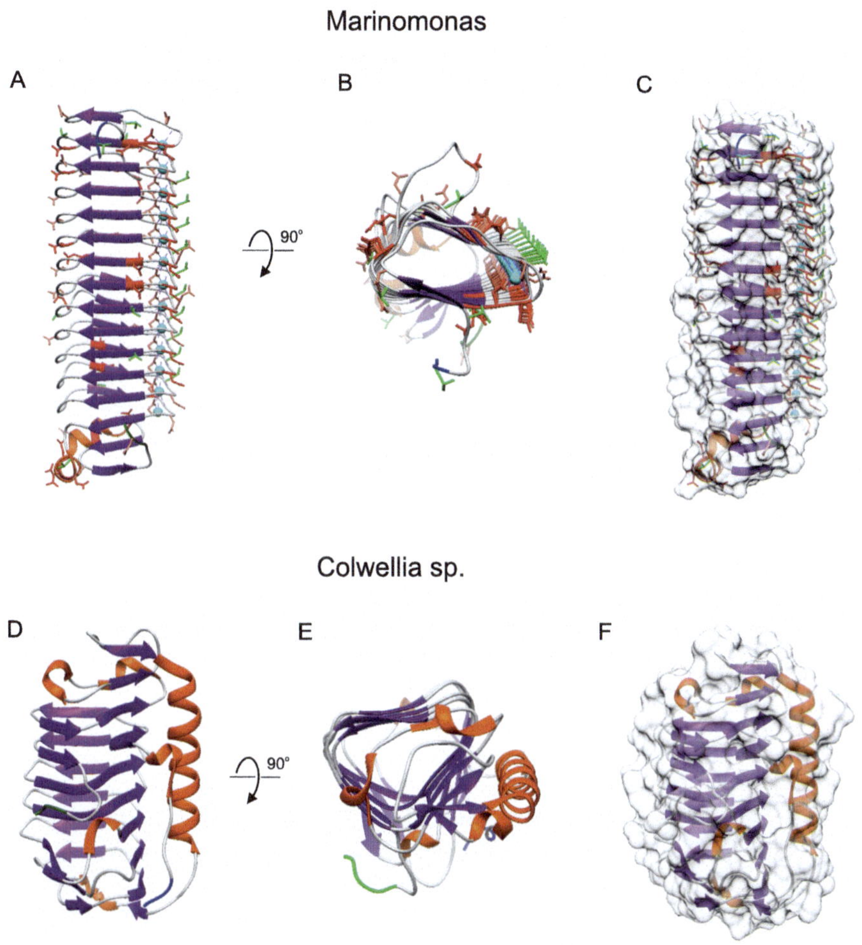

Fig. 8.4 Structures of two bacterial AFPs are presented. (**a–c**) Crystal structure of *Marinomonas protea* AFP, *Mp*AFP, (PDB 3P4G). (**a**) Secondary structure is shown with helices in orange, beta strands in purple and loops in gray. N- and C-termini are colored blue and green, respectively. Threonine residues are in green, aspartic acid in salmon and asparagine in red. Cyan spheres are calcium. (**b**) View in (**a**) is rotated 90° toward the reader. (**c**) Semitransparent molecular surface representation of the view in (**a**). (**d–f**) Crystal structure of *Colwellia* sp. AFP, *Col*AFP (PDB 3WP9). (**e**) View in (**d**) is rotated 90° toward the reader. (**f**) Semitransparent molecular surface representation of the view in (**d**)

A 26-kDa RIP (*Col*AFP) with 42% sequence identity to the *Typhula* fungus IBP (*Tis*AFP6) was identified in a species of Antarctic sea ice bacteria, *Colwellia* strain SLW05 (Raymond et al. 2007). The N-terminal sequence indicated *Col*AFP is secreted, suggesting it functions to keep open the ice channels that *Colwellia* inhabits, along with many other sea ice organisms. The crystal structure and other features of *Col*AFP were later determined (Hanada et al. 2014). *Col*AFP has an irregular β-helical structure (Fig. 8.4d–f) similar to the *Tis*AFP6 and similar IBPs of several other organisms (diatoms, fungi, copepod, etc.). It has a compound IBS encompassing the β-helix as well as the adjoining loop allowing it to bind to multiple ice planes, perhaps accounting for its hyperactive TH activity (~4 °C at 0.14 mM). Interestingly, *Col*AFP does not have the repetitive amino acid sequences usually present in other hyperactive AFPs.

A 54-kDa IBP (RIP) similar to the *Col*AFP and other related IBPs was identified from a bacterium taken from a 3500-m deep ice core just above the subglacial Lake Vostok in Antarctica (Raymond et al. 2008).

A screen of bacteria from multiple Antarctic sites identified several positives with TH in crude extracts of the bacteria ranging from 0.12 to 0.50 °C (Muñoz et al. 2017). These included isolates from five different genera—*Pseudomonas, Anthrobacter, Sporosarcina, Sphingomonas, and Plantibacter*. Genome analyses identified genes encoding putative AFPs, and the amino acid sequences of these were used for homology modeling. The modeled structures indicated a β-helix having a triangular cross section with a row of threonines along one side, that likely form the IBS. Once again these had high structural homology to the *Tis*AFP6 and *Col*AFP type IBPs known from other bacteria, fungi, etc.

Most IBP-secreting bacteria appear to secrete their IBPs, however, the Antarctic bacterium, *Flavobacterium xanthum* IAM12026, was reported to have intracellular IBP activity, low level TH and RI (Kawahara et al. 2007). The activity of the purified 59 kDa protein was increased by the addition of 0.5 M malate (Kawahara et al. 2007). Malate is also known to enhance the TH of the *Dendroides canadensis* beetle AFP (Li et al. 1998).

An AFP screen of 14 Arctic bacterial strains from five genera (*Cryobacterium, Leifsonia, Polaromonas, Pseudomonas*, and *Subtercola*) isolated from cryoconite holes in Arctic glaciers, identified eight strains with activity (Singh et al. 2014). RI activity was present in most. Of these eight, six strains of four species were further characterized with *Pseudomonas ficuserectae* having the highest TH. Cryoconite holes on the surface of glaciers initially form when dark accumulations of dust, small stones, soil, etc. absorb more sunlight than the surrounding ice, causing them to warm and melt into the ice. As the holes continue to melt they accumulate water and form small ecosystems of cryophilic bacteria, fungi, algae, ciliates, archaea, etc. Surface water freezing seals the wells while the subsurface water remains liquid. Secreted TH-producing IBPs from the contained organisms, such as *P. ficuserectae*, are likely to assist in keeping this water from freezing.

A bacterial IBP (*Ff*IBP) with considerable sequence homology to the yeast *Le*AFP and others was described from *Flavobacterium frigoris* PS1, isolated from ice in McMurdo Sound, Antarctica (Do et al. 2012a, b; Raymond and Kim 2012). While the amino acid sequence is very similar to that of *Le*IBP, *Ff*IBP lacks the C-terminal hydrophobic loop that connects the two homodimers that constitute *Le*IBP, and therefore, *Ff*IBPs exists as a monomer. However, *Ff*IBP reportedly has approximately 10X greater TH (2.5 °C at 50 μM) than *Le*IBP.

Two IBPs (labeled EFsymbIBP and EFsymbAFP) from an Antarctic microbiological consortium of a symbiotic psychrophilic uncultured bacterium and the psychrophilic ciliate *Euplotes focardii* were sequenced and found to be similar to previously described Antarctic bacterial IBPs, including that of Flavobacteriaceae bacterium from the glacial ice of Lake Vostok (i.e., EFsymbIBP = 53.4% identity) (Pucciarelli et al. 2014). This similarity indicated to the authors that the identified IBPs, which have RI activity, are likely to be produced by the bacterium rather than the ciliate. In a later study one of these IBPs, labeled as EfclIBP, was shown to have an N-terminal signal sequence, indicating that it is secreted. The authors speculate that the IBP either anchors the outer cell surface to the ice or is concentrated around cells to provide RI protection to the cell consortium (Mangiagalli et al. 2017).

A potential commensal relationship has been described between the Antarctic moss *Bryum argenteum* and multiple epiphytic bacteria (Raymond 2016). Metagenome sequencing identified hundreds of genes, mostly from *Actinobacteria* and *Bacteroidetes,* encoding a domain that is common in IBPs. The 42 encoded proteins had N-terminal signal peptides.

An interesting study employed an ingenious device, termed a "cryocycler," that used 48 alternating freeze–thaw cycles to isolate freeze-tolerant soil bacteria from a cold region of western Canada where warm Chinook winter winds periodically cause short periods of melting followed by refreezing (Walker et al. 2006). Of the five genera that survived this extreme freeze–thaw treatment, one of these, *Chryseobacterium* sp. strain C14, had RI activity. Also, cell-free media from the *Chryseobacterium* sp. culture enhanced the survival of another bacterial isolate, *Enterococcus* sp. strain C8.

8.5 Horizontal Gene Transfer of Ice-Binding Proteins

As mentioned multiple times, several bacterial TH and/or RI-producing IBPs are structurally similar to one another and to the IBPs of the snow mold *Typhula ishikariensis* (*Tis*AFP) and other fungi, diatoms, and even the copepod *Stephos longipes* (Compare Figs. 8.1a–c, 8.3, and 8.4d–f). This raises the possibility that these IBPs may have passed from one species to another by means of horizontal (lateral) gene transfer, most likely from bacteria to bacteria and from bacteria to eukaryotes. This point has been raised multiple times in the literature, providing convincing grounds for likely horizontal transmission, especially in the sea ice brine channels where IBP producing bacteria are in close proximity with co-inhabiting diatoms, yeasts, etc. (i.e., Raymond and Janech 2009; Raymond and Kim 2012; Raymond et al. 2017; Kiko 2010; Bayer-Giraldi et al. 2011). Similarity of the IBPs, in spite of many incongruities between the phylogenies of the organisms, provides strong grounds for this assertion. Phylogenetic analyses based on amino acid sequences of both verified and putative (based on biogenetic mining of public protein databases) AFPs illustrate both similarities and differences among these IBPs (Raymond and Janech 2009). The possible transfer of IBPs from viruses to bacteria or other organisms, especially in cryoconites where viruses are especially plentiful, has also been discussed (Singh et al. 2014). Either of these two potential scenarios, or convergent evolution, may have resulted in these similar IBPs in certain of these examples.

8.6 Ice-Nucleating Proteins

Ice nucleators organize water molecules on their surfaces into ice-like embryo crystals that, upon reaching a critical size, nucleate supercooled water. When temperature decreases, the size of the ice-like embryo crystals increases, while the critical size of embryo crystals decreases, thereby increasing the probability of ice nucleation. In the absence of INs small volumes of very pure liquid water can supercool to near −40 °C before an embryo crystal reaches the critical size at which ice nucleation occurs spontaneously (homogeneous nucleation). However, Zacchariassen and colleagues (Zachariassen et al. 2004, 2010) have argued that in larger volumes of water homogeneous nucleation is expected at higher temperatures (−18 °C to −22 °C) because the chance of the presence of that single embryo crystal reaching critical size is greater in larger volumes of water, including the hemolymph volumes of insects. (For the basic principles of supercooling and ice nucleations see Knight 1967; Vali 1995; Chap. 2.) In contrast, in biological systems the presence of ice-nucleating surfaces, most often proteins, that organize ice-like water aggregations on their surfaces can initiate freezing at much higher temperatures, with variations in the heterogeneous nucleation temperature dependant on the efficiency of the IN in organizing the ice-like aggregations to critical size. The most efficient INs, on the surface of certain Gram-negative bacteria, can nucleate at temperatures as high as −2 °C. Bacteria with IN capabilities were initially discovered and studied by atmospheric physicists searching for INs that initiate precipitation by nucleating ice formation on their surfaces from water vapor in the atmosphere. In addition to mineral dust and other nonbiological INs, some of the newly discovered INs were bacteria and some of these were well-known species such as *Pseudomonas syringae* and *Erwinia herbicola* that are common in soil, detritus, and as epiphytes on the surfaces of plants, etc. (Upper and Vali 1995). This led microbiologists and plant biologists to connect these bacteria to costly late spring and early autumn frosts that cause significant crop damage (history reviewed by Upper and Vali 1995). In the absence of the epiphytic bacterial ice nucleators the water that condenses on the surface of plants (dew) when temperature drops would not freeze and inoculate sensitive plants until approximately −5 °C to −6 °C, rather than −2 °C to −3 °C in their presence. The realization of the importance of bacterial INs, especially in agriculture (Lindow et al. 1978; Lindow 1983), stimulated considerable scientific interest leading to numerous studies on the ecology, biochemistry, molecular genetics, applied biology, etc. of bacterial and fungal INs. (These early studies are summarized in the edited book by Lee et al. 1995.)

IN in bacteria is produced by large INPs bound at the N-terminus to the outer bacterial membrane, and composed mostly of 16-mer tandem repeats (Green and Warren 1985; Wolber and Warren 1989; Warren and Corotto 1989) that form a β-solenoid region similar to those of certain insect AFPs (Graether and Jia 2001; Garnham et al. 2011). However, the ice-like water organizing regions of the INPs are much larger than those of the insect AFPs, and consequently they form the embryo crystal that nucleates with limited supercooling at a high subzero temperature where the embryo crystal reaches a critical size. In support of the theory that the water organizing mechanism in some AFPs and bacterial INPs, a truncated INP based on a 96-residue portion of the *P. syringae* INP was shown to have TH activity (Kobashigawa et al. 2005). Apparently, the embryo crystal formed by a single INP, or even by multiple INPs on a single bacterium is not sufficient to induce efficient ice nucleation. Multiple studies showed a requirement for cooperative action between multiple INPs that each add their embryo crystals to those of others to produce a larger combined embryo crystal capable of nucleating at higher temperature (Southworth et al. 1988; Wolber and Warren 1989; Mueller et al. 1990; Yeung et al. 1991).

While the function(s) of bacterial INPs is not completely understood or agreed upon, their most likely function is to injure, and perhaps kill, plants with which they are associated by initiating freezing at times when freezing would not otherwise occur, thereby creating abundant nutrients for the microbes.

An interesting example of the use of bacterial ice nucleators to prevent, rather than cause, frost damage occurs in the large Afro-alpine plant *Lobelia telekii* growing on Mount Kenya (Krog et al. 1979). In the center of a rosette of long leaves is a hollow cylindrical 2-meter tall inflorescence containing approximately 1 L of a viscous fluid. Subzero nighttime air temperatures are common at the high altitude where this plant grows. The fluid in the inflorescence freezes at approximately −4.4 °C (Zachariassen and Hammel 1988) due to the presence of potent ice nucleators. The heat of fusion given off by the freezing of the inflorescence fluid keeps the fluid temperature near 0 °C throughout the night, providing sufficient heat to protect nearby freeze-sensitive flower tissue from damage (Krog et al. 1979). While the ice nucleators were initially thought to be endogenous they are actually bacterial. Thereby, the bacteria and plant have a mutualistic symbiotic relationship, with the bacteria providing heat to the plant while the plant provides nutrition and shelter for the bacteria (Zachariassen and Kristiansen 2000).

8.7 Archaea

Although we often think of the archaea as thermophilic organisms, in fact, they are ubiquitous, and are common components of cold regions where they comprise a significant portion of the biomass (Cavicchioli 2006). Consequently, although exciting, it is not surprising to learn that putative IBPs are present in archaea. Genome searches based on the ice-binding domain of the *Colwellia* sp. SLW05 AFP (*Col*AFP) discussed earlier (Fig. 8.4d–f) identified orthologs from a number of organisms including archaea (Banerjee et al. 2015). Although the putative archaeal IBPs are structurally similar to those of *Colwellia* and certain eukaryotes such as the snow mold fungus *T. ishikaryensis* AFP (*Tis*AFP), there are differences. Molecular docking simulations indicated that the archaea and bacterial ice-binding sites bound more tightly to ice than those of the eukaryotes.

8.8 Conclusions

The majority of the studies of IBPs have been directed to the AFPs of fish, the AFPs and RIPs of insects, and the RIPs of plants (as evidenced by each being covered in a separate chapter of this book). The exception is the many studies of bacterial INPs. However, a considerable amount of information has now been accumulated concerning AFPs and RIPs in these many "other" organisms as well. This is especially true of fungi and bacteria. Some of these advances benefited from the ability to mine genomic databases (especially in bacteria) for potential IBP sequences and structures that could then be further investigated. However, in spite of this progress, there is still much to be learned that will enrich the studies of low-temperature researchers for many years. A few suggestions follow.

The physiological functions of the TH and RI IBPs of most of the organisms covered in this chapter merit considerable attention. This requires information on the in situ activity (TH and/or RI) of the IBPs, and on their specific locations—extracellular, on the cell membrane, intracellular, even outside the organism in the case of sea ice brine channel, and similarly located, species. Whether the proteins are best categorized as AFPs (functioning to prevent freezing) or RIPs is not always obvious. In this review, the authors have often differentiated AFPs from RIPs on the basis of two points (A) whether the organism is known to be freeze tolerant or freeze susceptible, and (B) the magnitude of the specific TH activity (TH/concentration) of the TH-producing IBPs. If the organism is freeze avoiding the assumption is that the IBP must be an AFP. This is probably correct. However, we now know that the distinction between freeze tolerance and avoidance is not always straightforward (Rozsypal and Kostal 2018), varying with the situation—freezing via inoculative freezing or after more significant supercooling, etc. In contrast, if the specific TH produced by the IBP is low and the organism is freeze tolerant, then we assume the protein must be a RIP. This is potentially misleading, for a number of reasons. (1) Expressed proteins are often used for the specific TH activity measurements, and whether all, or any, of these proteins are in their native state, properly folded, etc. is not always obvious. Also, do the concentrations used to reflect the actual in situ concentrations? (2) The TH and/or RI activities of the IBP are rarely measured in the presence of potential enhancers (polyols, sugars, etc.) present in the organism. Multiple enhancers of TH, supercooling promotion activity, etc. of the beetle *Dendroides canadensis* AFPs are known, including self-enhancement among certain AFP isomers (reviewed in Duman 2015; Chap. 6). We recall just one example where testing was conducted on the effect of an enhancer on the different AFPs and RIPs mentioned in this review. The *Flavobacterium xanthum* IAM12026 IBP increased TH and RI activity when 0.5 M maltose was added (Kawahara et al. 2007). (3) In some cases, the IBP may function as both an antifreeze and a RIP. If the organism is freeze tolerant and the IBP is extracellular, then the protein likely functions as a RIP in the extracellular fluid, including the periplasmic fluid of bacteria. However, the IBP may also be located on the cell membrane where it may inhibit inoculation of the cytoplasm by extracellular ice, and/or it may be intracellular and function as a cytoplasmic antifreeze. (4) In species with multiple IBP isoforms, the possibility of site-specific locations of the various isoforms should be investigated, as is the case with certain beetle AFPs (Duman et al. 2002; Nickell et al. 2013). The TH/RI activity, presence of local enhancers, and AFP/RIP function may vary with location.

In particular, the functions of TH/RI IBPs in bacteria, yeasts, and other single-celled organisms should be further investigated. Once again this requires improved information on the activity and location of the proteins. This includes the intriguing functions of IBPs of various single-celled sea ice organisms. Does the IBP function in the periplasmic space between the cell membrane and the cell wall and/or outside the cell wall where it may function to inhibit closure of the brine channels by further ice growth? While the *Marinomonas Mp*AFP functions as an attachment of the bacterium to the ice, this does not preclude it from functioning to inhibit growth of the ice. In fact, the site on the ice to which the *MP*AFP is attached is certainly precluded from adding water molecules. Also, the number of *Mp*AFPs on the surface of the bacterium is not known, but there may be many and their numbers may be sufficient to prevent the cell from being overgrown by ice. Also, if the *Mp*AFPs are distributed over the entire bacterial surface then there will likely be *Mp*AFPs projecting into the water as well as attaching to the sea ice. Recall that bacteria often form mats of cells on surfaces. Is this the case with *Marinomonas* and other brine channel bacteria, yeasts, etc.? If so, such a mat could form a significant barrier to channel closure. Of course, this does not preclude the secretion of free IBPs into the brine channel water as well.

In addition to sea ice, the potential importance, or even the existence, of a freshwater lake, river, etc. ice-associated IBP-producing organisms bear investigation. A recent review documents the dearth of winter information on freshwater lakes that periodically freeze (Hampton et al. 2017). While over half of the world's lakes freeze periodically, most for several months each year (Weyhenmeyer et al. 2008, 2011; Verpoorter et al. 2014), less than 2% of recent peer-reviewed freshwater literature included under-ice winter investigations (Hampton et al. 2015). In the Antarctic, the ice-associated algae account for 25%–30% of annual productivity, and these algae play an important role in determining the makeup of the plankton after the ice melts (Arrigo and Thomas 2004). Not only are large freshwater ice systems likely to have IBP-producing organisms, small shallow freshwater systems that freeze top to bottom should also be investigated. This includes small ephemeral seasonal (vernal) ponds. Freshwater animals, especially crustaceans such as certain *Daphnia* species, tadpole shrimp (Notostraca) that inhabit these waters are especially attractive candidates for study.

While many new IBP-producing organisms and their IBPs have been identified, there are certainly many more to be discovered and studied. At most, only a minor percentage of potential IBP-producing species of any group has been investigated, and of these the functions and structures of even fewer have been elucidated. Of particular interest among the "others" are spiders, mites, additional snow flea species, fungi, etc. In particular, the archaea are worthy of investigation.

The arctic yeast *Leucosporidium Le*AFP was reported as a homodimer where the monomers are connected by covalent bonds (Lee et al. 2010). To our knowledge, this is the only reported case of this sort. However, in the protein databases there are several examples where the structures of IBPs derived from crystal structures, etc. are shown as dimers or even tetramers bound together by standard hydrophobic interactions. This begs the question of the natural state of these IBPs in the organism. There are precedents for the formation of multimeric complexes of AFPs. Yeast-two hybrid studies showed that certain AFP isomers of the beetle *Dendroides canadensis* AFPs (DAFPs) form complexes composed of different isomers, and this has a synergistic effect on TH (Wang and Duman 2005). Another example is Maxi, a form of the winter flounder type-I fish AFP that forms a dimer that employs clathrate water molecules that project inward between the two monomers where they function to hold the two together (Sun et al. 2014).

Another area of interest should be the PPII peptide products of hydrolysis of collagens derived from the skin of certain endothermic vertebrates that have TH and RI activity. The possibility that these collagen peptides can inhibit or mitigate freezing of the skin of poorly insulated regions of mammals and birds should be determined. Alternatively, their TH/RI activity may be an artifact of the treatments and the peptides may not have these functions when in their natural form as components of the collagen.

Finally, these "other" organisms should be examined for the presence of AFGLs. To date, AFGLs have only been detected in several insects, two frogs, one plant, and one species of fungus. However, they have been searched for in only a few organisms.

As mentioned earlier, there is much to engage low-temperature physiologists and biochemists studying IBPs of "other" species far into the future.

References

Abraham NM, Liu L, Jutras BL, Yadav AK, Narasimhan S, Gopalakrishnan V, Ansari JM, Jefferson KK, Cava F, Jacobs-Wagner C, Fikrig E (2017) Pathogen-mediated manipulation of arthropod microbiota to promote infection. Proc Natl Acad Sci USA 114:E781–E790

Adhikari BN, Wall DH, Adams BJ (2009) Desiccation survival in an Antarctic nematode: molecular analysis using expressed sequence tags. BMC Genomics 10:69–85

Ali F, Wharton DA (2016) Ice-active substances from the infective juveniles of the freeze tolerant entomopathogenic nematode, *Steinernema feltiae*. PLoS One 11(5):e0156502. https://doi.org/10.1371/journal.pone.0156502

Anthony SE, Sinclair BJ (2019) Overwintering red velvet mites are freeze tolerant. Physiol Biochem Zool 92:201–205

Arrigo KR, Thomas DN (2004) Large scale importance of sea ice biology in the Southern Ocean. Antarct Sci 16:471–486

Ashworth EN, Kieth TL (1995) Ice nucleation activity associated with plants and fungi. In: Lee RE, Warren GJ, Gusta LV (eds) Biological ice nucleation and its applications. American Phytological Society Press, St Paul, MN, pp 137–162

Banerjee R, Chakraborti P, Bhowmick R, Mukhopadhyay S (2015) Distinct molecular features facilitating ice-binding mechanisms in hyperactive antifreeze proteins closely related to an Antarctic sea ice bacterium. J Biomol Struct Dyn 33:1424–4141

Bayer-Giraldi M, Weikusat I, Besir H, Dieckmann G (2011) Characterization of an antifreeze protein from the polar diatom Fragilariopsis cylindrus and its relevance in sea ice. Cryobiology 63(3):210–219

Biggar KK, Kolani E, Furusawa T, Storey KB (2013) Expression of freeze-responsive proteins, Fr10 and Li16, from freeze-tolerant frogs enhances freezing survival of BmN insect cells. FASEB J 27:3376–3383

Block W (1977) Oxygen consumption of the terrestrial mite *Alaskozetes antarcticus* (Acari: Cryptostigmata). J Exp Biol 68:69–87

Block W (1980) Survival strategies in polar terrestrial arthropods. Biol J Linn Soc 14:29–38

Block W (1982) Supercooling points of insects and mites on the Antarctic Peninsula. Ecol Entomol 7:1–8

Block W, Duman JG (1989) Presence of thermal hysteresis producing antifreeze proteins in the Antarctic mite, *Alaskozetes antarcticus*. J Exptl Zool 250:229–231

Block W, Zettel J (1980) Cold hardiness of some alpine Collembola. Ecol Entomol 5:1–9

Bryon A, Wybouw N, Dermauw W, Tirry L, Thomas Van Leeuwen T (2013) Genome wide gene-expression analysis of facultative reproductive diapause in the two-spotted spider mite *Tetranychus urticae*. BMC Genomics 14:815–835

Bryon A, Kurlovs AH, Van Leeuwen T, Clark RM (2017) A molecular-genetic understanding of diapause in spider mites: current knowledge and future directions. Physiol Entomol 42(3):211–224

Buzzini P, Margesin M (2014) Cold-adapted yeasts: biodiversity, adaptation strategies and biotechnological significance. Heidelberg, Springer, Germany, p 549

Cannon RJC, Block W (1988) Cold tolerance of microarthropods. Biol Rev 63:23–77

Cao H, Zhao Y, Zhu YB, Xu F, Yu JS, Yuan M (2016) Antifreeze and cryoprotective activities of ice-binding collagen peptides from pig skin. Food Chem 194:1245–1253

Carrasco MA, Duman JG (2011) A cross-species compendium of proteins related to cold stress identified by bioinformatic approaches. J Insect Physiol 57:1127–1135

Cavicchioli R (2006) Cold-adapted archaea. Nature Rev Microbiol 4:331–343

Chen L, DeVries AL, Cheng CHC (1997) Convergent evolution of antifreeze glycoproteins in Antarctic fish and Arctic cod. Proc Natl Acad Sci USA 94:3817–3822

Cheng J, Hanada Y, Miura A, Tsuda S, Kondo H (2016) Hydrophobic ice-binding sites confer hyperactivity of an antifreeze protein from a snow mold fungus. Biochem J 473:4011–4026

Costanzo (2019) Overwintering adaptations and extreme freeze tolerance in a subarctic population of the wood frog *Rana sylvatica*. J Comp Physiol B 189:1–5

Costanzo JP, Lee RE (2013) Avoidance and tolerance of freezing in ectothermic vertebrates. J Exp Biol 216:1961–1967

Crawford CS, Riddle WA (1974) Cold hardiness in centipedes and scorpions in New Mexico. Oikos 25(1):86

Crawford CS, Riddle WA, Pugach S (1975) Overwintering physiology of the centipede Scolopendra polymorpha. Physiol Zool 48(3):290–294

Damodaran S (2007) Inhibition of ice crystal growth in ice cream mix by gelatin hydrolysate. J Agric Food Chem 55:10918–10923

Danks HV (1981) Arctic arthropods. Entomological Society of Canada, Ottawa, p 608

Danks HV (1991) Winter habitats and ecological adaptations for winter survival. In: Lee RE, Denlinger DL (eds) Insects at low temperature. Chapman & Hall, London, pp 231–259

Dantel H, Knulle W (1996) The supercooling ability of ticks (Acari, Ixodoidea). J Comp Physiol B 166:517–524

Davies PL (2014) Ice-binding proteins: a remarkable diversity of structures for stopping and starting ice growth. Trends Biochem Sci 39:548–555

Dayton PK, Robilliard GA, DeVries AL (1969) Anchor ice formation in McMurdo Sound, Antarctica, and its biological effects. Science 163:273–274

Denlinger DL, Lee RE (2010) Low temperature biology of insects. Cambridge University Press, Cambridge, p 390

DeVries AL, Wohlschlag DE (1969) Freezing resistance in some Antarctic fishes. Science 163 (3871):1073–1075

DeVries AL (1971) Glycoproteins as biological antifreeze agents in Antarctic fishes. Science 172:1152–1155

DeVries AL, Cheng C-HC (2005) Antifreeze proteins in polar fishes. In: Farrell AP, Steffensen JF (eds) Fish physiology, vol XXII. Academic Press, San Diego, pp 155–201

Do H, Lee JH, Lee SG, Hak Jun Kim HJ (2012a) Crystallization and preliminary X-ray crystallographic analysis of an ice-binding protein (FfIBP) from *Flavobacterium frigoris* PS1. Acta Crystallogr Sect F Struct Biol Cryst Commun F68:806–809

Do H, Lee JH, Lee SG, Hak Jun Kim HJ (2012b) Crystallization and preliminary X-ray crystallographic analysis of an ice-binding protein (FfIBP) from *Flavobacterium frigoris* PS1. Addendum. Acta Crystallogr Sect F Struct Biol Cryst Commun F68:1418–1420

Doucet CJ, Byass L, Elias L, Worrall D, Smallwood M, Bowles DJ (2000) Distribution and characterization of recrystallization inhibitor activity in plant and lichen species from the UK and maritime Antarctic. Cryobiology 40:218–227

Du L, Betti M (2016) Identification and evaluation of cryoprotective peptides from chicken collagen: ice-growth inhibition activity compared to that of type I antifreeze proteins in sucrose model systems. J Agric Food Chem 64:5232–5240

Duman JG (1979) Subzero temperature tolerance in spiders: the role of thermal hysteresis factors. J Comp Physiol 131:347–352

Duman JG (2001) Antifreeze and ice nucleator proteins in terrestrial arthropods. Annu Rev Physiol 63:327–357

Duman JG (2015) Animal ice-binding (antifreeze) proteins and glycolipids: an overview with emphasis on physiological function. J Exp Biol 218:1846–1855

Duman JG, Olsen TM (1993) Thermal hysteresis activity in bacteria, fungi and phylogenetically diverse plants. Cryobiology 30:322–328

Duman JG, Verleye D, Li N (2002) Site specific forms of antifreeze proteins in the beetle *Dendroides canadensis*. J Comp Physiol B 172:547–552

Duman JG, Bennett V, Sformo T, Hochstrasser R, Barnes BM (2004) Antifreeze proteins in Alaskan insects and spiders. J Insect Physiol 50:259–266

Duman JG, Walters KR, Sformo T, Carrasco MA, Nickell P, Barnes BM (2010) Antifreeze and ice nucleator proteins. In: Denlinger DL, Lee RE (eds) Low temperature biology of insects. Cambridge University Press, Cambridge, pp 59–90

Evans CW, Hellman L, Middleditch M, Wojnar JM, Brimble MA, DeVries AL (2012) Synthesis and recycling of antifreeze glycoproteins in polar fishes. Antarct Sci 24:259–268

Garnham CP, Gilbert JA, Hartman CP, Campbell RL, Laybourn-Parry J, Davies PL (2008) A Ca^{2+}-dependent bacterial antifreeze protein domain has a novel beta-helical ice-binding fold. Biochem J 411:171–180

Garnham CP, Campbell RL, Davies PL (2011) Anchored clathrate waters bind antifreeze proteins to ice. Proc Natl Acad Sci USA 108:7363–7367

Gilbert JA, Hill PJ, Dodd CER, Laybourn-Parry J (2004) Demonstration of antifreeze protein activity in Antarctic lake bacteria. Microbiol 150:171–180

Gilbert J, Davies P, Laybourn-Parry J (2005) A hyperactive, Ca^{2+}-dependent antifreeze protein in an Antarctic bacterium. FEMS Microbiol Lett 245:67–72

Graether SP, Jia Z (2001) Modeling *Pseudomonas syringae* ice-nucleation protein as a beta-helical protein. Biophys J 80:1169–1173

Graham LA, Davies PL (2005) Glycine-rich antifreeze proteins from snow fleas. Science 310:461

Green RL, Warren GJ (1985) Physical and functional repetition in a bacterial ice nucleation gene. Nature 317:645–648

Griffith M, Yaish MW (2004) Antifreeze proteins in overwintering plants: a tale of two activities. Trends Plant Sci 9:399–405

Guo S, Garnham CP, Whitney JC, Graham LA, Davies PL (2012) Re-evaluation of a bacterial antifreeze protein as an adhesin with ice-binding activity. PLoS One 7(11):e48805. https://doi.org/10.1371/journal.pone.0048805

Hampton SE, Moore MV, Ozersky T, Stanley EH, Polashenski CM, Galloway AWE (2015) Heating up a cold subject: prospects for under-ice plankton research in lakes. J Plankton Res 37:277–284

Hampton SE, Galloway AWE, Powers SM, Ozersky T, Woo KH, Batt RD, Labou SG, O'Reilly CM, Sharma S, Lottig NR, Stanley EH, North RL, Stockwell JD, Adrian R, Wyhenmeyer GA, Arvola L, Baulch HM, Bertani I, Bowmn LL, Carey CC, Catalan J, Colom-Montero W, Domine LM, Felip M, Granados I, Gries C, Grossart H-P, Haberman J, Haldna M, Hayden B, Higgins SN (2017) Ecology under lake ice. Ecol Lett 20:98–111

Hanada Y, Nishimiya Y, Miura A, Tsuda S, Hidemasa Kondo H (2014) Hyperactive antifreeze protein from an Antarctic sea ice bacterium *Colwellia* sp. has a compound ice-binding site without repetitive sequences. FEBS J 281:3576–3590

Hargens AR, Shabica SV (1973) Protection against lethal freezing temperatures by mucus in an Antarctic limpet. Cryobiology 10:331–337

Hashim NH, Bharudin I, Nguong DL, Higa S, Bakar FD, Nathan S, Rabu A, Kawahara H, Illias RM, Najimudin N, Mahadi NM, Murad AM (2013) Characterization of Afp1, an antifreeze protein from the psychrophilic yeast *Glaciozyma antarctica* PI12. Extremophiles 17:63–73

Hashim NHF, Sulaiman S, Bakar FDA, Illias RM, Kawahara H, Najimudin N, Mahadi NM, Murad AMA (2014) Molecular cloning, expression and characterisation of Afp4, an antifreeze protein from *Glaciozyma antarctica*. Polar Biol 37:1495–1505

Hawes TC, Worland MR, Bale JS (2010) Freezing in the Antarctic limpet *Nacella concinna*. Cryobiology 61:128–132

Hawes TC, Marshall CJ, Wharton DA (2011) Antifreeze proteins in the Antarctic springtail, *Gressittacantha terranova*. J Comp Physiol B 181:713–719

Hawes TC, Marshall CJ, Wharton DA (2014) A 9kDa antifreeze protein from the Antarctic springtail *Gomphiocephalus hodgsoni*. Cryobiology 69:181–183

Heisig M, Abraham NM, Liu L, Neelakanta G, Mattessich S, Sultana H, Shang Z, Ansari JM, Killiam C, Walker W, Cooley L, Flavell RA, Agaisse H, Fikrig E (2014) Anti-virulence properties of an antifreeze protein. Cell Rep 9:417–424

Holmstrup M, Sømme L (1998) Dehydration and cold hardiness in the Arctic collembolan *Onychiurus arcticus* Tullberg 1876. J Comp Physiol B 168:197–203

Hopkin SP (1997) The biology of springtails. Oxford University Press, Oxford, p 344

Hoshino T, Matsumoto N, Araki T, Georges F, Goda T, Ohgiya S (1998) Freezing resistance among isolates of a psychrophilic fungus, *Typhula ishikariensis*, from Norway. Proc NIPR Symp Polar Biol 11:112–118

Hoshino T, Kiriaki M, Nakajima T (2003a) Novel thermal hysteresis proteins from low temperature basidiomycete, *Coprinus psychromorbidus*. Cryo Lett 24:135–142

Hoshino T, Kiriaki TM, Ohgiya S, Fujiwara M, Kondo H, Nishimiya Y, Yumoto I, Tsuda S (2003b) Antifreeze proteins from snow mold fungi. Can J Bot 8:1175–1181

Husby JA, Zachariassen ZE (1980) Antifreeze agents in the body fluid of winter active insects and spiders. Experientia 36:963–964

Kawahara H, Nakana Y, Omiya K, Muryoi N, Nishikawa J, Obata H (2004) Production of two types of ice crystal-controlling proteins in Antarctic bacterium. J Biosci Bioeng 98:220–223

Kawahara H, Iwanaka Y, Higa S, Muryoi N, Sato M, Honda M, Omura H, Obata H (2007) A novel, intracellular antifreeze protein in an antarctic bacterium, Flavobacterium xanthum. Cryo Letters 28:39–49

Kawahara H, Matsuda Y, Sakaguchi T, Arai N, Koide Y (2016) Antifreeze activity of xylomannan from the mycelium and fruit body of *Flammulina velutipes*. Biocontrol Sci 21:153–159

Khodayari S, Moharramipour S, Kamali K, Hidalgo K, Renault D (2012) Effects of acclimation and diapause on the thermal tolerance of the two-spotted spider mite *Tetranychus urticae*. J Thermal Biol 37:419–423

Khodayari S, Moharramipour S, Larvor V, Hidalgo K, Renault D (2013) Deciphering the metabolic changes associated with diapause syndrome and cold acclimation in the two-spotted spider mite *Tetranychus urticae*. PLoS One 8:e54025. https://doi.org/10.1371/journal.pone.0054205

Kiko R (2008) Ecophysiology of Antarctic sea-ice meiofauna. Doctoral Dissertation. Christian Albrechts University of Kiel, 116 p

Kiko R (2010) Acquisition of freeze protection in a sea-ice crustacean through horizontal gene transfer? Polar Biol 33:543–556

Kim HJ, Lee JH, Do H, Jung W (2014) Production of antifreeze proteins in cold-adapted yeasts. In: Buzzini P, Margesin M (eds) Cold-adapted yeasts: biodiversity, adaptation strategies and biotechnological significance. Springer, Heidelberg, pp 259–280

Kirchner W (1973) Ecological aspects of cold resistance in spiders (a comparative study). In: Weiser W (ed) Effects of temperature on ectothermic organisms. Springer, Berlin, pp 271–279

Kirchner W, Kestler P (1969) Untersuchungen zur chilfradpinne *Araneus cornutus* (Araneidae). J Insect Physiol 15:41–53

Knight CA (1967) The freezing of supercooled liquids. Van Nostrand, Princeton, p 246

Knight CA, DeVries AL (1989) Melting inhibition by fish antifreeze glycopeptides. Science 254:505–507

Knight CA, Duman JG (1986) Inhibition of recrystallization of ice by insect thermal hysteresis proteins: a possible cryoprotective role. Cryobiology 23:256–262

Knight CA, DeVries AL, Oolman LD (1984) Fish antifreeze protein and the freezing and recrystallization of ice. Nature 308:295–296

Kobashigawa Y, Nishimiya Y, Miura K, Ohgiya S, Miura A, Tsuda S (2005) A part of ice nucleation protein exhibits the ice-binding ability. FEBS Lett 579:1493–1497

Kolb JH (2013) Short circuit co-evolution by the perfect parasites: antifreeze glycoproteins in Antarctic fish leeches (Hirudinea Piscicolid). PhD dissertation, Massey University, New Zealand, 290 pp

Kondo H, Hanada Y, Sugimoto H, Hoshino T, Garnham CP, Davies PL, Tsuda S (2012) Ice-binding site of snow mold fungus antifreeze protein deviates from structural regularity and high conservation. PNAS USA 109:9360–9365

Kostal V (2010) Cell structural modifications in insects at low temperatures. In: Denlinger DL, Lee RE (eds) Low temperature biology of insects. Cambridge University Press, Cambridge, pp 116–140

Krog JO, Zachariassen KE, Larsen B, Smidsrød O (1979) Thermal buffering in afro-alpine plants due to nucleating agent-induced water freezing. Nature 282(5736):300–301

Larson D, Middle L, Vu H, Zhang W, Serianni AS, Duman JG, Barnes BM (2014) Wood frog adaptations to overwintering in Alaska: new limits to freezing tolerance. J Exp Biol 217:2193–2200

Layne JR (1991) External ice triggers freezing in freeze-tolerant frogs at temperatures above their supercooling point. J Herpetol 25:129–130

Layne JR, Lee RE (1987) Freeze tolerance and the dynamics of ice formation in wood frogs (*Rana sylvatica*) from southern Ohio. Can J Zool 65:2062–2065

Lee RE (2010) A primer on insect cold tolerance. In: Denlinger DL, Lee RE (eds) Low temperature biology of insects. Cambridge University Press, Cambridge, pp 3–34

Lee RE Jr, Costanzo JP (1998) Biological ice nucleation and ice distribution in cold-hardy ectothermic animals. Annu Rev Physiol 60:55–72

Lee RE, Denlinger DL (2010) Rapid cold hardening: ecological significance and underpinning mechanisms. In: Denlinger DL, Lee RE (eds) Low temperature biology of insects. Cambridge University Press, Cambridge, pp 35–58

Lee RE, Warren GJ, Gusta LV (eds) (1995) Biological ice nucleation and its applications. Americal Phytological Society Press, St Paul, MN, p 370
Lee JK, Park KS, Park S, Park H, Song YH, Kang SH, Kim HJ (2010) An extracellular ice-binding glycoprotein from an Arctic psychrophilic yeast. Cryobiology 60:222–228
Lee JH, Park AK, Do H, Park KS, Moh SH, Chi YM, Kim HJ (2012) Structural basis for antifreeze activity of ice-binding protein from Arctic yeast. J Biol Chem 287:11460–11468
Li N, Andorfer CA, Duman JG (1998) Enhancement of insect antifreeze protein activity by low molecular weight solutes. J Exp Biol 201:2243–2251
Lin Y, Duman JG, DeVries AL (1972) Studies on the structure and activity of low molecular weight glycoproteins from an Antarctic fish. Biochem Biophys Res Comm 46:87–92
Lin F-H, Graham LA, Campbell RL, Davies PL (2007) Structural modeling of snow flea antifreeze protein. Biophys J 92(5):1717–1723
Lindow SE (1983) The role of bacterial ice nucleation in frost injury to plants. Annu Rev Phytopathol 21:363–384
Lindow SE, Arny DC, Upper CD (1978) *Erwinia herbicola*: a bacterial ice nucleus active in increasing frost injury to corn. Phytopathology 68:523–527
Lundheim R, Zachariassen KE (1993) Water balance of overwintering beetles in relation to strategies for cold tolerance. J Comp Physiol B 163:1–4
Mangiagalli M, Bar-Dolev M, Tedesco P, Natalello A, Kaleda A, Brocca S, Pascale D, Pucciarelli S, Miceli C, Bravslavsky I (2017) Cryo-protective effect of an ice-binding protein derived from Antarctic bacteria. FEBS J 284:163–177
Marshall CJ (2003) Ice active proteins from an Antarctic nematode *Panogrolaimus davidii*. Cryo-Lett 24:404
Martof BS, Humphries RL (1959) Geographic variation in the wood frog, *Rana sylvatica*. Am Midl Nat 61:350–389
Meier P, Zettel J (1997) Cold hardiness in *Entomobrya nivalis* (Collembola, Entomobryidae): annual cycle of polyols and antifreeze proteins, and antifreeze triggering by temperature and photoperiod. J Comp Physiol 167:297–304
Michaud MR, Denlinger DL (2010) Genomics, proteomics and metabolomics: finding the other players in insect cold tolerance. In: Denlinger DL, Lee RE (eds) Low temperature biology of insects. Cambridge University Press, Cambridge, pp 91–115
Mock T, Thomas DN (2005) Recent advances in sea-ice microbiology. Environ Microbiol 7:605–619
Mok Y-F, Lin F-H, Laurie A, Graham LA, Celik Y, Braslavsky I, Davies PL (2010) Structural basis for the superior activity of the large isoform of snow flea antifreeze protein. Biochemistry 49:2593–2603
Mueller GM, Wolber PK, Warren GJ (1990) Clustering of ice nucleation protein correlates with ice nucleation activity. Cryobiology 27:416–422
Muñoz PA, Márquez SL, González-Nilo FD, Márquez-Miranda V, Blamey JM (2017) Structure and application of antifreeze proteins from Antarctic bacteria. Microb Cell Factories 16:138. Article number: 138
Neelakanta G, Sultana H, Fish D, Anderson JF, Fikrig E (2010) *Anaplasma phagocytophilum* induces *Ixodes scapularis* ticks to express an antifreeze glycoprotein gene that enhances their survival in the cold. J Clin Invest 120:3179–3190
Nickell PK, Sass S, Verleye D, Blumenthall EM, Duman JG (2013) Antifreeze proteins in the primary urine of larvae of the beetle *Dendroides canadensis* (Latreille). J Exp Biol 216:1695–1703
Olive LC, Meister K, DeVries AL, Duman JG, Guo S, Bakker HJ, Voets IK (2016) Blocking rapid ice crystal growth through non-basal plane adsorption of antifreeze proteins. Proc Natl Acad Sci USA 113:3740–3745
Park AK, Park KS, Kim HJ, Park H, Ahn IY, Chi YM, Moon JH (2011) Crystallization and preliminary x-ray crystallographic studies of the ice-binding protein from the Antarctic yeast *Leucosporidium* sp. AY30. Acta Cryststallogr Sect F Struct Biol Cryst Commun 67:800–802

Pentelute BL, Gates ZP, Tereshko V, Dashnau JL, Vanderkooi JM, Kossiakoff AA, Kent SB (2008) X-ray structure of snow flea antifreeze protein determined by racemic crystallization of synthetic protein enantiomers. J Am Chem Soc 130:9695–9701

Poinsot-Balaguer N, Barra JA (1983) Experimental and ultrastructural data on freezing resistance in *Folsomides angularis* (Insecta, Collembola). Pedobiologia 25:357–363

Priddle J, Heywood RB, Theriot E (1986) Some environmental factors influencing phytoplankton in the Southern Ocean around South Georgia. Polar Biol 5:65–79

Pucciarelli S, Chiappori F, Devaraj RR, Yang G, Yu T, Ballarini P, Miceli C (2014) Identification and analysis of two sequences encoding ice-binding proteins obtained from a putative bacterial symbiont of the psychrophilic Antarctic ciliate *Euplotes focardii*. Antarct Sci 26:491–501

Ramløv H (2000) Aspects of natural cold tolerance in ectothermic animals. Hum Reprod 15:26–46

Ramløv H, Wharton D, Wilson P (1996) Recrystallization in a freezing tolerant Antarctic nematode, *Panagrolaimus davidi*, and an alpine Weta, *Hemideina maori* (Orthoptera; Stenopelmatidae). Cryobiology 33:607–613

Raymond JA, Lin Y, DeVries AL (1975) Glycoprotein and protein antifreezes in two Alaskan fishes. J Exp Zool 193(1):125–130

Raymond JA (2016) Dependence on epiphytic bacteria for freezing protection in an Antarctic moss, *Bryum argenteum*. Environ Microbiol Rep 8:14–19

Raymond JA, Janech MG (2009) Ice-binding proteins from enoki and shiitake mushrooms. Cryobiology 58:151–156

Raymond JA, Kim KJ (2012) Possible role of horizontal gene transfer in the colonization of sea ice by algae. PLoS One 2012(7):e35968. https://doi.org/10.1371/journal.pone.0035968. pmid:22567121

Raymond JA, Knight C (2003) Ice binding, recrystallization inhibition, and cryoprotective properties of ice-active substances associated with Antarctic sea ice diatoms. Cryobiology 46:174–181

Raymond JA, Fritsen C, Shen K (2007) An ice-binding protein from an Antarctic sea ice bacterium. FEMS Microbiol Ecol 61:214–221

Raymond JA, Christner BC, Schuster SC (2008) A bacterial ice-binding protein from the Vostok ice core. Extremophiles 12:713–717

Raymond JA, Rachael M-K, Valentin K (2017) Multiple ICE-binding proteins of probable prokaryotic origin in an Antarctic lake alga, sp. ICE-MDV (Chlorophyceae). J Phycol 53(4):848–854

Robinson CH (2001) Cold adaptation in Arctic and Antarctic fungi. New Phytol 151:341–353

Rozsypal J, Kostal V (2018) Supercooling and freezing as eco-physiological alternatives rather than mutually exclusive strategies: a case study in *Pyrrhocoris apterus*. J Insect Physiol 111:53–62

Schmid WD (1982) Survival of frogs in low temperature. Science 215:697–698

Schnenker R (1983) Effects of temperature acclimation of cold-hardiness of alpine microarthropods. Rev Ecol Biol Sol 20:37–47

Schünemann H, Werner I (2005) Seasonal variations in distribution patterns of sympagic meiofauna in Arctic pack ice. Mar Biol 146:1091–1102

Shah SHH, Kar RK, Asmawi AA, Rahman MBA, Murad AMA, Mahadi NM, Basri M, Rahman RNZA, Salleh AB, Chatterjee S (2012) Solution structures, dynamics, and ice growth inhibitory activity of peptide fragments derived from an Antarctic yeast protein. https://doi.org/10.1371/journal.pone.0049788.s006

Shier WT, DeVries AL (1975) Carbohydrates of antifreeze proteins from an Antarctic fish. FEBS Lett 54:135–142

Shoulders MD, Raines RT (2009) Collagen structure and stability. Annu Rev Biochem 78:929–958

Sinclair BJ, Chown SL (2002) Haemolymph osmolality and thermal hysteresis activity in 17 species of arthropods from sub-Antarctic Marion Island. Polar Biol 25:928–933

Sinclair BJ, Sjursen H (2001) Cold tolerance of the Antarctic springtail *Gomphiocephalus hodgsoni* (Collembola, Hypogastruridae). Antarct Sci 13:271–279

Singh P, Hanada Y, Mohan Singh S, Tsuda (2014) Antifreeze protein activity in Arctic cryconite bacteria. FEMS Microbiol Lett 351:14-22

Sjursen H, Sømme L (2000) Seasonal changes in tolerance to cold and desiccation in *Phauloppia sp.* (Acari, Oribatida) from Finse, Norway. J Insect Physiol 46:1387–1396
Sømme L, Conradi-Larsen EM (1977) Cold hardiness of Collembolans and oribatid mites from windswept mountain ridges. Oikos 29:118–126
Sømme L, Stromme A, Zacchariassen KE (1993) Notes on the ecology and physiology of the Antarctic oribatid mite *Maudheimia wilsoni*. Polar Res 12:21–25
Southworth MW, Wolber PK, Warren GJ (1988) Nonlinear relationship between concentration and activity of a bacterial ice nucleation protein. J Biol Chem 263:15211–15216
Storey KB, Storey JM (1984) Biochemical adaption for freezing tolerance in the wood frog, Rana sylvatica. J Comp Physiol B 155(1):29–36
Storey KB, Storey JM (2004) Physiology, biochemistry, and molecular biology of vertebrate freeze tolerance: the wood frog. In: Fuller BJ, Lane N, Benson EE (eds) Life in the frozen state. CRC, Washington, DC, pp 243–274
Storey KB, Storey JM (2010) Oxygen: stress and adaptation in cold-hardy insects. In: Denlinger DL, Lee RE (eds) Low temperature biology of insects. Cambridge University Press, Cambridge, pp 141–165
Storey KB, Storey JM (2017) Molecular physiology of freeze tolerance in vertebrates. Physiol Rev 97:623–665
Sullivan KJ, Biggar KK, Storey KB (2015) Transcript expression of the freeze responsive gene fr10 in *Rana sylvatica* during freezing, anoxia, dehydration, and development. Mol Cell Biochem 399:17–25
Sun X, Griffith M, Pasternak J, Glick BR (1995) Low temperature growth, freezing survival, and production of antifreeze protein by the plant growth promoting rhizobacterium *Pseudomonas putida* GR12-2. Can J Microbiol 41:776–784
Sun T, Feng-Hsu Lin F-H, Campbell RL, Allingham JS, Davies PL (2014) An antifreeze protein folds with an interior network of more than 400 semi-clathrate waters. Science 343:795–798
Tereshina VM, Memorskaya AS (2005) Adaptation of *Flammulina velutipes* to hypothermia in natural environments: the role of lipids and carbohydrates. Microbiology 74:279–283
Theede HR, Schneppenheim R, Bevess L (1976) Frostschutz- glycoproteine bei *Mytilus edulis*? Mar Biol 36:183–189
Thomas DN, Dieckmann GS (2002) Antarctic Sea ice - a habitat for extremophiles. Science 295:641–644
Thorne M, Kagoshima H, Clark M, Marshall C, Wharton D (2014) Molecular analysis of the cold tolerant Antarctic nematode, *Panagrolaimus davidi*. PLoS One 9(8):e104526. http://doi.org/1371/journal.pone.0104526
Thorne MAS, Seybold A, Marshall C, Wharton D (2017) Molecular snapshot of an intracellular freezing event in an Antarctic nematode. Cryobiology 75:117–124
Todde G, Whitman C, Hovmoller S, Laaksonen A (2014) Induced ice melting by the snow flea antifreeze protein from molecular dynamics simulations. J Phys Chem B 118:13527–13534
Treonis AM, Wall DH, Virginia RA (2000) The use of anhydrobiosis by soil nematodes in the Antarctic dry valleys. Funct Ecol 14:460–467
Tursman D, Duman JG (1995) Cryoprotective effects of thermal hysteresis protein on survivorship of frozen gut cells from the freeze tolerant centipede *Lithobius forficatus*. J Exp Zool 272:249–257
Tursman D, Duman JG, Knight CA (1994) Freeze tolerance adaptations in the centipede *Lithobius forficatus*. J Exp Zool 268:347–353
Upper CD, Vali G (1995) The discovery of ice nucleation-active bacteria and its role in the injury of plants by frost. In: Lee RE, Warren GJ, Gusta LV (eds) Biological ice nucleation and its applications. Americal Phytological Society Press, St Paul, MN, pp 29–40
Vali G (1995) The principles of ice nucleation. In: Lee RE, Warren GJ, Gusta LV (eds) Biological ice nucleation and its applications. American Phytological Society Press, St Paul, MN, pp 1–28
VanVoorhies WV, Raymond JA, DeVries AL (1978) Glycoproteins as biological antifreeze agents in the cod, *Gadus ogac*. Physiol Zool 51:347–353

Verpoorter C, Kutser T, Seekell DA, Tranvik LJ (2014) A global inventory of lakes based on high-resolution satellite imagery. Geophys Res Lett 41:6396–6402

Walker VK, Palmer GR, Voordouw G (2006) Freeze-thaw tolerance and clues to the winter survival of a soil community. Appl Environ Microbiol 72:1784–1792

Waller CL, Worland MR, Convey P, Barnes DKA (2006) Ecophysiological strategies of Antarctic intertidal invertebrates faced with freezing stress. Polar Biol 29:1077–1083

Walsh LL, Tucker PK (2017) Contemporary range expansion of the Virginia opossum (*Didelphis virginiana*) impacted by humans and snow cover. Can J Zool 2018:107–115

Walters KR, Serianni AS, Sformo T, Barnes BM, Duman JG (2009) A non-protein thermal hysteresis-producing xylomannan antifreeze in a freeze tolerant Alaskan beetle. Proc Natl Acad Sci USA 106:20210–20215

Walters KR, Serianni AS, Voituron Y, Sformo T, Barnes BM, Doman JG (2011) A thermal hysteresis-producing xylomannan glycolipid antifreeze associated with cold-tolerance is found in diverse taxa. J Comp Physiol B 181:631–640

Wang SY, Damodoran S (2009) Ice-structuring peptides derived from bovine collagen. J Agric Food Chem 57:5501–5509

Wang L, Duman JG (2005) Antifreeze proteins of the beetle *Dendroides canadensis* enhance one another's activities. Biochemist 44:10305–10312

Wang S, Zhao J, Chen L, Zhou Y, Wu J (2014) Preparation, isolation and hypothermia protection activity of antifreeze peptides from shark skin collagen. Food Sci Technol 55:210–217

Warren G, Corotto L (1989) The consensus sequence of ice nucleation proteins from *Erwinia herbicola, Pseudomonas fluorescens* and *Pseudomonas syringae*. Gene 85:239–242

Weyhenmeyer GA, Westöö A-K, Willén E (2008) Increasingly ice-free winters and their effects on water quality in Sweden's largest lakes. Hydrobiology 599:111–118

Weyhenmeyer GA, Livingstone DM, Meili M, Jensen O, Benson B, Magnuson JJ (2011) Large geographical differences in the sensitivity of ice-covered lakes and rivers in the northern hemisphere to temperature changes. Glob Chang Biol 17(1):268–275

Wharton DA (1995) Cold tolerance strategies in nematodes. Biol Rev 70:161–185

Wharton DA (2003) The environmental physiology of Antarctic terrestrial nematodes: a review. J Comp Physiol B 173:621–628

Wharton D, Ferns D (1995) Survival of intracellular freezing by the Antarctic nematode *Panagrolaimus davidi*. J Exp Biol 198:381–387

Wharton D, Raymond MR (2015) Cold tolerance of the Antarctic nematodes *Plectus murrayi* and *Scottnema lindsayae*. J Comp Physiol B 185:281–285

Wharton DA, Goodall G, Marshall CJ (2003) Freezing survival and cryoprotective dehydration as cold tolerance mechanisms in the Antarctic nematode *Panagroulamis davidi*. J Exp Biol 206:215–221

Wharton D, Downes M, Goodall G, Marshall C (2005a) Freezing and cryoprotective dehydration in an Antarctic nematode (*Panagrolaimus davidi*) visualised using a freeze substitution techniques. Cryobiology 50:21–28

Wharton D, Barrett J, Goodall G, Marshall C, Ramløv H (2005b) Ice-active proteins from the Antarctic nematode, *Panagrolaimus davidi*. Cryobiology 51:198–207

Whitmore DH, Gonzalez R, Baust JG (1985) Scorpion cold hardiness. Physiol Zool 58:526–537

Wilkins SP, Blum AJ, Burkepile DE, Rutland TJ, Wierzbicki A, Kelly AM, Hamann MT (2002) Isolation of an antifreeze peptide from the Antarctic sponge *Homaxinella balfourensis*. Cell Mol Life Sci 59:2210–2215

Wolber PK, Warren GJ (1989) Bacterial ice nucleating proteins. Trends Biochem Sci 14:179–182

Worland MR, Block W (2003) Desiccation stress at subzero temperatures in polar terrestrial arthropods. J Insect Physiol 49:193–203

Worland MR, Grubor-Lajsik G, Montiel PO (1998) Partial desiccation induced by sub-zero temperatures as a component of the survival strategy of the Arctic collembolan *Onychiurus arcticus* (Tullberg). J Insect Physiol 44:211–219

Xiao N, Suzuki K, Nishimiya Y, Kondo H, Miura A, Tsuda S, Hoshino T (2010) Comparison of functional properties of two fungal antifreeze proteins from *Antarctomyces psychrotrophicus* and *Typhula ishikariensis*. FEBS J 277:394–403

Xu H, Griffith M, Patten CL, Glick BR (1998) Isolation and characterization of an antifreeze protein with ice nucleation activity from the plant growth promoting rhizobacterium *Pseudomonas putida* GR12-2. Can J Microbiol 44:64–73

Yamashita Y, Nakamura N, Omiya K, Nishikawa J, Kawahara H, Obata H (2002) Identification of an antifreeze lipoprotein from *Moraxella* sp. of Antarctic origin. Biosci Biotechnol Biochem 66:239–247

Yeung KL, Wolf EE, Duman JG (1991) A scanning tunneling microscopy study of an insect lipoprotein ice nucleator. J Vac Sci Technol B 9:1197–1201

Young SR, Block W (1980) Experimental studies of cold tolerance in an Antarctic terrestrial mite. J Insect Physiol 26:189–200

Zachariassen KE, Hammel HT (1988) The effect of ice-nucleating agents on ice-nucleating activity. Cryobiology 25(2):143–147

Zachariassen KE, Kristiansen E (2000) Ice nucleation and Antinucleation in nature. Cryobiology 41 (4):257–279

Zachariassen KE, Kristansen E, Pedersen SA, Hammel HT (2004) Ice nucleation in solutions and freezing in insects – homogeneous or heterogeneous? Cryobiology 48:309–321

Zachariassen KE, Duman JG, Kristiansen E, Pedersen S, Li N (2010) Ice nucleation and antifreeze proteins in animals. In: Graether S (ed) Biochemistry and function of antifreeze proteins. Nova Science, New York, pp 73–104

Zettel J (1984) Cold hardiness strategies and thermal hysteresis in Collembola. Rev Ecol Biol Sol 21:189–203

Zettel J, von Allmen H (1982) Jahresverlauf der Kalteresistenz zweir Collembola-Arten in den Berner Voralpen. Rev Suisse Zool 89:927–939

Zimmerman SL, Frisbie J, Goldstein DL, West J, Rivera JK, Krane CM (2007) Excretion and conservation of glycerol, and expression of aquaporins and glyceroporins, during cold acclimation in Cope's gray tree frog *Hyla chrysoscelis*. Am J Physiol 292:R544–R555

Chapter 9
Molecular Origins and Mechanisms of Fish Antifreeze Evolution

C.-H. Christina Cheng and Xuan Zhuang

9.1 Introduction

In the world of biological life, antifreeze proteins (AFPs) and ice-binding proteins (IBPs) occur only in a number of species that seasonally or perennially inhabit Earth's cryosphere. The encoding genes of these novel molecules, therefore, represent new genotypes, as they are absent in their temperate counterparts. Where from and how did these new genotypes evolve? Research into their evolutionary origins and mechanisms began about three decades ago, with the fish antifreeze proteins being the best characterized. A number of review articles on fish antifreeze protein evolution have been published over the years, with the most recent in 2018 (Davies and Graham 2018). Far less is known about the evolution of AFPs from insects (Doucet et al. 2009) and plants (Bredow and Walker 2017), and even less so for the plethora of IBPs from lower eukaryotes and prokaryotic organisms. To avoid repeating the syntheses already provided in these reviews (Doucet et al. 2009; Bredow and Walker 2017; Davies and Graham 2018), this chapter will emphasize on the fine level molecular processes of fish antifreeze evolution that are particularly well deduced, the associated environmental or evolutionary drivers, and the broader impact of the adaptive antifreeze trait on the fish group were known. As always in scientific inquiries, there are unsettled questions or hypotheses, and so it is the case for the evolutionary processes of antifreeze proteins; we will attempt to address some of them. Finally, in addition to understanding how life for teleosts in freezing

C.-H. C. Cheng (✉)
Department of Evolution, Ecology & Behavior, University of Illinois, Urbana-Champaign, IL, USA
e-mail: c-cheng@illinois.edu

X. Zhuang
Department of Ecology & Evolution, University of Chicago, Chicago, IL, USA
e-mail: xzhuang3@uchicago.edu

H. Ramløv, D. S. Friis (eds.), *Antifreeze Proteins Volume 1*,
https://doi.org/10.1007/978-3-030-41929-5_9

environment became possible, we will also illustrate how mechanistic elucidation of the evolutionary processes of fish antifreeze in the past three decades importantly contributed to the broader field of evolutionary biology, by providing clear examples that support conceptual models and hypotheses of the evolutionary process of new genes and functions.

9.1.1 How Do New Genes with New Function Arise?

Among the most fundamental questions in evolutionary and adaptational biology are how new genetic material emerge and evolve novel functions that confer phenotypic and fitness benefits in organisms. What are the ways? Research into the origins and mechanisms of fish antifreeze protein evolution began in the 1990s when the arrival of molecular methods first enabled inquiries at the DNA and RNA level, and has continued to the present with ever more powerful molecular approaches. In the broader field of molecular evolution, new conceptual models and hypotheses of how new genes evolve continue to be put forth, and various AFPs count as some of the best supporting examples. Until about two decades ago, the prevailing view of how new genes and functions arose was shaped by the landmark treatise "*Evolution by Gene Duplication*" by Ohno (1970) and proponents of the concept (Jacob 1977). The concept emphasized a dominant role of duplication of existing genes in generating functional novelty. The main thesis is that when an existing gene becomes duplicated, which occurs quite commonly, one copy continues with the original function, while the redundant copy either degenerates (non-functionalization), or is free for natural selection to act upon should the selective pressure for its maintenance happens to exist. If the latter occurs, selective tinkering of the gene sequence comes to bear and a new gene with a new expression pattern (sub-functionalization) or a novel function (neofunctionalization) may emerge (Jacob 1977; Lynch and Conery 2000).

The evolutionary relatedness of the novel gene and its putative ancestral homolog is embedded in sequence similarity they share. This relatedness can be readily revealed today by searching sequence databases using powerful Best Local Alignment Search Tools (BLAST) equipped with robust algorithms to determine statistically meaningful sequence similarities quickly. Ohno's thesis of the evolution of functional novelty by gene duplication plus sequence divergence proved to be prescient and one of the greatest insights in evolutionary thought, predating investigations and data resources in the molecular and genomics era that would prove him correct. To date, it is the most frequent mechanism of new gene evolution, with many supporting examples including fish antifreeze proteins (type II and type III).

The rapid advance of massively parallel DNA sequencing technologies in the recent two decades and the resulting stratospheric growth of genic, genomic, and transcriptomic data made possible comparative analyses of entire suits of coding genes and/or expressed genes across different lineages or populations of the same lineage. From these analyses, a much richer array of models of evolutionary origins and processes of new gene formation have emerged (Kaessmann 2010). The most recent is the growing evidence for de novo gene birth—the emergence of new coding genes "from scratch," i.e., from ancestral nonprotein coding sequences (Tautz 2014; McLysaght and Guerzoni 2015) that were vernacularly dubbed "junk DNA." De novo gene birth was previously thought to be improbable or impossible (Jacob 1977). However, advances within just the past 10 years brought to light that de novo gene birth may actually be an active generator of genetic novelty (Tautz 2014; McLysaght and Guerzoni 2015; Schlötterer 2015). The occurrence of de novo gene birth was usually detected through comparative analyses of genomes and transcriptomes on a phylogenetic framework, revealing lineage- or species-specific gene transcripts while the orthologous sequences in sister species are noncoding DNA. Ample evidence for lineage- or species-specific expressed transcripts have been reported [see review by McLysaght and Guerzoni (2015)]. However, in nearly all cases, questions abound as to what function the putative new genes may serve, what adaptive outcome they confer, what selective pressure drove their birth, and the molecular mechanism of its de novo formation (Tautz and Domazet-Lošo 2011; McLysaght and Hurst 2016). Putative new gene sequences in search of a purpose are very much the status quo and an ongoing challenge. Until functional and adaptive significance can be established, putative de novo genes would remain as orphan genes. In contrast, the function and adaptive significance of fish antifreeze proteins are fully established, and their encoding genes are decidedly novel. Here again, the recently deduced origin and mechanism of evolution of one of the fish antifreeze, the antifreeze glycoprotein (AFGP) of northern codfishes contributed the first clear example of the de novo gene birth from entirely noncoding DNA (Zhuang et al. 2019).

9.1.2 Selective Pressure in Marine Cryosphere for Antifreeze Evolution

What drove the evolution of antifreeze proteins in teleost fishes? Adaptive traits in higher organisms are almost always complexly multigenic, and the selective pressures that lead to their evolution are difficult to ascertain precisely. Evolution of fish antifreeze proteins is an exception in being unencumbered by many of these uncertainties. They represent the rare instances of a single gene producing a vital adaptive function that confers a clear and large fitness consequence. The selective pressure that drove their evolution is obvious and simple to comprehend. It stems from the mismatch between the colligative freezing points (fp) of marine water and teleost fish. Open ocean waters have salt (predominantly NaCl) concentrations of ~1030 mOsm, and would have an fp of −1.9 °C, following the physicochemical principle of colligative fp depression by dissolved solutes (−1.86 °C for 1000 mOsm dissolved solutes in water). Teleost fishes are hypoosmotic with respect to seawater, presumably an evolutionary holdover from their freshwater ancestry (Carrete Vega and Wiens 2013). Their blood and body fluids are more dilute, typically at about a third (~300 to 350 mOsm) of the osmotic concentrations of seawater (Prosser 1973), and thus would have a colligative fp of about −0.6 °C. Cold-adapted polar fishes have significantly more plasma electrolytes (mostly NaCl), with some species achieving osmotic concentrations up to ~580 mOsm (Enevoldsen et al. 2003; Jin and DeVries 2006) (except for rainbow smelt that has antifreeze, discussed later). This would depress colligative fp down to −1.0 °C, however, it is still nearly a full degree above the fp of seawater (−1.9 °C). Thus, teleost fishes are undercooled with respect to the fp of the marine water they live in, and are at risk of freezing when ambient water temperatures cool below their colligative fps.

The marine cryosphere where freezing conditions occur includes the high-latitude oceans at both polar regions and the nearshore waters of the cold temperate coasts of northern hemisphere continents. In the south polar region, the Southern Ocean (SO) surrounding Antarctica is by far the world's coldest and harshest marine environment for boney fishes. It is chronically at or near the fp of seawater (−1.9 °C) and laden with ice, reaching extremes of sustained cold and iciness at high latitude coastal embayments year round, exemplified by McMurdo Sound (78°S) (Hunt et al. 2003; Cziko et al. 2014), where the first discovery of fish antifreeze protein was made (DeVries 1968, 1971). The severe marine frigidity of the SO environment results from its isolation by the massive Antarctic Circumpolar Current (ACC)—the largest oceanic current on Earth. The ACC forms a giant revolving curtain of barrier reaching almost to the seafloor, decoupling thermal exchanges with the warmer oceans to its north. In contrast, the Arctic and northern oceans are not isolated bodies of water, but are in wide communications with each other. Fish habitats in these regions experience gradations of cold severity spatially and temporally, due to seasonal solar thermal influences, and heat transfer from poleward flow of warm southerly oceanic currents or southward flow of cold Arctic currents (DeVries and Steffensen 2005). Icy, freezing conditions comparable to the SO occur in the high Arctic (Scholander et al. 1957; Raymond et al. 1975; Denstad et al. 1987; DeVries and Steffensen 2005). The nearshore waters of northern cold temperate coasts could reach subzero temperatures during the winter months and become ice covered (Petzel et al. 1980; Fletcher et al. 1985, 2011). The daunting challenge to the hypoosmotic teleosts localized in these frigid conditions, either due to their life histories (northern species) or geographical isolation (fishes of the Southern Ocean), is the inescapable death from inoculative freezing of their undercooled body fluids when they encounter ice (Scholander et al. 1957; Denstad et al. 1987).

It is therefore remarkable that rich fish faunas thrive in the polar seas and cold temperate coasts. Some polar species even exploit crevices in ice formations to forage for food and escape predators (Andriashev 1970; DeVries 1974; Gradinger and Bluhm 2004). There are behavioral strategies to avoid freezing, including migrating to offshore warmer waters in the winter, or staying in deep waters where hydrostatic pressure depresses in situ freezing point below the ambient water temperature such that ice does not form, and thus fish do not encounter ice and can remain undercooled (Scholander et al. 1957). Endemic northern shallow water species and Antarctic notothenioids could not avoid ice in their subzero habitats, and thus are in constant threat of inoculating freezing. Compelled by this tremendous selective pressure to invent a life-preserving solution, absent which the lineage perishes, the evolutionary adaptive outcome was the emergence of novel antifreeze proteins that could bind internalized ice crystals and arrest ice growth in their undercooled fluids thereby prevent organismal freezing. The magnitude of antifreeze functional capacity in different fish lineages today relates to the severity of the environment they occupy, attesting to the causal relationship between the intensity of selective pressure and the strength of the adaptive trait.

The discovery of the first biological antifreeze, the antifreeze glycoprotein or AFGP in the Antarctic notothenioid fish in the late 1960 and early 1970 (DeVries 1968, 1971) laid the cornerstone of the field of macromolecular antifreeze research. Functional investigations confirmed that these macromolecular AFGPs, and no other small osmolytes, were responsible for further depressing the fish fp expected from colligative solutes to below that of the freezing seawater, enabling freezing avoidance. A flurry of surveys of northern cold water fishes for the presence of antifreeze proteins followed. Other types of antifreeze proteins were discovered in rapid succession, including a near-identical AFGP in the unrelated northern codfishes, and diverse antifreeze peptides (AFPs) in other cold-water species classified as type I, II, and III AFP based on their distinctive protein structures. The recognition that a protein molecule can interact with an unusual ligand—ice crystal, and impedes its growth down to temperatures below the expected colligative fp (Raymond and DeVries 1972, 1977), gave birth to the paradigm of nonequilibrium fp depression that has now found examples across all domains of life inhabiting the cryosphere, both marine and terrestrial.

In the following sections, the details of the evolutionary origins and mechanisms of types II and III AFPs, and of the near-identical AFGP in the unrelated Antarctic notothenioid fish and northern codfish will be discussed. The small alpha-helical type I AFP from the phylogenetically distant flounder, snailfish, sculpin, and cunner has been determined to have a polyphyletic origin, thus a fascinating case of convergent evolution (Graham et al. 2013). However, the ancestral source and mechanism of evolution are still unknown, and thus any discussion of the subject matter will have to await future discovery. Also, the putative antifreeze from the longhorn sculpin, previously named type IV AFP, is no longer considered as a physiologically relevant antifreeze due to its miniscule circulating concentrations. Thus, it will be excluded from the discussion.

9.2 Type II AFP in Herring, Rainbow Smelt and Sea Raven

Type II AFPs (or AFP II) are cysteine-rich globular proteins of about 15 kDa, first discovered in three northern cold-temperate fish species from three distantly related orders—sea raven *Hemitripertus americanus* (Scorpaeniformes) (Slaughter et al. 1981), rainbow smelt *Osmerus mordax* (Osmeriformes) (Ewart et al. 1992), and

Atlantic herring *Clupea harengus* (Clupeiformes) (Ewart and Fletcher 1993). The antifreeze activity or thermal hysteresis (TH) of the circulating levels of AFP II in these three fish range from 0.1 °C to a maximum of 0.48 °C in coldest months depending on species, population, and catch location (Duman and DeVries 1975; Ewart and Fletcher 1990; Raymond 1992). Compared to the high TH levels of 1.0 °C to >2.0 °C in the Antarctic notothenioid fishes contending with constant icy, −1.9 °C SO conditions (Duman and DeVries 1975; Cziko et al. 2014), the much lower AFP activities of these northern species reflect the less severe environments they inhabit. The typical mean plasma salt concentrations of these species (~450 mOsm) would depress plasma fp to −0.84 °C. Adding a maximum of 0.48 °C TH, the fish would be able to avoid freezing down to about −1.3 °C, slightly over half a degree above the fp of seawater. Regardless, when habitat temperatures chill below the fish colligative fp, even if only for part of the day, it only takes encountering a single ice crystal to rapidly nucleate the undercooled body fluids and kill the fish. Thus, the presence of AFP remains crucial in warding off that lethal circumstance to ensure organismal survival.

The adaptive role of AFP II in rainbow smelt is not entirely clear. This species has evolved a colligative fp depression strategy, by synthesizing an abundance of glycerol, reaching plasma concentrations as high as >400 mOsm in the winter (Raymond 1992, 1993). With the high glycerol level, in conjunction with plasma electrolytes and other osmolytes, the smelt becomes isosmotic with seawater. Thus, colligative fp depression alone would be sufficient to protect rainbow smelt from freezing. It is, therefore, puzzling why rainbow smelt employs both colligative (glycerol) and non-colligative (AFP II) freeze avoidance strategies, especially from the consideration of energetic costs involved in synthesizing both types of molecules. From an evolutionary standpoint, if the colligative freeze avoidance strategy evolved first and solved the freezing challenge, there would be no selective pressure to evolve non-colligative macromolecular antifreeze as a redundant solution. Or Vice versa. It would seem more evolutionarily efficacious to improve upon one emergent adaptive trait instead of two. Atlantic herring and sea raven ostensibly had assumed that path, with the resultant single phenotype—the AFP II and the TH it provides, which was comparable to those in smelt, apparently sufficient to preserve life. The responsible antifreeze genotype in rainbow smelt appears to consist of a single copy of AFP II gene, based on screening a large-insert genomic DNA BAC (Bacterial Chromosome Arm vector) library (Graham et al. 2012). It has not expanded to a multigene family typical of other fish antifreeze proteins that occur at high circulating levels such as AFP I of winter flounder (Scott et al. 1985), AFP III of zoarcid fishes (Hew et al. 1988; Wang et al. 1995b), and AFGPs of Antarctic notothenioids (Nicodemus-Johnson et al. 2011) as well as northern codfishes (Zhuang et al. 2012, 2018). Whether this implies a minor role of AFP II in freeze avoidance in rainbow smelt remains a query to grapple with.

9.2.1 Putative AFP II in Other Species: Physiological Relevance?

Besides herring, smelt, and sea raven, a type II AFP cDNA were more recently cloned from another osmerid, the Japanese smelt *Hypomesus nipponensis* (Yamashita et al. 2003). The translated protein sequence shares high sequence similarity with rainbow smelt AFP II, and the recombinant protein shows typical antifreeze behavior. The specimens were obtained from a fish market, and thus no native plasma thermal hysteresis values from circulating AFP II in the fish were reported. The environmental conditions where the smelt came from are also unknown. The species was called a "mid-latitude freshwater fish" (Yamashita et al. 2003). If its life history is completely freshwater, it would have no need for an antifreeze, as teleosts are hyperosmotic to freshwater with colligative fps lower than 0 °C, the fp of freshwater. Similarly lacking in physiological and environmental information is a cottoid species (family Agonidae), the longsnout poacher *Brachyopsis rostataus*. It was reported to have a type II AFP, for which the crystal structure of the recombinant protein was solved (Nishimiya et al. 2006, 2008), but without any information on the isolation and characterization of the native AFP II, or cDNA sequence.

Two other species unrelated to herring, smelt, and sea raven, the Chinese perch or mandarin fish *Siniperca chuatsi* (order Centrachiformes) and the American yellow perch *Perca flavescens* (order Perciformes) have unpublished mRNA sequences in Genbank annotated as "antifreeze protein" (EU719616) and "type II antifreeze protein" (FJ826540), respectively. The translated amino acid sequences of Chinese perch share 37%–43% (mandarin fish) with herring, smelt, and sea raven AFP II, while the yellow perch shares 44%–49% identity. Despite the absence of functional evidence, they were treated as AFPs in phylogenetic analyses to infer evolutionary origin and transmission of AFP II across teleost phylogeny (Sorhannus 2012). These two perch fishes are natives of freshwater rivers and lakes, thus there is no selective pressure for AFP to evolve, as they would not need AFP protection. We suggest that treating cloned sequences as actual antifreeze is inappropriate, and clouds the very evolutionary issue one tries to resolve. More broadly, we suggest it should be a prerequisite to ascertain functional, physiological, and organismal relevance before alluding to putative AFPs and AFP-like sequences as antifreeze proteins, which otherwise would imply function and an adaptive role where they may not exist. The exclusion of the longhorn sculpin AFP as a physiologically relevant AFP because of its extremely low (g/mL) circulating concentration is a case in point (Gauthier et al. 2008). On the other hand, if these sequences indeed prove to be biologically irrelevant for fish survival in their particular habitat, they become important from an evolutionary standpoint. They may in fact represent homologs in the evolutionary lineage possessing features of exaptation that predispose them to be the target of selection to achieve full function antifreeze genes.

9.2.2 Lectin Origin of AFP II: Which Lectin?

The evolutionary origin of type II AFP from sea raven, smelt, and herring was identified in the early 1990s, the first among fish antifreeze proteins. The rapid advances of molecular methods in the decade prior had contributed to an exponential growth of available gene sequences. Thus, once the AFP cDNA was cloned and sequenced, a database search readily revealed a Ca^{++}-dependent (C-type) lectin to be the closest evolutionary homolog (Ewart et al. 1992; Ewart and Fletcher 1993). Therefore, type II AFP is a classic case of evolution from a preexisting gene.

AFP II of herring and smelt share 83% amino acid identity, while sea raven AFP II is more divergent, sharing about 42% identity with herring and smelt. Ca^{++} is required for herring and smelt AFP II activity, while sea raven AFP II activity is Ca^{++}-independent (Ewart et al. 1992; Ewart and Fletcher 1993). All three have 10 conserved cysteines forming five positionally conserved disulfide bridges, while all other known C-type lectins have fewer cysteines.

C-type lectins (as opposed to Ca^{++}-independent lectins) are a large and diverse superfamily of animal lectins with reportedly over a thousand identified members even 14 years ago in an excellent review by Zelensky and Gready (Zelensky and Gready 2005). C-type lectins share a conserved structural module—the carbohydrate recognition domain (CRD), which in many lectin members binds a specific monosaccharide. Other members were discovered not to bind carbohydrates, or even Ca^{++}, but other ligands, thus the homologous CRD of these members is given the more general name of C-type lectin-like domain (CTLD), and the proteins that contain it are called CTLD-containing proteins (CTLDcps) (Zelensky and Gready 2005). Type II AFPs are themselves CTLDcps as they possess homologous CRDs that bind ice instead of carbohydrate. Collectively, the superfamily of C-type lectins represents a large and vastly functional fluid group of secreted proteins.

It remains uncertain to this day which CTLD or CTLDs gave rise to the type II AFPs. Definitive identification of the evolutionary precursor(s) is essential in resolving the lasting question of AFP II evolutionary history or histories. The large and ever expanding membership of the CTLD superfamily makes definitive identification both challenging and promising. The challenge is that as new putative members are identified, including bioinformatically annotated putative homologs from transcriptome and genome sequencing, the statistically closest (best hit) homologs of AFP II change with time. Being a soluble, extracellular, monomeric CTLD domain, AFP II most resembles Group VII CTLD, which comprises the members of the human REG (Regenerating) protein gene family including lithostathine (or PSP—pancreatic stone protein) (Zelensky and Gready 2005), now recognized as *Reg1A*. Amino sequence alignments at the time of AFP II discovery over two decades ago when the database was much smaller, identified lithostathine as the closest homolog (Ewart and Fletcher 1993; Ewart et al. 1999). However, the shared amino acid sequence identities are very low, at 23%–25%, indicating very remote relatedness. More recently, skin mucus mannose/galactose-specific lectins were used as representatives of the closest lectin relatives of type II AFPs in phylogenetic analyses to assess AFP II evolutionary history (Liu et al. 2007).

The influx of genome assemblies with entire suites of gene members in protein gene families on the other hand promises to increase the success of finding the most plausible (or closest) evolutionary ancestor.

A number of lower vertebrate CTLDcps have been recognized to have no human/mammalian homologs. In Neighbor-Joining phylogenetic inference they were recovered as distinct clades from human and other mammalian C-type lectins (Zelensky and Gready 2005). One clade-specific CTLDcps encompasses type II AFPs and a number of soluble serum lectins and predicted AFP-like sequences from bony fishes. To utilize the recent influx of genomic and transcriptomic sequences in search of the newest soluble serum lectins and predicted AFP-like homologs of AFP II, for this chapter, we performed BLAST (Best Local Alignment Search Tools) searches of the nr (nonredundant) database and Ensembl genomes and analyzed the results to considerable length. Using smelt, herring, and sea raven AFP II in separate BLASTP (protein BLAST) searches of the current nr (nonredundant) protein data base yielded best sequence matches (hits) on very recently submitted records. Nearly all the top one hundred non-self hits are deposited within the past 3 years (2016–2019). These non-self hits consist of two groups of sequences based on annotation labels—"ladderlectin/ladderlectin-like" and "type-2-ice structuring protein-like," plus a handful of "galactose-specific lectin nattectin-like." All hits are from teleost species, and except for one, they are sequences from across diverse taxa (28 species) in the Clupeocephalan cohort of Euteleosteomorpha. Amino acid sequence identities of these top non-self hits range between 36% and 43.7%, many of which are $\geq$40%, total positive residues are in the mid to upper 50%, and the alignments span the full length of the translated protein sequence. Searching Ensembl genomes yielded annotation matches in 39 species of teleost fishes, 19 of which are not represented in the nr BLASTP hit list. The genome of the yellow perch *Perca flavescens* with the reported putative AFP II-like sequence (FJ826540) has also been recently sequenced, and the assembly is available on NCBI, which we also utilized. Thus, based on the search results from the most recent versions of sequence data repositories, the closest evolutionary homologs of AFP II genes are ladderlectin-like, type-2-ice structuring protein-like, and/or galactose-specific lectin nattectin-like genes. They are ubiquitous in Euteleosteomorpha (47 species with sequence matches). Ladderlectin-like is also present Atlantic herring, which belongs to a different Clupeocephalan cohort—Otomorpha. Ladderlectin and nattectin are C-type lectins that have known function. Type-2-ice structuring protein-like annotation is strictly based on sequence similarities to known AFP II, and does not imply function, especially when all the species in the hit list are either freshwater or warm temperate and tropical marine fish. However, it may represent a ubiquitous CTLDcp in Euteleosteomorpha species, whose sequence happens to be AFP II-like, and potentially one or separate paralog might have been co-opted to form AFP II in smelt and sea raven, which belong to Euteleosteomorpha.

Ladderlectin is a multimeric Ca^{++}-dependent lectin first discovered in the plasma of rainbow trout *Oncorhynchus mykiss* (Jensen et al. 1997). On native gel electrophoresis, it migrated as a ladder of different sized multimers and thus was named ladderlectin by the discoverer. In reducing condition, it is a 16-kDa monomer, similar in size to AFP II. Subsequently, two ladderlectin variants were found in rainbow trout (Russell et al. 2008), one of which is longer than the other by one additional, small coding exon. The implication of this additional exon will be discussed in the next section on the mode of AFP II evolution. Nattectin is a C-type lectin found in fish venom and binds galactose, and is Ca^{++}-independent for its activity (Lopes-Ferreira et al. 2011).

9.2.3 *Mode(s) of AFP II Evolution*

The occurrence of AFP II with high degrees of sequence similarity in three phylogenetically distant species has invoked multiple hypotheses of the origin and mode of evolution. They include (1) multiple origins and parallel evolution, (2) single origin and varied retention and loss, and (3) lateral gene transfer (LGT).

The high protein sequence similarity inclusive of the identical galactose-type binding site and Ca^{++} requirement for activity suggested to Ewart et al. (1999) that an AFP-like lectin ortholog might have existed in smelt and herring, and evolved in parallel into AFP II in the respective species. Recognizing the antifreeze trait is likely a recent evolution, they reasonably inferred that evolution of an AFP II gene in the common ancestor and inheritance by descent in herring and smelt is unlikely because of the long species divergence time between the two lineages. The greater sequence and structural divergence of sea raven AFP II, on the other hand, suggested it evolved separately from a different C-type lectin.

Liu et al. (2007) proposed an alternate hypothesis of a single origin of type II AFPs based on phylogenetic reconstruction of the evolutionary relationship of type II AFPs and various C-type lectins and CTLDs, and mapping to species phylogeny. Currently known AFP II-bearing species belong to two cohorts within Clupeocephala, namely Otocehpahla and Euteleostei (revised as Otomorpha and Euteleosteomorpha in the current NCBI Taxonomy). Atlantic herring belongs to Otomorpha, while rainbow smelt and sea raven belong to Euteleosteomorpha. Thus, they inferred the AFP II gene originated from a duplication of mannose or galactose-binding lectin gene in the common Clupeocephalan ancestor of the two cohorts, and developed ice-binding ability. The emergent ice-binding gene subsequently evolved into a full-function ice-binding CTLDcp in the lineages leading to today's AFP II-bearing species in the two separate cohorts. There are two difficulties with this model. One, it requires AFP gene loss in a huge number of lineages in both the Otomorpha and Euteleosteomorpha cohort. Two, the common ancestor of Otomorpha and Euteleosteomorpha dated to an age of 230 Myr (million years) (TimeTree.org; time estimate derived from 35 studies). The planet then consisted of the supercontinent Pangea, polar regions had not differentiated, and world oceans were warm, thus no selective pressure existed for ice-binding ability to develop in the lectin ancestor in the first place.

LGT was proposed in two studies by Graham et al. (2008, 2012) to account for the acquisition of a very similar AFP II genotype in the phylogenetically divergent herring, rainbow smelt, and sea raven. Both studies rigorously tested multiple potential falsifications of the LGT hypothesis, and the results strongly supported LGT as the mechanism. The first study examined all three species. The strongest evidence is the high nucleotide sequence similarities not just in the protein-coding sequence but also in most of the intron sequences between herring and smelt, and to a lesser degree between sea raven and these two species. This contrasted sharply with exonic sequence conservation only, but not intronic sequences when comparing herring and smelt *Prp8* (Pre-RNA Processing factor 8), the highly conserved eukaryotic spliceosome protein. Also, southern blot hybridization of genomic DNA using smelt AFP II genomic sequence as probe detected no complementary sequences in a cross section of teleost fishes, validating AFP II gene is specific to herring, smelt, and sea raven, with no closely related CTLDcp homologs (which, if present, would hybridize to the probe) in the other fish. These evidences are consistent with the LGT hypothesis, and they proposed two instances of its occurrence—once to the ancestor of sea raven, and more recently to or between herring and smelt.

The second study (Graham et al. 2012) examined a longer distance genomic region encompassing the smelt AFP II gene, isolated from a large-insert genomic DNA BAC library. The AFP II genotype appears to consist of a single gene copy nested with 15 neighboring genes within a single BAC clone of 160 kbp insert length. Again the sequence conservation between smelt and herring AFP II genes exceeds that of the conserved syntenic regions between smelt and five other test species with the available genome. No AFP-like sequence was detected in the syntenic regions of these test species, indicating that AFP II was a new acquisition in the smelt. Tests of dN/dS (rate of non-synonymous to the rate of synonymous substitution) ratios also indicated that AFP II sequences have had little time to diverge as opposed to the test set of non-AFP protein genes, consistent with a recent transmission of the gene between the two species. No viral elements were found close to the smelt AFP II gene, supporting the mechanism of transmission was not mediated by such elements.

The Atlantic herring genome assembly was completed to chromosome level and released in April 2019, after the two Graham et al. studies (2008, 2012). In the current version of the assembly, we located one copy each of herring AFP II gene in two chromosomes, 15 and 26. The canonical AFP II gene is likely the copy in chromosome 15 as its structure closely resembles the previously cloned gene (Graham et al. 2008). The copy in chromosome 26 is atypical, spanning over 30 kbp in distance because of extremely long introns, and was annotated to consist of four exons instead of the known six, however, the amino acid sequence of one of the transcripts is near identical with the copy in chromosome 15. Neither chromosome 15 or 26 AFP genomic locus is syntenic with the smelt AFP II locus. The syntenic region is in herring chromosome 1, and it contains no AFP II gene.

The two Graham et al. studies (2008, 2012) were thorough and presented compelling molecular evidence supporting LGT. However, there are a number of issues that remain difficult to address. (1) LGT concerns how the gene was transmitted after it was formed. Thus, how it was formed in the first place and which was the lectin precursor remained unknown. (2) The timing of the LGT transmission, if it is indeed LGT, is difficult to reconcile. If one accepts the putative AFP II in the Japanese smelt *Hypomesus nipponensis* is physiologically relevant antifreeze, then the AFP II gene would have to be formed, or received from the donor herring species, in the common ancestor of the Japanese and rainbow smelts. The estimated age of their most recent common smelt ancestor is 28.8 Myr (TimeTree.org; estimate derived from 7 studies). This is much older than the onset of large-scale northern hemisphere glaciation in the late Pliocene [~3.5–3 Mya (million years ago)] (Maslin et al. 1998), i.e., before environmental selective pressure existed to drive AFP II evolution. (3) There is no current proven example of successful heterologous fertilization in vertebrate animals in the wild to support it as a plausible mechanism of LGT of *AFP II* as proposed by Graham and co-workers. The barrier against heterologous fertilization is near impenetrable. A recent definitive study on species-specific recognition in fish fertilization identified the gate-keeping receptor named "Bouncer" (Herberg et al. 2018). It is expressed on the oocyte membrane surface where it selects sperms of the same species for entry through the micropyle, and prevents entry by sperm of other species. Should the barrier be manipulated and bypassed, and heterologous union occurred, the progeny was sterile. Graham et al. (2008, 2012) cited various experimental heterologous fertilization for creating transgenic fish or other vertebrates (Patil and Khoo 1996; Venugopal et al. 2004; Smith and Spadafora 2005) to suggest the possibility of comparable processes happening in the wild. However, these experimental approaches involved drastic measures of high-voltage electroporation or other laboratory manipulations, which are unlikely to be occurring in the wild. (4) If LGT by heterologous fertilization were to occur, the donor and the recipient species would have to be sympatric species for at least part of their life history. In occupying the same environment, one expects that they would be subject to the same selective pressures, thus it is difficult to reconcile why one species would evolve antifreeze protection, and the other will not. Lastly, while there are overwhelming similarities between the AFP II sequences of herring and smelt, there are two distinct differences that are yet to be adequately addressed. The smelt AFP II gene has an additional, small cds exon—E3 (Fig. 9.1) than herring and sea raven AFP II. Also, much of the upstream and downstream intron (I2 and I3) sequences flanking this additional exon are not homologous with herring and sea raven intron 2. Despite this absence of homology, this distinct exon was considered as part of smelt E2 and named E2a by Graham et al. (2008). Its origin, and that of the nonhomologous intron sequences remain unaccounted for.

9.2.4 Ladderlectin/Nattectin-Like Origin(S) of AFP II?

Figure 9.1 shows the structural alignment of smelt, herring, and sea raven *AFP II*, and the closest lectin and CTLDcps homologs we identified from our BLAST

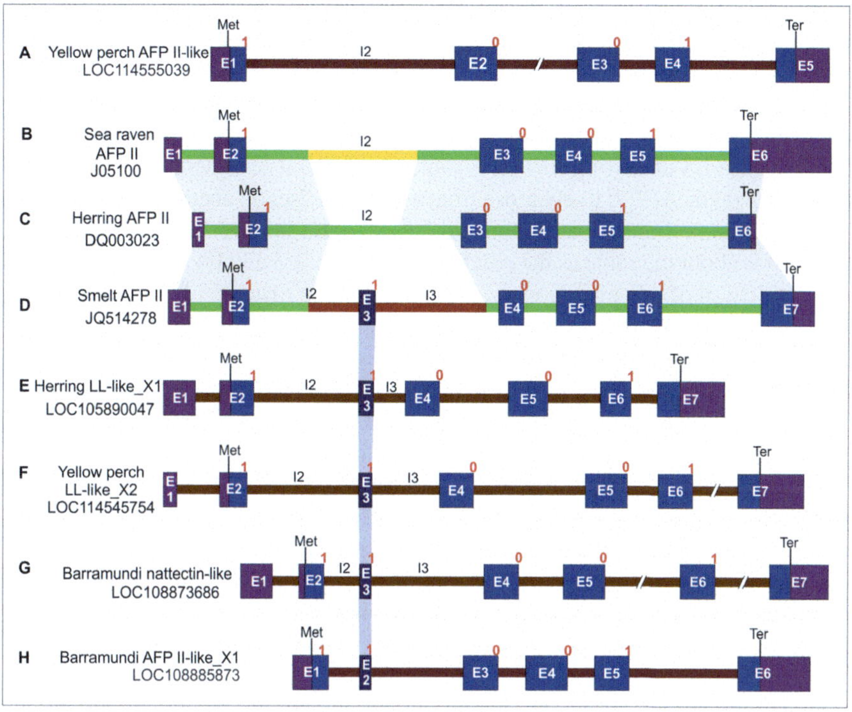

Fig. 9.1 Alignment of gene structure of herring, smelt, and sea raven *AFP II*, and representatives of top matching CLTD homologs including ladderlectin (LL)-like, Nattectin-like, and type-2 ice structuring-like (*AFP II*-like) identified from current (2019) sequence databases of teleost fishes. Exons and introns are drawn to scale, except for the introns that contain //, which indicates a longer actual length. Purple boxes are 5′UTR and 3′UTR sequences. 5′UTR is split between the first two exons in most homologs, while a few (yellow perch *AFP II*-like (top), and Barramundi *AFP II*-like (bottom)) do not have the first 5′UTR exon. Navy and deep blue boxes are protein-coding exons, with Met start and Ter (stop) sites indicated. The red numbers, 1 and 0, indicate the intron phase of the particular exon. Gray blocks connecting sea raven, herring and smelt AFPII genes are regions of sequence similarity. Smelt AFP II has an additional exon (E3) that is absent in herring and sea raven AFP II. Also, substantial portions of smelt AFP I2 and I3 are not homologous to I2 of either herring or sea raven. The seven exon structure inclusive of the small E3 of smelt AFP II gene is conserved in various ladderlectin (LL)-like, nattectin-like, and AFP II-like homologs across a wide phylogenetic representation of teleosts, raising the possibility that one of these homologs in smelt could be the evolutionary precursor of smelt *AFP II*

searches of the current databases as described above. Smelt AFP II gene consists of seven exons (Fig. 9.1d), while herring and sea raven *AFP II* consist of six exons (Fig. 9.1b, c). The additional smelt exon (E3) is shared by the ladderlectin (LL)-like, nattectin-like and AFP II-like homologs, and all have the same phase 1 intron junction for the downstream intron (I3) (Fig. 9.1e–h). All other corresponding cds exons of the depicted genes share the same intron phases. Conservation of gene structure, i.e., number of exons and introns, and intron phases are characteristics of evolutionary homologs. This structural analysis opens an alternate possibility that smelt *AFP II* could have evolved from a preexisting 7-exon ladderlectin-like/nattectin-like/AFP II-like gene in its genome, instead of acquiring it exogenously from herring. The 6-exon herring *AFP II* could arise in two different ways—from a 7-exon ancestor and subsequently lost the small E3, or from a 6-exon precursor similar to the yellow perch *AFP-II like*. The 7-exon ancestor hypothesis is more likely because E3 of herring *AFP II* is short, similar to its counterpart (E4) in smelt *AFP II*. Sea raven *AFP II* likely arose from a different 7-exon ancestor, and involved fusion of the ancestral E3 and E4 to form the extant E3 in the AFP gene, because its E3 length is about the same as the combined lengths of E3 and E4 in smelt *AFP II*. A major limitation of this evolutionary scheme is that the matching LL-like, nattectin-like, and AFP II-like homologs, while having the 7-exon structure and are significantly similar in translated amino acid sequences, 36%–42% (similar level between sea raven AFP II and herring/smelt AFP II), are quite dissimilar at the nucleotide sequence level. This applies to the herring LL-like X1 (Fig. 9.1e) and herring AFP II (Fig. 9.1c). Thus, whether a preexisting 7-exon ladderlectin-like/nattectin-like/AFP II-like ancestor had given rise to the incipient AFP II gene, which then diverged quickly to its extant sequence, remains to verified. One approach that may provide resolution is sequencing the genomes of rainbow smelt and sea raven, as well as the Japanese smelt and the longsnout poacher that have putative AFP II. Identifying and comparing the full suite of lectin and CTLDcp homologs and the genomic context they occur in these species may eventually lead to more definitive identification of the genetic origin or origins of the type II AFPs.

9.3 Type III AFP of Zoarcoid Fishes

Type III AFP is a small globular protein of about 64 amino acids and 7 kDa in mass, found in fishes in the Perciformes suborder of Zoarcoidei. AFP III has been characterized for species from two Zoarcoid families—Zoarcidae and Anarhichadidae. Zoarcid species include the Atlantic ocean pout *Macrozoarces americanus* (Hew et al. 1984; Liu et al. 2007), the European life-bearer eelpout *Zoarces viviparous* (Albers et al. 2007), the Alaskan eelpout *Lycodes polaris* (Schrag et al. 1987), and two Antarctic eelpouts, *Lycodichthys dearborni* (formerly *Rhigophila dearborni*) (Wang et al. 1995a, b), and *Pachycara brachycephalus* (formerly *Austrolycichthys brachycephalum*) (Cheng and DeVries 1989). Anarhichadids include two wolffishes, *Anarhichas lupus* and *A. minor* (Desjardins et al. 2012). Type III AFP has been found only in these related zoarcoid species, indicating it evolved once in a common ancestor of these particular taxa. As such it represents the only instance of a monophyletic origin among fish antifreeze proteins thus far.

Anarhichadids occur along coasts on both sides of the northern Atlantic down to Cape Cod on the west and northwestern Mediterranean on the east, and on the coasts of landmasses in between, i.e., Greenland, Svalbard, and Iceland (Barsukov 1986; Robins and Ray 1986). Of the two Newfoundland wolffishes examined, the Atlantic wolffish *A. lupus* lives in shallow water and encounters greatly more severe cold and icy conditions. Accordingly, it has evolved a much larger AFP III gene dosage providing for a high plasma TH of ≈1 °C. In contrast, the spotted wolffish *A. minor* inhabits deeper water (>100 m) that may be relatively ice free, correlating with a smaller AFP III gene family and much lower plasma TH of 0.13–0.24 °C (Desjardins et al. 2012).

In contrast to anarhichadids' exclusive northern distribution, zoarcid fishes are globally distributed. As of 1994, there are 220 recognized species (with new species being described since), most of which are rare and deep-water (Anderson 1994). Zoarcids are most abundant in the boreal seas, and their origin was considered to be in the North Pacific ocean. It was hypothesized a pre-Miocene radiation had taken place along the western coasts of the Americas, with subsequent endemism establishing in the Megellanic Province of S. America and Antarctica (Anderson 1994). The two Antarctic eelpouts with described AFP III are deep water (≥600 m) residents of McMurdo Sound. Their plasma TH levels are about 0.9 °C, substantially less than the notothenioid fishes inhabiting icy, freezing shallow waters of McMurdo Sound. When raised to surface during capture, the eelpouts would freeze if the ice hole has not been cleared of floating ice (personal observation). Along the lower latitude West Antarctic Peninsula (WAP) coast, our deepwater trawling (>600 m) in the past decade often brought up large biomass of zoarcid fishes, with species identity yet to be definitively determined. Most of these do not have detectable plasma TH, while some have low levels (Cheng, unpublished data). Thus, despite being polar inhabitants, with sympatric exposure to similar environmental selective pressure, not all Antarctic eelpouts possess the AF phenotype. The species without TH suggests they may be evolutionarily recent arrivals from northerly non-AFP bearing population through the Antarctic bottom water, which forms part of the global oceanic conveyor belt. The Antarctic bottom water is a distinctive nonfreezing water layer, in which antifreeze-lacking fish could exploit.

The divergence time for Anarhichadidae and Zoarcidae is estimated to be about 13.0 Myr (TimeTree.org; estimate derived from five studies). For AFP III to be present in various species within the two families, the logical inference is that the trait is inherited from a common ancestor as mentioned above. However, given the huge species diversity of zoarcidae, most of which inhabit warm boreal seas and are not expected to have AFP, and including WAP species lacking AFP trait, the current pattern of AFP III distribution in Zoarcoidei implies a loss of the trait in the great majority of zoarcid lineages post divergence from anarhichadids. It is unknown whether a parallel pattern of loss occurred in Anarhichadidae since only two species have been studied. A robust phylogeny of Zoarcoidei, with simultaneous examination for the presence/absence of the AFP III genotype and phenotype to map on the phylogeny are needed to resolve the evolutionary history of AFP III. A divergence time of 13 Myr is very recent in evolutionary time scale. If AFP III genotype evolved in the most recent common ancestor of anarhichadids and zoarcids, AFP III-like sequences should be quite recognizable if present.

9.3.1 Sialic Acid Synthase Origin of AFP III Gene

The first AFP III sequence, that of the ocean pout *M. americanus,* was determined in the early 1980s, and those of its relatives not long after. However, the genetic origin of AFP III remained elusive until the landmark human draft genome was published in February 2001. One key finding reported in the paper from the publically funded sequencing effort (International Human Genome Sequencing Consortium) (Lander et al. 2001) was two examples of human homologs in fish. One of them is human sialic acid synthase (BAA91818.1), found to contain a domain homologous to polar fish antifreeze III protein, suggesting that "fish created the antifreeze function by adaptation of this domain." The alignment of sialic acid synthase (SAS) amino acid sequence with type III AFPs and inference of SAS ancestry followed 6 months later (Baardsnes and Davies 2001). Thus, the emergence of AFP III is another classic case of evolution from a preexisting gene, but with an added complexity of co-opting only one domain from SAS, the small C-terminal domain (Fig. 9.2, left panel), which

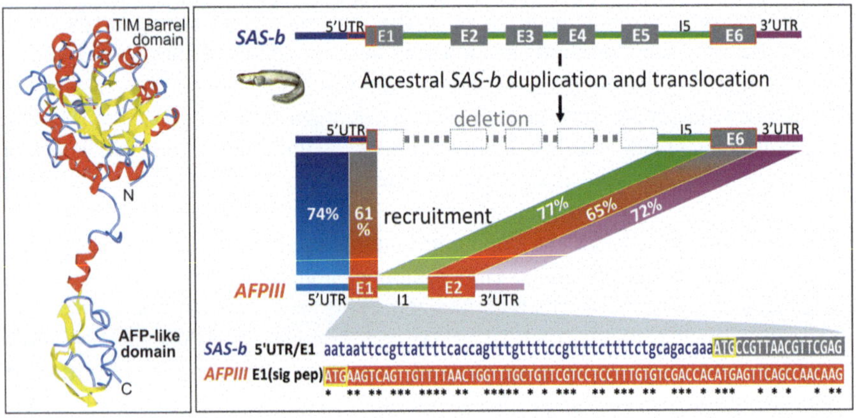

Fig. 9.2 (Left panel) Crystal structure of sialic acid synthase (NeuB) from *Neisseria meningitidis* (PDB 1XUU). The large N-terminal domain has a TIM (triosephosphate isomerase) barrel fold. The small C-terminal domain, encoded by exon 6 in zoarcid SAS (right panel), was co-opted and neofunctionalized into AFP III. (Right panel) Molecular mechanism of *SAS* to *AFP III* evolution deduced for AFP III of the Antarctic eelpout *Lycodichthys dearborni* (image). A duplicated ancestral copy of the *SAS-b* paralog was translocated from the SAS locus to a different chromosome. A deletion event removed the N-terminal domain coding sequence (from seventh codon of E1 through E5—gray dashed line and boxes). The remaining *SAS-b* structure (colored), i.e., the front end (5′UTR and partial E1), and the tail end (intron5, E6 and 3′UTR) were recruited forming the two-exon AFP III gene. Part of the nonprotein coding 5′UTR of *SAS-b* (red framed navy bar segment) was exonized to become the signal peptide cds (E1) of the new AFP III gene, where a 2-nt change (aat to ATG) established the start codon. Yellow bracketed ATG indicates the start codon of and *AFP III* and *SAS-b*. *L. dearborni AFP III* and *SAS-b* share high nucleotide sequence identities (61%–77%) in their homologous regions (indicated by gradient colored blocks), supporting the *SAS-b* ancestry and the depicted mechanism of evolution of *AFP III* from *SAS-b*

neofunctionalized into a potent antifreeze protein under natural selection (Deng et al. 2010).

SAS is a cytoplasmic enzyme conserved across domains of life. It is known as NeuB (prokaryotic N-acetylneuraminic acid synthase) in bacteria, and SAS in eukaryotes, or as the more generalized N-Acetylneuraminic Acid Synthase (NANS) across taxa (Betenbaugh et al. 2014). It catalyzes intracellular synthesis of sialic acids (9-carbon 2-keto-3-deoxy sugars) from N-acetylmannosamine or from mannose-N-acetyl-6-phosphate and phosphoenolpyruvate. The SAS protein ranges from 346 to 359 amino acids in length, encoded by a six exon gene in eukaryotes (Fig. 9.2, right panel). The first five exons encode a large N-terminal domain that displays a TIM barrel fold plus a linker to the small C-terminal domain (66 residues) encoded by the sixth exon, the homolog of zoarcoid AFP III gene (Fig. 9.2).

The molecular process of zoarcid *SAS* to *AFP III* evolution was deduced by characterizing the *SAS* and *AFP III* genomic regions isolated from a large-insert genomic DNA BAC library generated for the Antarctic eelpout *Lycodichthys dearborni* (Deng et al. 2010) (Fig. 9.2, right panel). *L. dearborni* was found to have two *SAS* paralogs, *SAS-a, and SAS-b.* Both share high nucleotide (nt) identities with *AFP III* in their homologous protein coding regions and intervening introns. However, only *SAS-b* additionally shares nt identity in the 5′untranslated region (UTR) and 5′ flanking sequences with *AFP III* (Fig. 9.2, right panel). The proximal 5′UTR immediately upstream of the ATG start codon in *SAS-b* (54 nt, nt in navy) is 64% identical to *AFP III* signal peptide cds (red background highlight), indicating it was an ancestral *SAS-b* copy that evolved into *AFP III.* The *SAS* and *AFP III* genomic regions occupy distinct chromosomes in *L. dearborni*, indicating that a translocation of the *SAS-b* ancestor had occurred, likely aided by a transposable element found to flank the extant *AFP III* locus.

The transformation of a 6-exon *SAS-b* to a 2-exon *AFP III* required deletion of the bulk of the SAS gene structure (seventh codon of exon 1 through exon 5), and recruitment of the upstream (5′UTR and partial E1) and downstream sequences (I5, E6, and 3′UTR). The deletion removed the coding sequence of the large N-terminal domain and linker in the SAS protein, and the new 2-exon gene would produce only the C-terminal domain, which evolved into a full-function AFP III protein under natural selection. AFP III is a secreted protein, requiring a signal peptide, which was absent in its cytoplasmic SAS-b ancestor. The signal peptide cds was acquired by exonization of the 54-nt proximal 5′UTR of the ancestral *SAS-b,* i.e., adding this previously noncoding sequence to the first seven codons in forming a large E1 in the emerging AFP III gene (Fig. 9.2, right panel, sequence alignment at the bottom). Under selective tinkering, adaptive nucleotide changes occurred and established an in-frame start codon (aat to ATG), endowing the new *AFP III* with a signal peptide coding E1.

9.3.2 Neofunctionalization of AFP III Under Escape from Adaptive Conflict

A central question in the evolution of functional novelty is why certain ancestral sequences became targets of selection leading to subfunctionalization and in rare instances neofunctionalization. In the case of AFP II, it is intuitively fitting that a member of the vastly diverse and functionally fluid C-type lectins, which are themselves secreted proteins, with a CRD that has proved to be amenable to tinkering to bind an array of ligands, became the target of selection and shaping into an extracellular ice-binding protein. In contrast, SAS, the precursor of AFP III, is encoded by a single copy gene in most teleosts, and two copies in a few others (Deng et al. 2010), and serves a highly conserved enzymatic function in sugar biosynthesis within cells (Betenbaugh et al. 2014). How did this old cytoplasmic enzyme become selected upon to become an extracellular antifreeze protein is more difficult to reconcile. The prevailing conceptual models on the fate of duplicated genes (Lynch and Conery 2000; Lynch and Katju 2004; Conant and Wolfe 2008; Hahn 2009) assume gene duplication happens at random, and functional divergence occurs after the duplication event. These models therefore do not address what may predispose genes to duplication and adaptive change.

Escape from Adaptive Conflict (EAC) is a more recent model that attempts to fill this conceptual gap. The model recognizes many molecules may have more than one function, or possess an incidental side function besides its primary function. There are ample examples of multi- or bi-functional proteins. Prominent multifunctional proteins are the taxon-specific ocular crystalline proteins, which are recruited from metabolic or stress-related enzymes in non-ocular tissues to become abundantly expressed as the refractive proteins of the eye lens or cornea, through a process called "gene sharing" (de Jong et al. 1994; Piatigorsky 1998). In cold hardy winter rye, three pathogenesis-related enzyme proteins exhibit low level of thermal hysteresis indicative of an ice growth inhibition side function (Hon et al. 1995; Griffith et al. 2005).

The EAC model makes the following predictions: (1) The ancestral molecule is bifunctional. (2) The side function could be subject to selection when relevant selective regime comes to bear, commencing adaptive sequence change before gene duplication. (3) This may lead to structural conflict that would constrain improvement of both the ancestral and the selected functions. (4) Gene duplication would resolve the adaptive conflict, whereupon both the ancestral and selected functions could continue to evolve and improve. These predictions are found to be fulfilled in the evolution of Antarctic eelpout AFP III.

In the Deng et al. study (2010), recombinant eelpout SAS-b protein was verified to be bifunctional, having both enzymatic activity and rudimentary ice growth inhibition ability (faceting of single-crystal ice) specifically in the small C-terminal domain (prediction 1). In addition, two of the six key residues that form the flat ice-binding surface in ocean pout AFP III pre-existed in SAS-b. They may have constituted the ancestral structural basis for rudimentary ice affinity in the SAS protein, priming it as a target for further selective tinkering. SAS-b gene (but not its paralog SAS-a) was tested to evolve at a fast rate (significant large dN/dS substitution rate ratio), and so were AFP III genes. Seven residues in SAS-b N-terminal domain, and an astounding 63% of AFP III were detective to be under positive selection. These two observations support the continued improvement of the ancestral function and at an accelerated rate of the new (AFP) function after gene duplication (prediction 4). Three residues in the SAS-b C-terminal domain also experience positive selection, including one (K351) whose ortholog (K61) in AFP III plays a key role in stabilizing the flat ice-binding surface. No such adaptive changes occurred in the SAS-a, and the orthologous position is D351. The specific adaptive changes in SAS-b, including D351 to K351, must have resulted from positive selection on the ancestral SAS-b prior to gene duplication for improving the incipient ice affinity (prediction 2). The prediction of structural conflict (3) was supported by the loss of SAS activity when SAS-b was mutated to contain four residues of the AFP III ice-binding surface in their orthologous positions in the C-terminal domain to mimic adaptive changes in the evolving molecule. It is known that SAS activity requires dimerization of two SAS monomers, juxtaposed N to C, to form the substrate-binding site (Huang et al. 2005). The inactive fourfold SAS-b mutant indicates adaptive changes making the C-terminal domain more AFP-like adversely affect the structural co-ordination of SAS monomers to form an active holoenzyme. Taken together, a cytoplasmic SAS contained a structural aspect that approximated the requirement for rudimentary ice affinity. This could explain why it became selected upon to become a secreted antifreeze when relevant selective pressure came to bear. The structural, experimental, and evolutionary analyses collectively strongly support the operation of EAC in the evolution of eelpout *AFP III*, and provided the first clear example of neofunctionalization of a preexisting protein under the EAC model for the broader field of evolutionary biology.

9.4 AFGP of Antarctic Notothenioids

Among cold water fishes, the fitness and ecological consequences of the evolution of an ice-growth inhibiting protein are particularly profound for the Antarctic notothenioid fishes endemic to the isolated icy, freezing Southern Ocean (SO). The isolation of the SO began with tectonic movements of Gondwana landmasses and opening of seaways leading to unrestricted circumpolar oceanic flow in the early Oligocene (40–35 Mya) (Kennett 1977; Livermore et al. 2005). The confinement of Antarctic notothenioids within the SO has no parallel in any other marine fish group elsewhere. Unable to escape what would have become a frigid marine grave, evolution of the AFGP gene in the Antarctic notothenioid ancestor was an adaptive innovation that singularly preserved the lineage. The impact on the SO fish fauna was dramatic. The Oligocene and subsequent marine cooling and glacial destruction of habitats massively extirpated the rich late Eocene ($\approx$40 Myr) fish lineages that were cold-labile (Eastman and McCune 2000; Eastman 2005). The AFGP fortified ancestral Antarctic notothenioid was able to diversify into all ice-laden water column niches, forming an adaptive radiation and the only known marine species flock today (Eastman and McCune 2000; Eastman 2005; Matschiner et al. 2011). At 130 recognized species and representing over 90% of fish biomass on continental shelves and slopes, Antarctic notothenioids became the predominant fish group (Eastman 2005, 2017), and crucially sustain the SO marine food web (La Mesa et al. 2004). Had AFGP protection not evolved and enabled the ecological success of the Antarctic notothenioids, the modern SO marine macrofauna would have become far more depauperate or assumed an entirely different character.

9.4.1 Simple Tripeptide Repeat Unit, Daunting Repetitive Coding Sequence

Unlike the great majority of known proteins, which are composed of many of the 22 proteogenic amino acids, the AFGP at the protein level appears extremely simple. It is a protein "polymer" consisting of repeats of a glycotripeptide unit. The peptide backbone contains only two amino acids, Ala and Thr, in the repeating tripeptides $(\text{AlaAlaThr})_n$ (DeVries et al. 1970, 1971) with an occasional substitution of the first Ala with Pro (Lin et al. 1972). The carbohydrate moiety is a disaccharide, galactose-N-acetylgalactosamine, O-linked to each Thr (Shier et al. 1972; Shier and DeVries 1975). The apparent complexity lies in its occurrence as a family of different length isoforms, however, they all share the same simple tripeptide repeat. AFGP is not the only protein with a repetitive sequence. Well-known proteins with short repeats include the diverse collagens with repeating $(\text{Gly-X-Y})_n$ (Kadler et al. 2007), and the extraordinary plethora of insect and spider silk proteins that are rich in Gly, Ser, and Ala (Sutherland et al. 2009; Chaw et al. 2017), but present in the repeats are various other amino acids, essential for secondary structure formation. Thus, the AFGPs are by far the most minimalist in amino acid composition and peptide backbone structure. The functional implication of this simple repeat structure, in that it can project the regularly spaced disaccharide side chains in the AFGP molecule to hydrogen bond with regularly spaced water molecules in the ice crystal was recognized early on (DeVries and Lin 1977; Raymond and DeVries 1977). The evolutionary implication of the simple repeat sequence, however, was not fully appreciated till later, when the evolutionary origin and mechanism of the encoding gene was solved.

Molecular technology of amplifying genes and cDNA by PCR (polymerase chain reaction) readily led to the discovery of the C-type lectin ancestry of type II AFPs (Ewart and Fletcher 1993; Ewart et al. 1998). In contrast, early attempts using PCR failed to amplify the AFGP gene sequence from genomic DNA of Antarctic notothenioids, nor could AFGP cDNA be amplified from tissue RNA by RT-PCR, despite the protein being synthesized at high circulating concentrations, at 30–35 mg/mL in many Antarctic notothenioid species (Jin and DeVries 2006), suggesting high levels of AFGP mRNA transcription. The reasons for the inexplicable recalcitrance to amplification would only become clear later, and it was because highly repetitive sequences are not amenable to PCR amplification.

Absent amplifiable AFGP gene or cDNA sequences in the early investigations, construction of a genomic DNA library and isolation of AFGP-positive clones for sequencing became the only avenue for determining the AFP gene sequence. The first AFGP gene was obtained from a genomic DNA library of the Antarctic notothenioid *Notothenia coriiceps* constructed with a phage vector (Hsiao et al. 1990). It is a partial gene lacking the coding sequence (cds) for the secretory signal, but the AFGP tripeptide repeat coding region—a single exon of about 2.5 kbp was complete, and it is tremendously repetitive. It encodes a long AFGP polyprotein precursor containing 46 copies of the two smallest length isoforms (AFGP8 and AFGP7, with 4 and 5 tripeptide repeats, respectively), linked in tandem by conserved three-residue linkers of primarily Leu-Ile/Asn-Phe. The linker residues are absent in the mature protein, thus it was proposed that they are posttranslationally cleaved, producing 46 mature AFGP molecules per round of transcription and translation. The polyprotein gene structure thus effectively boosts the gene dosage and abundance of the gene product (Hsiao et al. 1990).

At a glance, the uninterrupted 9-nucleotide tripeptide coding repeats may be mistaken for noncoding SSRs (simple sequence repeats) or microsatellite DNA. This high degree of repetitiveness, and the long run of it, were the root cause of earlier failures to PCR amplify discrete genomic AFGP sequence, as there are no unique primer sites, and the DNA template or any amplicons generated can effectively self-prime along any stretch of the repetitive sequence. The highly repetitive tripeptide cds also presented challenges in the sequencing of AFGP clones. The lack of unique primer sites in a 2.5 kbp run of 9-nt repeats precluded primer walking. It necessitated generating a large nested set of unidirectional deletion subclones from the parent clone for sequencing, identifying reliable nucleotide overlaps, and reconstructing the contiguous sequence.

9.4.2 Partial De Novo Evolution of Antarctic Notothenioid AFGP Gene

Database search using the partial *N. coriiceps* AFGP gene sequence identified a seemingly unlikely evolutionary homolog—a fish trypsinogen in the GenBank database, with sequence similarity located in the nonprotein coding 3′ flanking sequence of the *AFGP* gene. The AFGP protein coding sequence and the translated tripeptide AFGP repeats bear no resemblance to the trypsinogen sequence that contains all 20 amino acids encoded in the genetic code. This counters the criterion that evolutionary homologs share statistically significant sequence similarity, as would be expected on the principle of evolution by gene duplication and sequence divergence (Ohno 1970). To understand how a trypsinogen could be transformed to

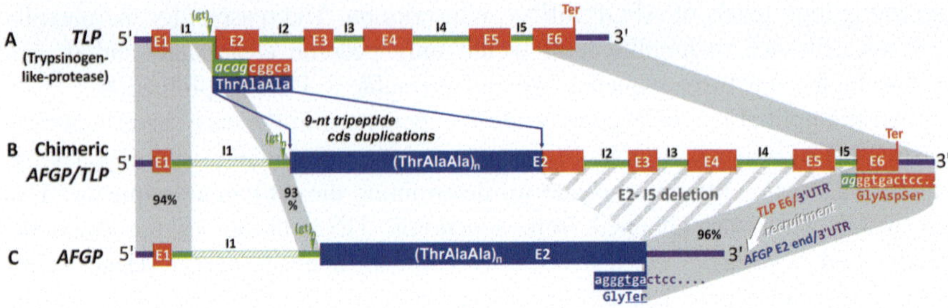

Fig. 9.3 Proposed evolutionary process of an ancestral TLP gene to an AFGP gene in Antarctic notothenioid fish. (**a**) TLP gene structure showing the 9-nt ThrAlaAla coding element spanning I1/E2 (intron 1/exon 2) junction, and the $(gt)_n$ immediately upstream that likely facilitated the first duplication of the 9-nt coding element. (**b**) Chimeric AFGP/TLP genes occur in notothenioid genomes validating the expansion of the 9-nt coding ThrAlaAla element in the formation of a long AFGP coding region leading to a hybrid *TLP* E2. (**c**) Deletion of the TLP E2 through I5 of the chimeric AFGP/TLP gene, and linking of the front and tail portions resulted in the new AFGP gene. The inherited *TLP* E1 provides the secretory signal, and *TLP* E6 provided the stop codon through a 1-nt reading frame shift plus the 3′UTR of the new AFGP gene. The proposed process is supported by the regions of very high nucleotide sequence identities (93%–96%, gray blocks) between *TLP* and *AFGP* determined for the notothenioid *Dissostichus mawsoni*

AFGP, complete AFGP genes inclusive of the signal peptide cds were isolated from a genomic library for the giant Antarctic toothfish *Dissostichus mawsoni*, and the trypsinogen gene and cDNA were obtained by PCR amplification of genomic DNA, and RT-PCR amplification of pancreatic tissue RNA, respectively, from the same species (Chen et al. 1997a).

The complete AFGP gene consists of two exons—the first exon encodes the signal peptide, and the second exon encodes the AFGP polyprotein (Fig. 9.3c), while the trypsinogen-like protease gene consists of six exons (Fig. 9.3a). The trypsinogen gene was subsequently named *trypsinogen-like protease* (*TLP*) since whether it is an ortholog of the digestive trypsins was yet to be experimentally tested. The evolutionary relatedness between *AFGP* and *TLP* became evident when the full-length gene sequences were aligned. They share >90% nt sequence identity at their 5′ and 3′ ends (Fig. 9.3, gray shaded regions) (Chen et al. 1997a). The 5′ region of sequence similarity encompasses 5′UTR, exon1(E1), and intron1 (I1) in *TLP*, and these are conserved as the same gene components in the new *AFGP*, with an additional gain of I1 sequence. The 3′ region of sequence similarity encompasses the 3′splice sequence of intron5 (I5), exon6 (E6), and 3′UTR in *TLP*, which in *AFGP* became the 3′ end of its E2 and 3′UTR. The origin of the long *AFGP* E2 that contains the repetitive $(\text{ThrAlaAla})_n$ cds was not initially obvious, and only on close scrutiny was it mapped to 9 nt that span the *TLP* I1/E2 splice junction. This 9-nt, partly intronic sequence comprises the three codons "*aca*-gca-gca" (I1 sequence in italic) for one (ThrAlaAla) tripeptide unit, the building block of AFGPs (Fig. 9.3a).

These evidences suggested a molecular mechanism of *TLP* to *AFGP* evolutionary transformation as follows (Fig. 9.3) (Chen et al. 1997a). The 5′ and 3′ ends (the regions of sequence similarity in extant genes) in an ancestral copy of *TLP* were recruited and joined upon the deletion of the intervening sequence—most of E2 up to 3′ splice sequence of I5, positioning the 9-nt partly intronic (ThrAlaAla) coding element in the middle. The inherited *TLP* E1, which encodes the signal peptide, preserved the secretory nature of the emergent new gene. The initial duplication of the 9-nt (ThrAlaAla) coding element was most likely random, which could result from slippage DNA replication at the stretch of dinucleotide $(gt)_n$ microsatellite DNA immediately ahead in I1 (Cheng 1998) (Figs. 9.3 and 9.4). The duplication had not affected the reading frame of *TLP* E2, preserving the I1 3′ splice sequence and only adding a small number of nucleotides. The resulting three additional residues in the TLP (Fig. 9.4) would lengthen the signal peptide slightly but was inconsequential to *TLP* activity, and thus could be tolerated. Further tripeptide cds duplications would require selective pressure to come to bear. In the depicted slippage replication (Fig. 9.4), only three duplications will produce four tripeptide repeats, the peptide backbone of AFGP8. This is the shortest AFGP isoform that possesses robust antifreeze activity, and could immediately contribute a fitness advantage. The extant long cds of the AFGP polyprotein (E2) (Fig. 9.3c) indicates further tripeptide duplications had occurred under strong selective pressure from the advent of freezing conditions in the SO.

A chimeric AFGP/TLP gene was subsequently discovered in the Antarctic toothfish (Cheng and Chen 1999) (Fig. 9.3b). It was a transcribed gene and thus presumed functional. The structure is a complete TLP gene, but its E2 is a hybrid composed of a 1.3-kbp long AFGP polyprotein coding sequence ahead of and in-frame with the cds in *TLP* E2 (Fig. 9.3b). The chimeric gene structure validated the *TLP* ancestry of the new *AFGP*, and was inferred to be an evolutionary intermediate form. It also suggests that the *TLP* to *AFGP* evolutionary process might have involved a substantial expansion of the tripeptide cds first, before the deletion that removed the bulk of *TLP* (Fig. 9.3b, c). This would form a much more stable intermediate gene structure than a minimal primordial *AFGP* if the order of the two events was reversed. The complete genomic region containing the AFGP gene family and neighboring genes was subsequently characterized from the Antarctic toothfish (Nicodemus-Johnson et al. 2011), and the AFGP genotype was nested among three paralogous trypsinogen gene families, all within less than 400 kbp total distance. Surprisingly a total of three chimeric TLP/AFGP genes were found in the genomic region. It called to question why multiple putative evolutionary intermediates would persist in the extant notothenioid genome when a large family of independent AFGP genes already formed. The chimeric genes all encode copies of the long AFGP isoforms, up to 82 tripeptide repeats, while all independent AFGP genes encode the short isoforms, predominantly AFGP7 (5 repeats) and AFGP8 (4 repeats) (Nicodemus-Johnson et al. 2011). The reason for this distinctive partitioning of coding content remains unknown. Another lasting question is the origin of the cds of the conserved three-residue linker sequence (Leu-Ile/Asn-Phe) in the AFGP polyprotein. Sequencing and scouring through the orthologous region of the AFGP genomic locus in basal non-Antarctic sister notothenioids may be one approach to pin down potential candidates and mechanism.

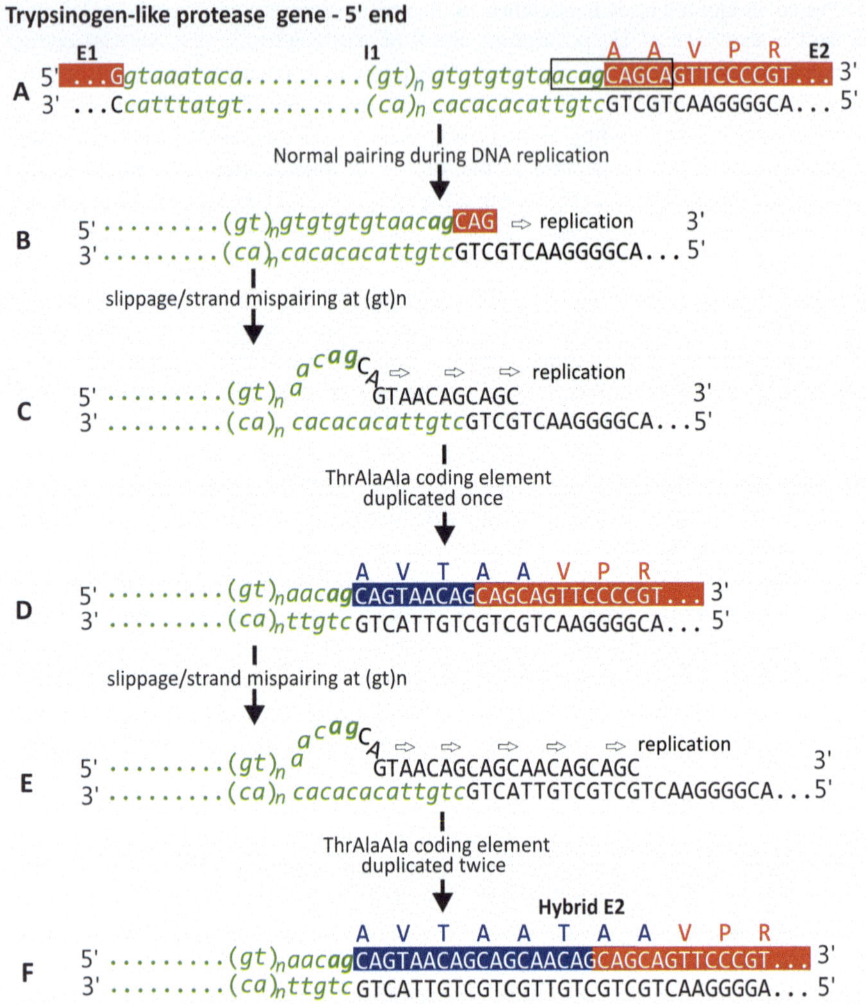

Fig. 9.4 Proposed process of the first duplications of the 9-nt ThrAlaAla coding element in an ancestral TLP gene. (**a**) 5′ end of TLP gene—E1, I1 and E2. The 9-nt *acag*CAGCA ThrAlaAla coding element straddles the I1/E2 junction (rectangular box; green italic nt—intron sequence, red shaded uppercase nt—E2 cds). (**b**) Normal strand paring during DNA replication leads to unaltered TLP gene. (**c**, **d**) Strand slippage and mispairing at the $(gt)_n$ stretch immediately ahead of the ThrAlaAla coding element would lead to its first duplication (navy boxed nt sequence, and navy amino acid translation). (**e**, **f**) The second round of slippage replication would lead to 2 tripeptide repeats. A third round would lead to 4 tripeptide repeats, resulting in AFGP 8, the smallest functional AFGP isoform

In sum, the evolutionary process of the Antarctic notothenioid AFGP gene is an eminent example of partial de novo gene evolution. The process utilized a preexisting copy of *TLP*, however, the new function was not derived from sequence tinkering of the co-opted ancestor. Instead, its origin was in a 9-nt snippet of *TLP*

DNA that was half noncoding intron sequence but comprised the three codons for one (ThrAlaAla) tripeptide. And an entirely new coding region for AFGP molecules arose de novo by microsatellite DNA like duplications when selective pressure from the chilling Antarctic marine environment came to bear, producing the ice growth-inhibiting protein that preserved life and empowered the ecological success of the Antarctic notothenioids.

9.4.3 Non-Hepatic Origin of Notothenioid AFGPs

The vertebrate liver is well known as the major synthesis organ for plasma proteins with the hepatic artery serving as the direct route of deployment to the circulatory system. Indeed hepatic synthesis provided type I, II, and III AFPs in the blood of the respective fish species (Ewart et al. 1992; Gong et al. 1992; Ewart and Fletcher 1993; Wang et al. 1995b) as well as the northern cod AFGPs (Chen et al. 1997b). A priori, the liver would also be the logical site of synthesis for the notothenioid AFGPs, however, RT-PCR amplification of liver RNA failed to produce AFGP cDNA in early attempts. Having solved that a pancreatic *TLP* gave rise to the notothenioid *AFGP* led to the realization that the liver may not be the tissue site of AFGP synthesis, but that the pancreas may be. A thorough study determined that AFGPs are indeed not synthesized in the liver in any stage of the notothenioid's life history, instead, they are synthesized at high levels in the pancreas, and secreted into the intestine (Cheng et al. 2006). Sequence alignment of AFGP and TLP genes from the complete trypsinogens/AFGP genome region (Nicodemus-Johnson et al. 2011) show they share 300 nt of 5′ sequence upstream of the start codon with 99% identity, thus AFGP genes very likely share *cis* promotors that direct their expression in the pancreas. In the persistently ice-laden condition of the Southern Ocean, where ingestion of icy food and water that would quickly freeze the hypoosmotic gastro-intestinal fluid and thus the whole fish is a constant danger, natural selection appeared to have acted first and foremost upon gastrointestinal (GI) freeze avoidance as the adaptive outcome. The pancreatic TLP to AFGP gene evolution, and the absence of hepatic AFGP expression are consistent with this hypothesis. AFGPs are also found to be expressed by the mucosa of the anterior stomach (Cheng et al. 2006), adding support to GI freeze avoidance as the prime target of selective pressure in the icy SO environment. The two amino acids Ala and Thr that make up the tripeptide repeats are not known substrates for any digestive protease, and thus AFGPs are immune to proteolysis in the GI fluids, and can effectively serve their ice growth inhibition function. A survey of other antifreeze-bearing fish species also found pancreatic expression of their respective AF type (Cheng et al. 2006), in addition to expression in the liver. The universal pancreatic expression in antifreeze-bearing fishes underscores GI freeze avoidance is a common need for these hypoosmotic teleosts in icy freezing habitats. The lasting mystery is why evolution has not driven the hepatic expression of AFGPs in the notothenioids to enable expeditious delivery to the blood circulation. There is no direct anatomical route for large-scale secretion of AFGPs made by the pancreas into blood circulation. Much of the pancreatic AFGPs are destined for the intestinal tract as high concentrations are found there (Cheng et al. 2006), thus there will be little to spare for transport to the blood to begin with. While some experimental evidence suggests that AFGPs from the GI fluid can be reabsorbed through the rectal epithelium and returned to the bloodstream (Evans et al. 2012), it seems too inefficient and cumbersome a process to achieve and maintain the very high concentrations in the blood. While not entirely improbable, rectal recycling of AFGPs to high blood levels is counter to physiological and evolutionary parsimony.

9.4.4 Broader Evolutionary Significance and Trade-Off of Notothenioid AFGP

When the mechanism of Antarctic notothenioid *AFGP* evolution was described in the late 1990s (Chen et al. 1997a; Cheng and Chen 1999), it was the first clear example of partial de novo gene evolution at the time. It brought to light that there is a larger creative repertoire of molecular evolutionary processes beyond the prevailing model of new gene evolution by gene duplication followed by sequence divergence driven by natural selection envisioned by Ohno and other proponents (Ohno 1970; Jacob 1977). The broader evolutionary implication of the simple tripeptide repeat structure of the AFGP protein also came to be appreciated in retrospect. That is, a functional protein gene of crucial fitness consequence could evolve by repeated duplication of a simple short DNA sequence, if the resulting protein structure has the right properties to interact with its target. In this regard, other proteins composed of short repeats, including the type I AFPs (Ala-rich 11-amino acid repeats), and various insect AFPs, especially the snow flea AFP (Gly-X-X repeats) (Graham and Davies 2005), and structural proteins like collagens and arthropod silks, could have evolved similarly, and not necessarily requiring a protein-coding DNA precursor sequence. The field of de novo gene creation advanced substantially in the ensuing decade. The evolution of the near-identical AFGP gene in the unrelated northern codfish (family Gadidae), to be described below, proved to be an example of the de novo gene evolution of highly satisfying clarity (Zhuang et al. 2019).

Since the discovery of macromolecular AFGP in the Antarctic notothenioid fish, and its crucial role in preventing death from freezing, the evolution of *AFGP* is regarded, and justifiably so, as one of the best example of adaptive evolution compelled by environmental change. Without AFGPs, death from inoculative freezing is for certain for these fish sequestered in the perennially icy freezing SO environment. This crucial evolutionary adaptive trait, however, has a trade-off—antifreeze protein is also an antimelt protein (Knight and Devries 1989; Celik et al. 2010; Cziko et al. 2014). The tight binding of AFGPs to ice crystals that deters water molecules from joining the ice lattice thereby inhibits ice growth would also hinder water molecules from leaving the ice lattice thereby inhibits ice melting (Knight and Devries 1989; Cziko et al. 2014). Temperatures substantially warmer than the expected equilibrium melting point of ice in the fish body fluids are needed to melt the ice. In the high-latitude McMurdo Sound, a 11-year high-resolution temperature record showed nearshore fish habitat waters hovered close to freezing almost year-round, and austral summer warming never raised temperatures above the experimentally determined melting temperature of ice in the presence of AFGPs (Cziko et al. 2014). Thus, the local notothenioid fishes are destined to accumulate internalized and AFGP-stabilized ice crystals through a substantial part or all of their life span (~20 years) potentially to injurious levels. This evolutionary pleiotropy is a unique challenge to the Antarctic notothenioids due to their isolation in the unwavering icy, freezing SO marine environment as a consequence of the unique geological evolutionary history of the Antarctic region. For antifreeze endowed fishes in the open sub-Arctic and Arctic marine systems, which experience large seasonal thermal variations, melting inhibition would be readily overcome to render the fish ice-free.

9.5 Convergent Evolution of AFGP in Unrelated Northern Codfish

The Arctic and sub-Arctic regions have different geography and a much more recent glacial history compared to the Antarctic. Thus, the evolution of antifreeze gene and function in various northern and Arctic teleost species would be much more recent events. The Arctic consists of a polar ocean surrounded by continental landmasses. The northern landmasses reached their present-day positions during the Miocene (~10–15 Mya), but Arctic sea temperatures remained mild and did not reach freezing till the Pliocene (Kennett 1977; Barry 1989). Thus, Arctic glaciation lagged the Antarctic by about 10 Myr, with the onset of glacial conditions around 3.2 Mya (Dunton 1992; Zachos et al. 2001), the formation of permanent ice cover on the Arctic Ocean at ~2 to 0.7 Mya (Dunton 1992), and cyclical glacial advances and retreats in the Northern Hemisphere during the Pleistocene.

The northern codfishes (family Gadidae) comprise a large assemblage of about 54 species including important commercial species, and form a vital component of the Arctic and sub-Arctic fish fauna (Cohen et al. 1990). The ancestral codfish is believed to be of boreal Atlantic origin (Howe 1991) and had dispersed and diversified in boreal-Arctic and circum-Arctic directions over evolutionary time (Coulson et al. 2006; Carr and Marshall 2008). The open marine circulation systems in the northern oceans means northern cod species are not restricted hydrographically to freezing bodies of water, and many high-latitude species do not experience freezing conditions in their habitats because of heat transfer from poleward flowing warm currents from lower latitudes, such as the Atlantic Current along the northeast Atlantic coast that is rich in cod grounds. Some cod populations are known to migrate across large geographic distances in the open northern oceanic system (Howe 1991; Kurlansky 1997), and presumably could avoid freezing conditions by migration to warmer climes. These geographic/oceanographic characteristics and life history traits correlate with the absence of antifreeze proteins in most cod species. Only six species, all in the gadid subfamily Gadinae are known to have antifreeze proteins. They are the saffron cod *(Eleginus gracilis*) from the Bering Sea (Raymond et al. 1975), Atlantic tomcod (*Microgadus tomcod*) (Reisman et al. 1984), Atlantic cod (*Gadus morhua*) (Hew et al. 1981), Greenland cod (*Gadus ogac*) (O'Grady et al. 1982), and the Arctic cod (*Boreogadus saida,* now known as polar cod) (Osuga and Feeney 1978; Chen et al. 1997b) and the ice cod *Arctogadus glacialis* (Præbel and Ramløv 2005) from the Arctic seas. The presence of antifreeze protection in these six species is consistent with their domicile in perennially or seasonally (winter) icy and freezing waters.

It was a great surprise when the antifreeze protein in the cods was first found to be near-identical in protein sequence and glycosylation to the AFGPs in the Antarctic notothenioid fishes. Gadids belong to the older teleost order Gadiformes while notothenioids are modern perciform fishes (order Perciformes), and thus the two groups are widely separated phylogenetically, besides geographically. The selective pressure for cod AFGP evolution likely arose from the onset of frigid marine conditions associated with northern hemisphere glaciation in Pliocene/Pleistocene only ~3.2 Mya. Thus, the timing of AFGP gene evolution in the gadids would be much more recent than that in Antarctic notothenioids. These considerations indicate inheritance of the AFGP trait from a common ancestor is highly unlikely. Molecular evidence showing distinct characteristics in gene sequence and structure, and distinct genomic origins had established *AFGP* in the unrelated codfish and Antarctic notothenioid evolved through convergent evolution in 1997 (Chen et al. 1997b). However, the precise genomic origin and details of the evolutionary process of the cod *AFGP* were only recently discovered (Zhuang et al. 2019). The belated discovery is due to the challenge that cod *AFGP* has no homologs even in the huge universe of known sequences in databases to provide a starting point. Alternative strategies had to be employed, ultimately leading to clear evidence of its evolution from entirely nonprotein coding DNA.

9.5.1 Gene Structure of AFGP in Northern Gadids

The gadid AFGP gene structure in Chen et al. (1997b) was not accurately delineated at the 5′ end. It has been corrected in the recent study (Zhuang et al. 2019). AFGP genomic regions and the resident genes from the polar cod, Atlantic tomcod and Atlantic cod were characterized. A typical gadid AFGP gene consists of three exons (Fig. 9.5f, functional AFGP gene). Two small exons at the 5′ end code for a signal peptide that directs the secretion of AFGPs into the bloodstream. The long exon 3 contains the coding sequence of many tripeptide repeats (Thr-Ala/Pro-Ala)$_n$. Between the signal peptide and the AFGP tripeptide repeats is a short stretch of

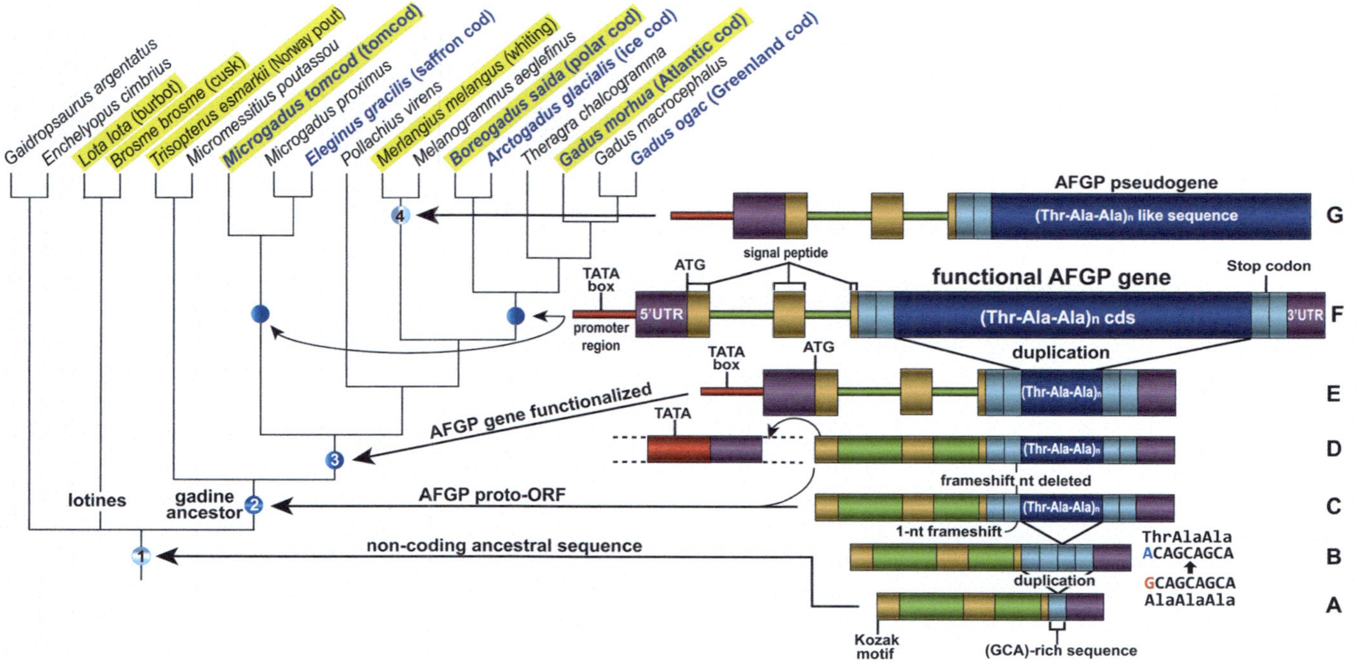

Fig. 9.5 Gadid phylogeny and evolutionary process of AFGP. AFGP-endowed and AFGP-lacking species in the tree are labeled in blue and black text, respectively. The seven yellow highlighted species were characterized for their AFGP genomic loci or AFGP orthologous regions. A typical functional AFGP gene and its component parts are represented by different colors: red—putative core promoter region, purple—UTR, gold—signal peptide cds, navy blue—AFGP tripeptide repeats cds, cyan—(GCA)-rich repeats, and green—introns. Structures (**a**–**g**) from bottom to top shows the evolutionary process of AFGP gene, which started from an ancestral noncoding sequence (**a**) that evolved to a proto-ORF through duplication and mutation (**c**, **d**), and became functionalized by obtaining a transcriptional regulatory sequence (**d**, **e**). These evolutionary stages are supported by the presence of the evolutionary sequence structures in the relevant cod lineage, indicated by the arrows pointing to the internodes (gradient blue circles), each representing the common ancestor of corresponding clades. See main text for detailed descriptions of the evolutionary process

Gln(Q)-rich sequence, which is presumably removed posttranslationally since they are not in mature AFGPs. The encoded long, AFGP polypeptide contains an occasional Arg or Lys in place of a Thr. Some of these sites likely serve as posttranslational cleavage sites, leading to multiple mature AFGP molecules of different sizes from the long protein precursor. Besides these, the gadid AFGP gene contains the typical components of a functional gene, including the core promoter TATA box upstream from the putative transcription start site (TSS), followed by 5′ UTR, a Kozak motif with the embedded Met translation start codon, followed by the signal peptide cds. The AFGP locus of each of the three AFGP-bearing gadid species encompass an AFGP gene family of different gene copy numbers, greatest in the polar cod, and least in Atlantic tomcod, correlating with the level environmental demand of gene dosage. In addition, the length of the (Thr-Ala/Pro-Ala)$_n$ tripeptide cds also varies substantially among members within the AFGP gene family (Zhuang et al. 2019).

9.5.2 De Novo Evolution of Gadid AFGP Gene

The evolution of gadid *AFGP* proved to be more creative than the partial de novo evolution of notothenioid *AFGP*. Not only did the coding sequence of the AFGP tripeptide repeats evolved from nonprotein coding DNA, all other essential components of the gene were assembled out of "genomic scraps," unassociated with any preexisting genes. The evolution of gadid *AFGP* offered the field of evolutionary biology a clear case of de novo origination of a crucial adaptive functional gene (Zhuang et al. 2019). All other known mechanisms of new gene evolution (reviewed earlier in the overview section) all involve duplication of existing materials in a genome to some extent, either preexisting genes or mobile elements, in generating genetic novelty. In contrast, de novo genes originate from previously noncoding DNA. The very slim odds of de novo gene evolution had been famously declared by Jacob (1977), "The probability that a functional protein would appear de novo by random association of amino acids is practically zero." How could evolution produce a functional gene from a random DNA sequence? For a new protein-coding gene to form de novo, the minimal essential components required include a proper start and stop codons, a coding sequence in between for a meaningful protein that will fold properly to execute a biological function, plus a transcription start site and transcription factor binding sites for the synthesis of its mRNA. How could these genic components be created through the random processes of genomic changes to produce a read through gene sequence that must also be sufficiently useful for selection to take hold?

Absence of evolutionary homologs in sequence databases, which indicated to us that cod *AFGP* might not have a protein-coding ancestor, we took the alternate approach of re-tracing the evolutionary history of cod *AFGP* within the gadid phylogeny. Three AFGP-endowed species (derived) and four AFGP-lacking species (more basal) were chosen to span the progression of evolutionary stages in the formation of the new AFGP gene (Fig. 9.5, yellow highlighted). We constructed genomic DNA libraries, isolated and characterized the *AFGP* genomic loci of the AFGP-endowed gadids and the orthologous loci from the AFGP-lacking species. Through fine-scale comparative analyses, and experimental verifications, we identified the noncoding origin of the cod *AFGP*, and reconstructed the steps and their evolutionary order by which all the essential genic components were assembled from entirely nonsense DNA to form the functional new gene (Zhuang et al. 2019).

The evolutionary process of the gadid *AFGP* was deduced as follows (Fig. 9.5) (Zhuang et al. 2019). An ancestral, small noncoding region in the genome of the last common ancestor of Gadidae (Fig. 9.5, node 1) contained a short (27-nt) GCA-rich (>3 tandem GCA triplets) sequence (Fig. 9.5a, light blue segment). Immediately ahead was a sequence that included a Kozak motif (translation activation), which could code for a signal peptide if properly spliced (Fig. 9.5a, gold segments—latent signal peptide cds; green segments—latent introns). The GCA-rich sequence duplicated forming four contiguous copies (Fig. 9.5b), the evidence for which is the two pairs of this GCA-rich sequence, now flanking extant AFGP coding region (Fig. 9.5f). Neutral sequence drift likely led to a 1-nt substitution (G to A) of a GCAGCAGCA in the middle of the ancestral four copies, leading to ACAGCAGCA, which would become the 9-nt coding element for one Thr-Ala-Ala tripeptide (Fig. 9.5, indicated next to b). The first duplications of the 9-nt Thr-Ala-Ala coding element occurred, very likely due to random microsatellite-style expansion because of the ACA/GCA repetitiveness, and incidentally formed a small open reading frame (ORF) for a few AFGP tripeptide repeats (Fig. 9.5c, navy segment). This early expansion began spreading the two pairs of GCA-rich copies apart to flanking positions. We found the first 3′ GCA-rich copy contains a preexisting in-frame TAG codon, which became a logical stop (indicated in Fig. 9.5f) for the emerging AFGP tripeptide coding region. The latent signal peptide cds upstream of the budding AFGP tripeptide ORF could not yet contribute a secretory signal because the two regions are not in the same reading frame, due to a 1-nt reading frameshift in the second 5′ GCA-rich copy (Fig. 9.5c). This frameshift was deduced from the presence of a conserved extra nt at the orthologous sites in the sequenced basal AFGP-lacking burbot and cusk, as well as in AFGP pseudogenes of AFGP-endowed species. Upon deletion of this frameshift nt (sometime between structures c and d, Fig. 9.5), the signal peptide and AFGP tripeptide coding regions became assembled into a single read-through ORF. Formation of this proto-ORF occurred in an ancestor of the gadines (Fig. 9.5, node 2) because the orthologs in extant basal lotines had not developed an equivalent AFGP tripeptide ORF. The birth of the proto-ORF, however, could not lead to transcription without a cis-acting minimal promotor. A putative translocation event joined the proto-ORF with another noncoding genomic segment, which provided a 5′UTR sequence and a nearby TATA motif (Fig. 9.5d). This translocation event is deduced based on the finding that the basal gadine Norway pout, while having the orthologous AFGP proto-ORF structure, does not have the same upstream flanking sequence or TATA box. Thus, a recombinant event must have either relocated the proto-ORF elsewhere, or moved in a noncoding genomic segment. And the event occurred in the most recent common ancestor of the AFGP-endowed gadines (Fig. 9.5, node 3), after the divergence of the Norway pout. Acquisition of this 5′ UTR/flanking region functionalized the AFGP proto-ORF, enabling its transcription, and translation from the preexisting Kozak motif. The de novo AFGP gene could produce a nascent ice-binding protein, which could become secreted into the blood and confer a fitness benefit. The expansion of the AFGP repetitive tripeptide repeats in Atlantic tomcod, polar cod, and Atlantic cod (Fig. 9.5, the two unnumbered nodes) would have been the result of the intensification of selective pressure from Northern hemisphere marine glaciation. Gadine species that no longer face freezing pressures, such as the whiting (from the ice-free Norway fjords) witnessed AFGP cds degeneration into pseudogenes (Fig. 9.5, node 4).

9.5.3 Proto-ORF Model of De Novo Gene Birth of Gadid AFGP

In order for a noncoding DNA sequence to evolve into a protein-coding gene, two fundamental steps must occur in the evolutionary process. They are (a) the DNA sequence must be able to recruit the transcription apparatus for mRNA synthesis, and (b) the noncoding sequence must achieve an ORF recognizable by the translation machinery. Two conceptual models have been proposed to recapitulate the evolutionary mechanism of a de novo gene birth. Should step (a) occurs before step (b), it fits the transcription-first model, while if the order is reversed, it fits the proto-ORF model (McLysaght and Guerzoni 2015). The gadid AFGP gene birth is a supporting example of the proto-ORF model. A nascent AFGP ORF complete with proper translational signals had developed before the promotor region was put in place (Fig. 9.5d). Evidence for the AFGP proto-ORF still persists in the Norway pout (*Trisopterus esmarkii*), the outgroup species of the AFGP-bearing gadine clade (Fig. 9.5) (Zhuang et al. 2019). Thus, the de novo birth of cod AFGP gene is one of the clearest cases if not the only clear case validating the proto-ORF model to date. The difficulty in detecting proto-ORF thus far is likely due to setting in of mutations interrupting the non-transcribed ORF before selective pressure came to bear, rendering intermediate stages undiscernible. The case of gadid AFGP gene birth differs in that the repetitive tripeptide ORF could form relatively quickly and easily through microsatellite-like expansion. In addition, the selective pressure to avoid death from freezing is so strong that any newly formed protein with the incipient ice-binding function will be quickly fixed in the population. The speed of these processes was likely the key factor that prevented the AFGP proto-ORF from degenerating back into nonsense DNA.

The birth and death of new genes are now recognized to happen dynamically, with duplications, consecutive point mutations, or genomic rearrangements generating random patterns that could almost make functional sense, but these gene-like sequences then dissolve and disperse again before natural selection could act to hold and conserve them. Thus, the recently evolved gadid AFGP gene family, compelled by the strong selection, provided us a rare instance where multiple intermediary developing stages are still traceable in members of the gene family and among lineages of different phylogenetic age, such that each step of the evolutionary processes could be pinpointed.

9.6 Conclusions

The strong selective pressure from episodes of marine glaciation in the history of Earth has driven independent evolution of a life-preserving solution in different lineages of hypoosmotic cold-water bony fishes in the form of distinctive types of macromolecular antifreeze proteins. It is one of the most remarkable adaptive evolutionary innovations in the visible fitness benefit to the organism, and the clear causal relationship between genotype and phenotype, which are often elusive in other biological systems. Understanding the diversity of genetic and genomic origins, and the evolutionary mechanisms and processes that gave rise to this crucial adaptive trait has powerfully contributed to the broader field of evolutionary biology in providing clear examples that would validate conceptual models and hypotheses of how genetic novelties arose. Molecular sleuthing of the origins and evolution of novel proteins with definitive evidential support proved to be an arduous task. The four fish antifreezes discussed in this chapter took three decades of investigations, but represent only a small fraction of the world of antifreeze proteins and the fast accruing plethora of ice-binding proteins. Even within teleost antifreezes, the evolutionary origin(s) and mechanism(s) of type I AFP, which was first discovered in the winter flounder three decades ago, remain unknown to this date. Neither are those of the insect antifreezes known. The microbial IBPs form a large, new unknown group. Thus, evolutionary studies of antifreeze proteins and IBPs will continue to be a robust field, that no doubt promises more fascinating discoveries.

References

Albers CN, Bjørn-Mortensen M, Hansen PE, Ramløv H, Sørensen T (2007) Purification and structural analysis of a type III antifreeze protein from the european eelpout *Zoarces viviparus*. Cryo Lett 28:51–60

Anderson ME (1994) Systematics and osteology of the Zoarcidae (Teleostei: Perciformes). In: Icthyological bulletin, vol 60. J.L.B. Smith Institute of Ichthyology, Grahamstown, pp 1–120. http://vital.seals.ac.za:8080/vital/access/manager/Repository/vital:15033

Andriashev AP (ed) (1970) Cryopelagic fishes in the Arctic and Antarctic and their significance in polar ecosystems. Academic Press, London

Baardsnes J, Davies PL (2001) Sialic acid synthase: the origin of fish type III antifreeze protein? Trends Biochem Sci 26:468–469

Barry RG (1989) The present climate of the Arctic Ocean and possible past and future states. In: Herman Y (ed) The Arctic seas: climatology, oceanography, geology, and biology. Van Nostrand Reinhold, New York

Barsukov VV (1986) Anarhichadidae. In: Whitehead PJP, Bauchot M-L, Hureau J-C, Nielsen J, Tortonese E (eds) Fishes of the North-eastern Atlantic and the Mediterranean. UNESCO, Paris, pp 1113–1116

Betenbaugh MJ, Yin B, Blake E, Kristoffersen L, Narang S, Viswanathan K (2014) N-Acetylneuraminic acid synthase (NANS). In: Taniguchi N, Honke K, Fukuda M, Narimatsu H, Yamaguchi Y, Angata T (eds) Handbook of glycosyltransferases and related genes. Springer, Tokyo

Bredow M, Walker VK (2017) Ice-binding proteins in plants. Front Plant Sci 8:2153

Carr SM, Marshall HD (2008) Intraspecific phylogeographic genomics from multiple complete mtDNA genomes in Atlantic cod (*Gadus morhua*): origins of the "Codmother," transatlantic vicariance and midglacial population expansion. Genetics 180:381–389

Carrete Vega G, Wiens JJ (2013) Why are there so few fish in the sea? Proc R Soc B 279:2323–2329

Celik Y, Graham LA, Mok Y-F, Bar M, Davies PL, Braslavsky I (2010) Superheating of ice crystals in antifreeze protein solutions. Proc Natl Acad Sci 107:5423

Chaw RC, Saski CA, Hayashi CY (2017) Complete gene sequence of spider attachment silk protein (PySp1) reveals novel linker regions and extreme repeat homogenization. Insect Biochem Mol Biol 81:80–90

Chen L, DeVries AL, Cheng C-HC (1997a) Evolution of antifreeze glycoprotein gene from a trypsinogen gene in Antarctic notothenioid fish. Proc Natl Acad Sci USA 94:3811–3816

Chen L, DeVries AL, Cheng C-HC (1997b) Convergent evolution of antifreeze glycoproteins in Antarctic notothenioid fish and Arctic cod. Proc Natl Acad Sci USA 94:3817–3822

Cheng C-HC (1998) Origin and mechanism of evolution of antifreeze glycoproteins in polar fishes. In: Di Prisco G, Pisano E, Clarke A (eds) Evolution of the Antarctic Ichthyofauna. Springer, Berlin, pp 311–328

Cheng C-HC, Chen L (1999) Evolution of an antifreeze glycoprotein. Nature 40:443–444

Cheng C-HC, DeVries AL (1989) Structures of antifreeze peptides from the antarctic eel pout, *Austrolycichthys brachycephalus*. Biochim Biophys Acta 997:55–64

Cheng C-HC, Cziko PA, Evans CW (2006) Nonhepatic origin of notothenioid antifreeze reveals pancreatic synthesis as common mechanism in polar fish freezing avoidance. Proc Natl Acad Sci USA 103:10491–10496

Cohen DM, Inada T, Iwamoto T, Scialabba N, Whitehead PJP (1990) FAO species catalogue: vol. 10 gadiform fishes of the world (order gadiformes), an annotated and illustrated catalogue of Cods. Hakes, grenadiers and other gadiform fishes known to date. FAO

Conant GC, Wolfe KH (2008) Turning a hobby into a job: how duplicated genes find new functions. Nat Rev Genet 9:938–950

Coulson MW, Marshall HD, Pepin PC, Carr SM (2006) Mitochondrial genomics of gadine fishes: implications for taxonomy and biogeographic origins from whole-genome data sets. Genome Biol 49:1115–1130

Cziko PA, DeVries AL, Evans CW, Cheng C-HC (2014) Antifreeze protein-induced superheating of ice inside Antarctic notothenioid fishes inhibits melting during summer warming. Proc Natl Acad Sci 111:14583–14588

Davies PL, Graham LA (2018) Protein evolution revisited. Syst Biol Reprod Med 64:403–416

de Jong WW, Lubsen NH, Kraft HJ (1994) Molecular evolution of the eye lens. Prog Retin Eye Res 13:391–442

Deng C, Cheng C-HC, Ye H, He X, Chen L (2010) Evolution of an antifreeze protein by neofunctionalization under escape from adaptive conflict. Proc Natl Acad Sci 107:21593–21598

Denstad J-P, Aunaas T, Borseth JF, Aaset AV, Zachariassen KE (1987) Thermal hysteresis antifreeze agents in fishes from Spitsbergen waters. Polar Res 5:171–174

Desjardins M, Graham LA, Davies PL, Fletcher GL (2012) Antifreeze protein gene amplification facilitated niche exploitation and speciation in wolffish. FEBS J 279:2215–2230
DeVries AL (1968) Freezing resistance in some Antractic fishes (PhD thesis). PhD thesis, Stanford University
DeVries AL (1971) Glycoproteins as biological antifreeze agents in Antarctic fishes. Science 172:1152–1155
DeVries AL (1974) Survival at freezing temperatures. In: Malins DC, Sargent JR (eds) Biochemical and biophysical perspectives in marine biology. Academic Press, London, pp 289–330
DeVries AL, Lin Y (1977) The role of glycoprotein antifreezes in the survival of Antarctic fishes. In: Llano GA (ed) Adaptations within Antarctic ecosystems. Gulf, Houston, pp 439–458
DeVries AL, Steffensen JF (2005) The Arctic and Antarctic polar marine environments. In: Farrell AP, Steffensen JF (eds) Fish physiology. Academic Press, San Diego, pp 1–24
DeVries AL, Komatsu SK, Feeney RE (1970) Chemical and physical properties of freezing point depressing glycoproteins from Antarctic fishes. J Biol Chem 245:2901–2908
DeVries AL, Vandenheede J, Feeney RE (1971) Primary structure of freezing point-depressing glycoproteins. J Biol Chem 246:305–308
Doucet D, Walker VK, Qin W (2009) The bugs that came in from the cold: molecular adaptations to low temperatures in insects. Cell Mol Life Sci 66:1404–1418
Duman JG, DeVries AL (1975) The role of macromolecular antifreezes in cold water fishes. Comp Biochem Physiol 52A:193–199
Dunton K (1992) Arctic biogeography: the paradox of the marine benthic fauna and flora. Trends Ecol Evol 7:183–189
Eastman JT (2005) The nature of the diversity of Antarctic fishes. Polar Biol 28:93–107
Eastman JT (2017) Bathymetric distributions of notothenioid fishes. Polar Biol 40:2077–2095
Eastman JT, McCune AR (2000) Fishes on the Antarctic continental shelf: evolution of a marine species flock? J Fish Biol 57:84–102
Enevoldsen LT, Heiner I, DeVries AL, Steffensen JF (2003) Does fish from the Disko Bay area of Greenland possess antifreeze proteins during the summer? Polar Biol 26:365–370
Evans CW, Hellman L, Middleditch M, Wojnar JM, Brimble MA, Devries AL (2012) Synthesis and recycling of antifreeze glycoproteins in polar fishes. Antarct Sci 24:259–268
Ewart KV, Fletcher GL (1990) Isolation and characterization of antifreeze proteins from smelt (*Osmerus mordax*) and Atlantic herring (*Culpea harengus harengus*). Can J Zool 68:1652–1658
Ewart KV, Fletcher GL (1993) Herring antifreeze protein: primary structure and evidence for a C-type lectin evolutionary origin. Mol Mar Biol Biotechnol 2:20–27
Ewart KV, Rubinsky B, Fletcher GL (1992) Structural and functional similarity between fish antifreeze proteins and calcium-dependent lectins. Biochim Biophy Res Commun 185:335–340
Ewart KV, Li Z, Yang DSC, Fletcher GL, Hew CL (1998) The ice-binding site of Atlantic herring antifreeze protein corresponds to the carbohydrate-binding site of C-type lectins. Biochemist 37:4080–4085
Ewart KV, Lin Q, Hew CL (1999) Structure, function and evolution of antifreeze proteins. Cell Mol Life Sci 55:271–283
Fletcher GL, Hew CL, Li X, Haya K, Kao MH (1985) Year-round presence of high levels of plasma antifreeze peptides in a temperate fish, ocean pout (*Macrozoarces americanus*). Can J Zool 63:488–493
Fletcher G, Kao M, Haya K (2011) Seasonal and phenotypic variations in plasma protein antifreeze levels in a population of marine fish, sea raven (*Hemitripterus americanus*). Can J Fish Aquat Sci 41:819–824
Gauthier SY, Scotter AJ, Lin F-H, Baardsnes J, Fletcher GL, Davies PL (2008) A re-evaluation of the role of type IV antifreeze protein. Cryobiology 57:292–296
Gong Z, Fletcher GL, Hew CL (1992) Tissue distribution of fish antifreeze protein mRNAs. Can J Zool 70:810–814

Gradinger RR, Bluhm BA (2004) In-situ observations on the distribution and behavior of amphipods and Arctic cod (*Boreogadus saida*) under the sea ice of the high Arctic Canada Basin. Polar Biol 27:595–603

Graham LA, Davies PL (2005) Glycine-rich antifreeze proteins from snow fleas. Science 310:461–461

Graham LA, Lougheed SC, Ewart KV, Davies PL (2008) Lateral transfer of a lectin-like antifreeze protein gene in fishes. PLoS One 3(7):e2616

Graham LA, Li J, Davidson WS, Davies PL (2012) Smelt was the likely beneficiary of an antifreeze gene laterally transferred between fishes. BMC Evol Biol 12:190

Graham LA, Hobbs RS, Fletcher GL, Davies PL (2013) Helical antifreeze proteins have independently evolved in fishes on four occasions. PLoS One 8(12):e81285

Griffith M, Lumb C, Wiseman SB, Wisniewski M, Johnson RW, Marangoni AG (2005) Antifreeze proteins modify the freezing process in planta. Plant Physiol 138:330–340

Hahn MW (2009) Distinguishing among evolutionary models for the maintenance of gene duplicates. J Hered 100:605–617

Herberg S, Gert KR, Schleiffer A, Pauli A (2018) The Ly6/uPAR protein bouncer is necessary and sufficient for species-specific fertilization. Science 361:1029–1033

Hew CL, Slaughter D, Fletcher GL, Joshi S (1981) Antifreeze glycoproteins in the plasma of Newfoundland Atlantic cod (*Gadus morhua*). Can J Zool 59:2186–2192

Hew CL, Slaughter D, Joshi S, Fletcher GL, Ananthanarayanan VS (1984) Antifreeze polypeptides from the Newfoundland Ocean pout, *Macrozoarces americanus*: presence of multiple and compositionally diverse components. J Comp Physiol B155:81–88

Hew CL, Wang N-C, Joshi S, Fletcher GL, Scott GK, Hayes PH, Buettner B, Davies PL (1988) Multiple genes provide the basis for antifreeze protein diversity and dosage in the ocean pout, *Macrozoarces americanus*. J Biol Chem 263:12049–12055

Hon WC, Griffith M, Mlynarz A, Kwok YC, Yang DS (1995) Antifreeze proteins in winter rye are similar to pathogenesis-related proteins. Plant Physiol 109:879–889

Howe GJ (1991) Biogeography of gadoid fishes. J Biogeogr 18:595–622

Hsiao K-C, Cheng C, Fernandes IE, Detrich HW, DeVries AL (1990) An antifreeze glycopeptide gene from the Antarctic cod *Notothenia coriiceps* neglecta encodes a polyprotein of high peptide copy number. Proc Natl Acad Sci 87:9265–9269

Huang HH, Liao HK, Chen YJ, Hwang TS, Lin YH, Lin CH (2005) Structural characterization of sialic acid synthase by electrospray mass spectrometry - a tetrameric enzyme composed of dimeric dimers. J Am Soc Mass Spectrom 16:324–332

Hunt BM, Hoefling K, Cheng C-HC (2003) Annual warming episodes in seawater temperatures in McMurdo Sound in relationship to endogenous ice in notothenioid fish. Antarct Sci 15:333–338

Jacob F (1977) Evolution and tinkering. Science 196:1161–1166

Jensen LE, Thiel S, Petersen TE, Jensenius JC (1997) A rainbow trout lectin with multimeric structure. Comp Biochem Physiol B Biochem Mol Biol 116:385–390

Jin Y, DeVries AL (2006) Antifreeze glycoprotein levels in Antarctic notothenioid fishes inhabiting different thermal environments and the effect of warm acclimation. Comp Biochem Physiol 76B:560–600

Kadler KE, Baldock C, Bella J, Boot-Handford RP (2007) Collagens at a glance. J Cell Sci 120:1955

Kaessmann H (2010) Origins, evolution, and phenotypic impact of new genes. Genome Res 20:1313–1326

Kennett JP (1977) Cenozoic evolution of Antarctic glaciation, the circum-Antarctic Ocean, and their impact on global paleoceanography. J Geophys Res 82:3843–3860

Knight CA, Devries AL (1989) Melting inhibition and superheating of ice by an antifreeze Glycopeptide. Science 245:505–507

Kurlansky M (1997) Cod: a biography of the fish that changed the world. Penguin Books, New York

La Mesa M, Eastman JT, Vacchi M (2004) The role of notothenioid fish in the food web of the Ross Sea shelf waters: a review. Polar Biol 27:321–338

Lander ES, Linton LM, Birren B, Nusbaum C, Zody MC, Baldwin J, Devon K, Dewar K, Doyle M, FitzHugh W, Funke R, Gage D, Harris K, Heaford A, Howland J, Kann L, Lehoczky J, LeVine R, McEwan P, McKernan K, Meldrim J, Mesirov JP, Miranda C, Morris W, Naylor J, Raymond C, Rosetti M, Santos R, Sheridan A, Sougnez C, Stange-Thomann Y, Stojanovic N, Subramanian A, Wyman D, Rogers J, Sulston J, Ainscough R, Beck S, Bentley D, Burton J, Clee C, Carter N, Coulson A, Deadman R, Deloukas P, Dunham A, Dunham I, Durbin R, French L, Grafham D, Gregory S, Hubbard T, Humphray S, Hunt A, Jones M, Lloyd C, McMurray A, Matthews L, Mercer S, Milne S, Mullikin JC, Mungall A, Plumb R, Ross M, Shownkeen R, Sims S, Waterston RH, Wilson RK, Hillier LW, McPherson JD, Marra MA, Mardis ER, Fulton LA, Chinwalla AT, Pepin KH, Gish WR, Chissoe SL, Wendl MC, Delehaunty KD, Miner TL, Delehaunty A, Kramer JB, Cook LL, Fulton RS, Johnson DL, Minx PJ, Clifton SW, Hawkins T, Branscomb E, Predki P, Richardson P, Wenning S, Slezak T, Doggett N, Cheng JF, Olsen A, Lucas S, Elkin C, Uberbacher E, Frazier M, Gibbs RA, Muzny DM, Scherer SE, Bouck JB, Sodergren EJ, Worley KC, Rives CM, Gorrell JH, Metzker ML, Naylor SL, Kucherlapati RS, Nelson DL, Weinstock GM, Sakaki Y, Fujiyama A, Hattori M, Yada T, Toyoda A, Itoh T, Kawagoe C, Watanabe H, Totoki Y, Taylor T, Weissenbach J, Heilig R, Saurin W, Artiguenave F, Brottier P, Bruls T, Pelletier E, Robert C, Wincker P, Smith DR, Doucette-Stamm L, Rubenfield M, Weinstock K, Lee HM, Dubois J, Rosenthal A, Platzer M, Nyakatura G, Taudien S, Rump A, Yang H, Yu J, Wang J, Huang G, Gu J, Hood L, Rowen L, Madan A, Qin S, Davis RW, Federspiel NA, Abola AP, Proctor MJ, Myers RM, Schmutz J, Dickson M, Grimwood J, Cox DR, Olson MV, Kaul R, Raymond C, Shimizu N, Kawasaki K, Minoshima S, Evans GA, Athanasiou M, Schultz R, Roe BA, Chen F, Pan H, Ramser J, Lehrach H, Reinhardt R, WR MC, de la Bastide M, Dedhia N, Blöcker H, Hornischer K, Nordsiek G, Agarwala R, Aravind L, Bailey JA, Bateman A, Batzoglou S, Birney E, Bork P, Brown DG, Burge CB, Cerutti L, Chen HC, Church D, Clamp M, Copley RR, Doerks T, Eddy SR, Eichler EE, Furey TS, Galagan J, Gilbert JG, Harmon C, Hayashizaki Y, Haussler D, Hermjakob H, Hokamp K, Jang W, Johnson LS, Jones TA, Kasif S, Kaspryzk A, Kennedy S, Kent WJ, Kitts P, Koonin EV, Korf I, Kulp D, Lancet D, Lowe TM, McLysaght A, Mikkelsen T, Moran JV, Mulder N, Pollara VJ, Ponting CP, Schuler G, Schultz J, Slater G, Smit AF, Stupka E, Szustakowki J, Thierry-Mieg D, Thierry-Mieg J, Wagner L, Wallis J, Wheeler R, Williams A, Wolf YI, Wolfe KH, Yang SP, Yeh RF, Collins F, Guyer MS, Peterson J, Felsenfeld A, Wetterstrand KA, Patrinos A, Morgan MJ, de Jong P, Catanese JJ, Osoegawa K, Shizuya H, Choi S, Chen YJ, Szustakowki J, International Human Genome Sequencing Consortium (2001) Initial sequencing and analysis of the human genome. Nature 409:860–921

Lin Y, Duman JG, DeVries AL (1972) Studies on the structure and activity of low molecular weight glycoproteins from an antarctic fish. Biochem Biophys Res Commun 46:87–92

Liu Y, Li Z, Lin Q, Kosinski J, Seetharaman J, Bujnicki JM, Sivaraman J, Hew C-L (2007) Structure and evolutionary origin of Ca^{2+}-dependent herring type II antifreeze protein. PLoS One 2:e548

Livermore R, Nankivell A, Eagles G, Morris P (2005) Paleogene opening of Drake passage. Earth Planet Sci Lett 236:459–470

Lopes-Ferreira M, Magalhães GS, Fernandez JH, de Junqueira-de-Azevedo ILM, Le Ho P, Lima C, Valente RH, Moura-da-Silva AM (2011) Structural and biological characterization of Nattectin, a new C-type lectin from the venomous fish *Thalassophryne nattereri*. Biochimie 93:971–980

Lynch M, Conery JS (2000) The evolutionary fate and consequences of duplicate genes. Science 290:1151–1155

Lynch M, Katju V (2004) The altered evolutionary trajectories of gene duplicates. Trends Genet 20:544–549

Maslin MA, Li XS, Loutre MF, Berger A (1998) The contribution of orbital forcing to the progressive intensification of northern hemisphere glaciation. Quat Sci Rev 17:411–426

Matschiner M, Hanel R, Salzburger W (2011) On the origin and trigger of the Notothenioid adaptive radiation. PLoS One 6:e18911

McLysaght A, Guerzoni D (2015) New genes from non-coding sequence: the role of *de novo* protein-coding genes in eukaryotic evolutionary innovation. Philos Trans R Soc B 370:20140332

McLysaght A, Hurst LD (2016) Open questions in the study of *de novo* genes: what, how and why. Nat Rev Genet 17:579–579

Nicodemus-Johnson J, Silic S, Ghigliotti L, Pisano E, Cheng C-HC (2011) Assembly of the antifreeze glycoprotein/trypsinogen-like protease genomic locus in the Antarctic fish *Dissostichus mawsoni* (Norman). Genomics. https://doi.org/10.1016/j.ygeno.2011.06.002

Nishimiya Y, Kondo H, Yasui M, Sugimoto H, Noro N, Sato R, Suzuki M, Miura A, Tsuda S (2006) Crystallization and preliminary X-ray crystallographic analysis of Ca^{2+}-independent and $Ca2^{+}$-dependent species of the type II antifreeze protein. Acta Crystallogr Sect F Struct Biol Cryst Commun 62:538–541

Nishimiya Y, Kondo H, Takamichi M, Sugimoto H, Suzuki M, Miura A, Tsuda S (2008) Crystal structure and mutational analysis of Ca^{2+}-independent type II antifreeze protein from Longsnout poacher, *Brachyopsis rostratus*. J Mol Biol 382:734–746

O'Grady SM, Schrag JD, Raymond JA, DeVries AL (1982) Comparison of antifreeze glycopeptides from Arctic and Antarctic fishes. J Exp Zool 224:177–185

Ohno S (1970) Evolution by gene duplication. Springer, Berlin

Osuga DT, Feeney RE (1978) Antifreeze glycoproteins from Arctic fish. J Biol Chem 253:5338–5343

Patil JG, Khoo HW (1996) Nuclear internalization of foreign DNA by zebrafish spermatozoa and its enhancement by electroporation. J Exp Zool 274:121–129

Petzel DH, Reisman HM, DeVries AL (1980) Seasonal variation of antifreeze peptide in the winter flounder, *Pseudopleuronectes americanus*. J Exp Zool 211:63–69

Piatigorsky J (1998) Multifunctional lens crystallins and corneal enzymes. More than meets the eye. Ann N Y Acad Sci 842:7–15

Præbel K, Ramløv H (2005) Antifreeze activity in the gastrointestinal fluids of *Arctogadus glacialis* (Peters 1874) is dependent upon food type. J Exp Biol 208:2609–2613

Prosser CL (1973) Water: osmotic balance; hormonal regulation. In: Prosser CL (ed) Comparative animal physiology. Saunders, Philadelphia, pp 1–78

Raymond JA (1992) Glycerol is a colligative antifreeze in some northern fishes. J Exp Zool 262:347–352

Raymond JA (1993) Glycerol and water balance in a near-isosmotic teleost, winter-acclimatized rainbow smelt. Can J Zool 71:1849–1854

Raymond JA, DeVries AL (1972) Freezing behavior of fish blood glycoproteins with antifreeze properties. Cryobiology 9:541–547

Raymond JA, DeVries AL (1977) Adsorption-inhibition as a mechanism of freezing resistance in polar fishes. Proc Natl Acad Sci USA 74:2589–2593

Raymond JA, Lin Y, DeVries AL (1975) Glycoprotein and protein antifreezes in two Alaskan fishes. J Exp Zool 193:125–130

Reisman HM, Kao MH, Fletcher GL (1984) Antifreeze glycoprotein in a southern population of Atlantic tomcod, *Microgadus tomcod*. Comp Biochem Physiol 78A:445–447

Robins CR, Ray GC (1986) A field guide to Atlantic coast fishes of North America. Houghton Mifflin Company, Boston

Russell S, Young KM, Smith M, Hayes MA, Lumsden JS (2008) Cloning, binding properties, and tissue localization of rainbow trout (*Oncorhynchus mykiss*) ladderlectin. Fish Shellfish Immunol 24:669–683

Schlötterer C (2015) Genes from scratch–the evolutionary fate of *de novo* genes. Trends Genet 31:215–219

Scholander PF, van Dam L, Kanwisher JW, Hammel HT, Gordon MS (1957) Supercooling and osmoregulation in arctic fish. J Cell Comp Physiol 49:5–24

Schrag JD, Cheng C-HC, Panico M, Morris HR, DeVries AL (1987) Primary and secondary structure of antifreeze peptides from arctic and antarctic zoarcid fishes. Biochim Biophys Acta 915:357–370

Scott GK, Hew CL, Davies PL (1985) Antifreeze proteins genes are tandemly linked and clustered in the genome of the genome of the winter flounder. Proc Natl Acad Sci USA 82:2613–2617

Shier WT, DeVries AL (1975) Carbohydrate of antifreeze glycoproteins from an Antarctic fish. FEBS Lett 54:135–138

Shier WT, Lin Y, DeVries AL (1972) Structure and mode of action of glycoproteins from Antarctic fishes. Biochim Biophys Acta 263:406–413

Slaughter D, Fletcher GL, Ananthanarayanan VS, Hew CH (1981) Antifreeze proteins from the sea raven, *Hemitripterus americanus*. J Biol Chem 256:2022–2026

Smith K, Spadafora C (2005) Sperm-mediated gene transfer: applications and implications. BioEssays 27:551–562

Sorhannus U (2012) Evolution of type II antifreeze protein genes in teleost fish: a complex scenario involving lateral gene transfers and episodic directional selection. Evol Bioinform 8:535–544

Sutherland TD, Young JH, Weisman S, Hayashi CY, Merritt DJ (2009) Insect silk: one name, many materials. Annu Rev Entomol 55:171–188

Tautz D (2014) The discovery of *de novo* gene evolution. Perspect Biol Med 57:149–161

Tautz D, Domazet-Lošo T (2011) The evolutionary origin of orphan genes. Nat Rev Genet 12:692–702

Venugopal T, Anathy V, Kirankumar S, Pandian TJ (2004) Growth enhancement and food conversion efficiency of transgenic fish *Labeo rohita*. J Exp Zool A Comp Exp Biol 301A:477–490

Wang X, DeVries AL, Cheng C-HC (1995a) Antifreeze peptide heterogeneity in an Antarctic eel pout includes an unusually large major variant comprised of two 7 kDa type III AFPs linked in tandem. Biochim Biophys Acta 1247:163–172

Wang X, DeVries AL, Cheng C-HC (1995b) Genomic basis for antifreeze peptide heterogeneity and abundance in an Antarctic eel pout: gene structures and organization. Mol Mar Biol Biotechnol 4:135–147

Yamashita Y, Miura R, Takemoto Y, Tsuda S, Kawahara H, Obata H (2003) Type II antifreeze protein from a mid-latitude freshwater fish, Japanese smelt (*Hypomesus nipponensis*). Biosci Biotechnol Biochem 67:461–466

Zachos J, Pagani M, Sloan L, Thomas E, Billups K (2001) Trends, rhythms, and aberrations in global climate 65 Ma to present. Science 292:686–693

Zelensky AN, Gready JE (2005) The C-type lectin-like domain superfamily. FEBS J 272:6179–6217

Zhuang X, Yang C, Fevolden S-E, Cheng CC (2012) Protein genes in repetitive sequence-antifreeze glycoproteins in Atlantic cod genome. BMC Genomics 13:293

Zhuang X, Murphy KR, Ghigliotti L, Pisano E, Cheng CC, (2018) Reconstruction of the repetitive antifreeze genomic loci in the cold-water gadids *Boreogadus saida* and *Microgadus tomcod*. Mar Genomics 39:73–84

Zhuang X, Yang C, Murphy KR, Cheng CHC (2019) Molecular mechanism and history of non-sense to sense evolution of antifreeze glycoprotein gene in northern gadids. Proc Natl Acad Sci 116:4400–4405

N cturer's authorised representative in the EU is Springer
Ge mer Service Centre GmbH, Europaplatz 3, 69115 Heidelberg,
con ou have any concerns regarding our products, please
ctSafety@springernature.com

Printe nd by CPI Group (UK) Ltd, Croydon, CR0 4YY
30/09/2026
02225231-0001